Introduction to Forestry Science

L. DeVere Burton

Delmar Publishers

an International Thomson Publishing company I(T)P®

Albany • Bonn • Boston • Cincinnati • Detroit • London • Madrid
Melbourne • Mexico City • New York • Pacific Grove • Paris • San Francisco
Singapore • Tokyo • Toronto • Washington

NOTICE TO THE READER

Publisher does not warrant or guarantee any of the products described herein or perform any independent analysis in connection with any of the product information contained herein. Publisher does not assume, and expressly disclaims, any obligation to obtain and include information other than that provided to it by the manufacturer.

The reader is expressly warned to consider and adopt all safety precautions that might be indicated by the activities herein and to avoid all potential hazards. By following the instructions contained herein, the reader willingly assumes all risks in connection with such instructions.

The publisher makes no representation or warranties of any kind, including but not limited to, the warranties of fitness for particular purpose or merchantability, nor are any such representations implied with respect to the material set forth herein, and the publisher takes no responsibility with respect to such material. The publisher shall not be liable for any special, consequential, or exemplary damages resulting, in whole or part, from the readers' use of, or reliance upon, this material.

Cover design: Carolyn Miller

Cover photo courtesy of Images © 1998 PhotoDisc.

Delmar Staff:

Publisher: Susan Simpfenderfer
Acquisitions Editor: Jeff Burnham
Developmental Editor: Andrea Edwards Myers

Production Manager: Wendy Troeger
Production Editor: Carolyn Miller
Marketing Manager: Katherine M. Hans

COPYRIGHT © 2000
By Delmar Publishers
an International Thomson Publishing company I**T**P®

The ITP logo is a trademark under license
Printed in the United States of America

For more information contact:

Delmar Publishers
3 Columbia Circle, Box 15015
Albany, New York 12212- 5015

International Thomson Publishing Europe
Berkshire House
168-173 High Holborn
London, WC1V 7AA
United Kingdom

Nelson ITP, Australia
102 Dodds Street
South Melbourne,
Victoria, 3205 Australia

Nelson Canada
1120 Birchmont Road
Scarborough, Ontario
M1K 5G4, Canada

International Thomson Publishing France
Tour Maine-Montparnasse
33 Avenue du Maine
75755 Paris Cedex 15, France

International Thomson Editores
Seneca 53
Colonia Polanco
11560 Mexico D. F. Mexico

International Thomson Publishing GmbH
Königswinterer Strasße 418
53227 Bonn
Germany

International Thomson Publishing Asia
60 Albert Street
#15-01 Albert Complex
Singapore 189969

International Thomson Publishing Japan
Hirakawa-cho Kyowa Building, 3F
2-2-1 Hirakawa-cho, Chiyoda-ku,
Tokyo 102, Japan

ITE Spain/Paraninfo
Calle Magallanes, 25
28015-Madrid, Espana

6 7 8 9 10   XXX   03 02

Library of Congress Cataloging-in-Publication Data

Burton, L. DeVere
  Introduction to forestry science / L. DeVere Burton.
    p.   cm.
  Includes index.
  Summary: A senior high textbook focusing on the North American forest regions; classification, anatomy, and diseases of trees; forest management and products; and urban forestry.
  ISBN 0-8273-8010-0
  1. Forests and forestry—North America—Juvenile literature.
2. Forest management—United States—Juvenile literature.
3. Forests and forestry—Juvenile literature. 4. Forest management—Juvenile literature. [1. Forests and forestry. 2. Forest management.] I. Title
SD376.B87   1999
634.9—dc21   98-36342
                                               CIP
                                               AC

Contents

v

Preface

The forests of North America are among our most valuable and treasured natural resources. They contribute wealth to the economy, wood products for the benefit of society, outdoor environments for our enjoyment, and living environments for animals and birds. Improved watersheds and fresh, clean air are other positive benefits derived from forests. Because forest resources are in high demand, they are also vulnerable to abuses. They are vulnerable because some human activities are capable of upsetting the delicate relationships that exist between living organisms such as trees, animals, and forest plants, and nonliving forest resources such as soil and water.

This textbook is about the principles of science that contribute to healthy forests. Science is the foundation upon which the management of forest environments should be based. It is unfortunate that science is sometimes ignored, and that political considerations are allowed to shape and control forest management policies. It is important that a balance be struck that allows our forests to be used for productive purposes, while fostering forest management practices that maintain and improve the environment.

A serious attempt has been made in this textbook to present both sides of major environmental issues. It is the contention of the author that we should seek "middle ground" in resolving environmental conflicts, and that it is wise to avoid radical positions on either side of these issues. It is usually wise to consider the long-term effects that some management strategies are likely to have on humans. It is probably unrealistic to expect that people will abruptly cease to use resources that have historically provided income and security for their families.

This textbook is divided into six sections.

Section 1—Introduction: Getting Acquainted with the Forest
Section 2—Dendrology: The Scientific Study of Trees
Section 3—Forest Management
Section 4—Forest Technologies
Section 5—Forest Products
Section 6—New Directions and Technologies in Forestry

The first section discusses the importance of forests, and it names and describes the different forest regions. Section 2 presents scientific information about the structure and life processes of trees. It also includes the study of relationships between trees and their environments, and the effects of insects, diseases, and pests on forest health. Section 3 deals with forest management practices, and Section 4 is a study of the technologies that are used to manage forests. Section 5 deals with producing and processing forest products. Finally, Section 6 discusses the new directions and technologies in forest management. It includes a chapter on the relatively new branch of forestry known as urban forestry.

Included in each chapter are features entitled Objectives, Terms to Know, Forest Profiles, Career Options, Profiles on Forest Safety, Looking Back, Questions for Discussion and Review, and Learning Activities. Each chapter is filled with photographs and illustrations that will aid students as they seek understanding of the concepts that are presented.

Current efforts at school reform stress integration of technical and academic curricula. The School-to-Work and Tech Prep initiatives are two examples of career education movements that take advantage of opportunities to teach science, math, and other subjects in the context of a career. The forestry curriculum helps students to know how they will use what they have learned. This textbook is intended to facilitate integration of science in curricula that deal with principles of forestry.

Acknowledgments

This book is dedicated to students and teachers who enjoy learning and working in the woods and forests of North America. It is also written in tribute to the individuals and organizations who use and manage our forest resources in a responsible manner to ensure forests for future generations. The decisions of forest managers will have lasting impacts on our forests for good or ill far into the future. It is to the foresters of tomorrow that these pages are dedicated, in the hope that those who manage the forests in the twenty-first century will learn from the successes and mistakes of the past.

Special thanks are offered to the people who have contributed resources, pictures, expertise, and encouragement during the writing of this textbook. Gratitude is offered to my family and friends, and appreciation is extended to colleagues and all who contributed to this project. Thanks to those who have reviewed the text and edited the manuscripts. Likewise, thanks to those writers whose written works have provided information of a technical nature, and thanks to the individuals who have contributed photographs:

American Forest & Paper Association
Washington, DC

Boise Cascade Corporation
Burley, ID

Boise National Forest
Boise, ID

Forestry Suppliers, Inc.
Jackson, MS

Clare Harkins
Boise, ID

Idaho Cedar Sales
Troy, ID

Gary Moen
Boise, ID

National Interagency Fire Center
Boise, ID

Denise Ortiz
Moscow, ID

Marilyn Parker
Rupert, ID

Potlatch Corporation
Lewiston, ID

Robert Pratt
Nez Perce, ID

Savannah Park Department
Savannah, GA

United States Department of
Agriculture
Washington, DC

United States Department of Energy
Washington, DC

United States Forest Service
Washington, DC

Wendy Wedum
Lewistown, ID

Woody Contracting
Council, ID

J. W. Buck Wright
Genesee, ID

Agricultural Mechanics: Fundamentals and Applications, 3rd Edition, by Elmer L. Cooper, copyright 1997 by Delmar Publishers.

Agriscience and Technology, by L. DeVere Burton, copyright 1998 by Delmar Publishers.

Ecology of Fish and Wildlife, by L. DeVere Burton, copyright 1996 by Delmar Publishers.

Introductory Horticulture, 5th Edition, by H. Edward Reiley and Carroll L. Shry, Jr., copyright 1997 by Delmar Publishers.

Managing Our Natural Resources, 3rd Edition, by William G. Camp and Thomas B. Daugherty, copyright 1997 by Delmar Publishers.

Soil Science & Management, 3rd Edition, by Edward J. Plaster, copyright 1997 by Delmar Publishers.

The services of the following reviewers of this textbook are gratefully acknowledged:

Douglas Brown
Central Columbia High School
Nescopec, PA

E. F. Cowen
Mitchell-Baker High School
Camilla, GA

Terry Crawford
Cul de Sac High School
Cul de Sac, ID

Bill Schrum
Seminole Vocational Education Center
Seminole, FL

Larry Stevenson
Tulelake High School
Tulelake, CA

Michael Waltersheidt
Texas A&M
College Station, TX

About the Author

L. DeVere Burton, author of *Introduction to Forestry Science,* is Director of Research, and former State Supervisor, for Agricultural Science & Technology with the Idaho State Division of Vocational Education. He has served as President of the National Association of Supervisors of Agriculture Education, and he has participated as a member of several national curriculum-related task forces for the National Agricultural Education Council.

The author was a high school agriculture teacher for fifteen years, and has been involved as a professional educator in agricultural education since 1967. He has experienced teaching assignments in both large and small schools, and in both single- and multiple-teacher departments. He has taught at four different schools, and at a major land grant university. He was involved in agriculture program supervision from 1987 to 1997. All of these experiences have contributed to his philosophy that "education must be fun and exciting for those who learn and for those who teach."

A wide range of experiences have prepared the author for his career as an educator in agriculture and natural resources. He was raised on a farm in western Wyoming that bordered on forest lands, and he experienced many pleasant hours in the canyons and along the streams that were part of the forests of his youth. During his years as a university student, he worked in the forest industry

as a logger and sawmill worker. Other jobs held by the author include testing milk for butterfat content; caring for livestock on a combination beef, swine, and trout ranch; maintenance/warehouse worker in a feed mill; manager of a dairy; finish carpenter; and animal research assistant. He has also worked in the food processing, metal fabrication, and concrete construction industries, and he owned and managed a purebred sheep and row crop farm for several years.

Dr. Burton earned his B.S. degree in Agricultural Education from Utah State University in 1967. He was awarded an M.S. degree in Animal Science from Brigham Young University in 1972. His Ph.D. degree was earned at Iowa State University in 1987 where he was also an instructor in the Agricultural Engineering Department.

This is the third textbook that Dr. Burton has authored. His first textbook is entitled *Agriscience & Technology,* and his second textbook is *Ecology of Fish and Wildlife.* All three textbooks have been written in a serious attempt to strengthen the science content and expand the breadth of the curriculum in the nation's agriculture and natural resource education programs.

INTRODUCTION

Getting Acquainted with the Forest

Chapter

1

Terms to Know

forestry
forest
strata
canopy
understory
shrub
shrub layer
herb layer
forest floor
biological value
transpiration
watershed
renewable resource
nonrenewable resource
gross national product (GNP)
coke
particulate matter
biomass
biomass power
short rotation woody crops (SRWC)
short rotation intensive culture
  (SRIC)
multiple use
riparian zone
silt
silt load

Introduction to Forestry

Objectives

After completing this chapter, you should be able to

❋ identify important forest products that contribute to the comfort and health of people and to the economies of nations

❋ describe the kinds of plants that compose the vegetative strata that are found in a forest environment

❋ list the major life forms that contribute to the biological value of a forest

❋ suggest some natural functions of a forest that affect its biological value

❋ describe how a watershed functions, and explain why a forested watershed is superior to a watershed that lacks forest plant cover

❋ identify ways that forest environments contribute to stable populations of wild animals

❋ distinguish between renewable resources and nonrenewable resources

❋ account for the major uses of forest resources in the United States

❋ list ways that forest products such as wood and other biomass materials are used as sources of energy

❋ explain the multiple-use concept of management for public lands

Management of the forest ecosystem is a complicated and controversial profession in our politically charged world (Figures 1–1, 1–2). The social and political sciences have become as important in forest management as the biological sciences that are the basis for modern forest management practices. Biological, political, and social sciences along with business and management skills contribute to the field of study known as **forestry.**

IMPORTANCE OF FORESTS

Thirty percent of the land area in the world is forest land, and forest products are very important to the economies of the developed countries of the world. The importance of the forests of North America and the world goes far beyond the production of wood products such as paper, cardboard, lumber, plywood, and structural beams. Forests also provide solvents, medicines, fuels, and many other products that are important for our health and comfort (Figure 1–3). It is forests and other forms of plant life that restore oxygen in our atmosphere through photosynthesis. Forests function as huge biological filter systems that clean the environment by removing impurities from air and water. They also function in the elemental cycles and in the water cycle.

Biological Value of Forests

A **forest** consists of an area where trees are the most dominant living organisms. The kinds of trees that are found in forests sometimes consist of a single species, but many forests are made up of more than one kind of tree (Figures 1–4, 1–5). Forest environments also include many plants other than trees. Several layers of vegetation called **strata** are found in a forest (Figure 1–6). The tall broadleaf trees form the ceiling or **canopy** at the highest levels. The area beneath the

Figure 1–1 Most Americans believe that it is important to manage forests in ways that will ensure that they will be available to future generations.

Figure 1–2 Forest management has become much more difficult in recent years due to conflicts over appropriate uses of forest resources.

Forest Products

Atmospheric oxygen

Medicines

Solvents

Fuels

Paper

Cardboard

Lumber and wood

Figure 1–3 Forests are important sources of some of our most useful products. Many products are made or extracted from wood.

Figure 1–4 Some forests consist almost entirely of a single species of trees.

Figure 1–5 The majority of forests are made up of different kinds of trees.

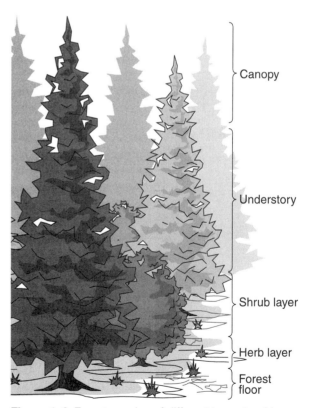

Canopy

Understory

Shrub layer

Herb layer

Forest floor

Figure 1–6 Forest species of different types tend to become dominant in different strata or vegetative forest layers.

Figure 1–7 Forests provide living environments for many different kinds of animals. *(Photo courtesy of Clare Harkins)*

canopy is filled in with smaller trees that make up the **understory** of the forest. The lower strata of a forest is often inhabited by small woody plants called **shrubs.** This layer of the forest strata is called the **shrub layer.** The shortest plants such as ferns, grasses, and flowering plants are collectively called the **herb layer.** The **forest floor** is composed of a layer of decaying plant materials that act as a mulch to preserve soil moisture. Each of these strata is inhabited by living organisms that are adapted to live in that particular environment.

Living organisms besides plants that are found in forests include insects, mammals, birds, amphibians, reptiles, and fish (Figures 1–7, 1–8). All of the

Figure 1–8 A forest provides ideal living environments for birds of many kinds.

Figure 1–9 Trees are important to fish habitat in streams. They provide places for fish to hide, and trees contribute to cool water temperatures by shading the water from direct sunlight.

Figure 1–10 A forested watershed is ideal because much of the seasonal moisture tends to be absorbed rather than running off the land.

living organisms that are found in forest environments contribute to the value of a forest. The **biological value** of a forest takes into account the worth of all of the life forms that are found there. It also includes the value of important natural functions of forests such as effects on climate, watersheds, water temperature, soil erosion, and wildlife.

Forests have some influence on the climate in the local area. They provide shade and cooler temperatures inside forest environments during summer months than in surrounding areas (Figure 1–9). Forest environments also contribute to comparatively warmer temperatures in winter within the confines of the forest. The movement of air is restricted by dense vegetation in the forest. This reduces the chill factors and restricts evaporation of moisture from the forest floor. In contrast, large amounts of water are lost from plant leaves to the surrounding area, resulting in increased humidity in the air. This loss of water from leaf surfaces is called **transpiration.**

A **watershed** is a region where water from rain and snow is absorbed into the soil (Figure 1–10). It emerges later from springs that feed into streams and rivers (Figure 1–11). A watershed that is forested is superior to other watersheds because trees tend to slow the melting of snow, allowing it to be absorbed into the soil instead of running off the soil surface. Trees aid in water absorption by reducing the depth to which soils become frozen. This allows water to seep into the soil sooner than it does in unprotected areas. The snow usually lasts several

Figure 1–11 A clean, pure water supply is available only when the watershed from which it flows is well managed.

weeks longer in forested areas than it does in open areas. This helps to maintain constant stream flows (Figure 1–12).

Flooding of lowland areas becomes much more severe when forests have been cleared from watersheds. Flooding occurs when melted snow or runoff water from rain exceeds the rate at which the water can be absorbed into the soil. Frozen soil prevents water from being absorbed down through the soil profile. Deep-rooted plants also contribute to the infiltration of water into the soil by breaking up hard layers of soil. Smaller fibrous roots tend to hold soil particles in place when water moves across the soil surface. Healthy forests help provide protection against erosion and flooding.

Figure 1–12 A healthy watershed is important in stabilizing stream flows.

Figure 1–13 Vegetation along stream banks provides protective cover for fish behind the overhanging plant materials.

Streams that flow through forests are protected from the heat of the sun by the shade of trees and brush growing along their banks. This is important to the survival of some kinds of fish such as trout that require cool water temperatures. Trees and brush also contribute to fish survival by providing cover on and beneath the surface of the water (Figure 1–13). Fish need places in the water where they can go to escape predators (Figure 1–14). Cover plants also provide insect habitat, and the fish feed on insects that drop into the water from these plants.

Many species of wild animals and birds make their homes in forest environments. They use the materials found in forests to provide dens and nesting

Figure 1–14 Fish habitat is enhanced by logs and fallen trees that become submerged in streams where they provide cover and protection for fish.

Figure 1–15 Trees are sources of both food and shelter for birds such as woodpeckers and flickers.

places (Figure 1–15). Forests provide food for many of these creatures, and the animals and birds that live in forests find protection from their enemies among the trees, shrubs, and other forest plants. Forests tend to isolate animals and birds from humans and human civilization. This allows them to avoid disturbances during critical periods in their lives. Examples of critical periods for birds include the nesting and fledgling periods. The eggs and young birds are quite vulnerable to predators. Young mammals are also vulnerable to predators due in part to their natural curiosity. Some of them also lack the mobility and caution that they develop later in life (Figure 1–16).

Figure 1–16 A newborn elk calf is most vulnerable to predators in the first few hours after it is born, and its safety depends on forest cover and camouflage coloring. *(Photo courtesy of Robert Pratt)*

Many of the medicines that are in use today are derived from plant materials. Many more plants with medicinal value may still be identified. Forests offer the potential for new medical cures that have not yet been explored. We are likely to find many new plant materials in our forests that are valuable to society.

Commercial Value of Forests

Forest products are important natural resources to the United States economy. A forest is a **renewable resource,** meaning that it is capable of regrowth following use (Figure 1–17). A **nonrenewable resource** such as coal is permanently used up when it is consumed. Many North American forests have been harvested at least twice since European settlers arrived. Good management practices should make it possible to extend the production of our forests well into the future.

Approximately 42% of the timber harvested in the United States is used for lumber (Figures 1–18, 1–19). Pulpwood for production of paper products ranks second in consumption at 28%. Fuelwood consumption accounts for 20% of all timber harvested. A growing segment of the timber industry is the production of plywood and veneers. Approximately 8% of all timber products is consumed for

Figure 1–17 A forest is a renewable resource that replaces itself when conditions are favorable to new growth.

Figure 1–18 Nearly half of all harvested trees are transported to sawmills where they are processed into lumber products.

this purpose. Other uses for wood products account for 2% of forest product consumption (Figure 1–20).

Fuelwood consumption tends to increase or decrease in response to the cost of other heating fuels such as coal and heating oil. The dramatic increase in oil prices in the 1970s was accompanied by a sharp increase in the use of wood for fuel (Figure 1–21). This occurred because the cost of wood became favorable in comparison with other fuels.

Lumber production has tended to remain constant since the beginning of this century at approximately 150 million cubic meters. The value of timber production accounts for about 6% of the **gross national product (GNP).** The

Figure 1–19 Lumber is an important forest product due to its widespread use in the construction of homes and commercial buildings.

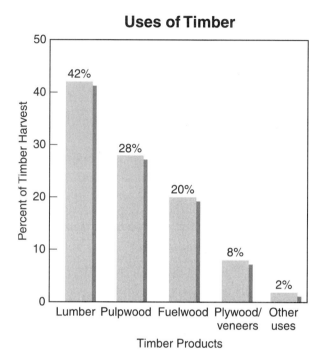

Uses of Timber

Figure 1–20 Timber harvested in the United States provides raw materials that are used to produce many different kinds of products.

Figure 1–21 Wood is the most important fuel in the world for heating homes and cooking. Even in the United States, 20% of the wood that is harvested is still used for fuel.

GNP is a measurement of the strength of the national economy. Since the passage of the North American Free Trade Agreement (NAFTA) legislation in 1993, Canadian timber has entered United States markets in increasing volume, while domestic timber sales have declined. A growing number of United States sawmills have closed due in part to a lack of consistent and dependable supplies of timber.

Energy Source

Wood is the most important source of heat in many of the countries of the world. It is used for cooking and for heating homes and other buildings that require supplemental heat. Energy is released from wood in the form of heat when combustion or burning occurs. Most of the wood that is used to heat homes is harvested and dried before it is used. Some wood fuel is obtained as a by-product of wood manufacturing in the form of wood pellets (Figure 1–22). This product consists of waste lumber that has been ground into small chips and extruded through a pellet mill for use in wood-burning stoves and furnaces. This is an efficient use of wood products that otherwise would be wasted.

Some commercial uses are made of wood as a heat source. The most common of these are in industries that convert wood to coke. **Coke** is a wood product that is obtained by heating wood to temperatures in excess of its combustion temperature using large ovens from which oxygen is excluded. The product that is obtained from this process is capable of burning at very high temperatures. Charcoal briquettes that are used in home barbecue grills are a type of coke (Figure 1–23). This fuel is used in processes that require high temperatures and clean-burning fuels.

The various species of wood are similar in chemical composition, but as the density of fuelwood increases, the amount of heat produced increases. For this

Figure 1–22 A recent development in woodstoves is the use of wood pellets for fuel. Wood pellets are made from wood scraps, sawdust, shavings, and other waste materials.

Figure 1–23 Charcoal and some kinds of coke are special fuels that are produced from wood for industrial and home uses.

reason, the high-density hardwoods are considered to be better sources of fuel than low-density softwoods. There is very little difference in the amount of heat that is produced per pound of wood regardless of the species. High-density woods are more efficient than low-density woods because they produce more heat per cord. Only a small part of the total harvest of softwood is used for fuel, but nearly half of the hardwood harvest is used for fuel.

One advantage of using wood for fuel is that it is a renewable resource. It tends to be low in sulfur compared to coal, so very little pollution such as acid precipitation (in the form of rain or snow) is produced. It does, however, produce pollution from **particulate matter** consisting of tiny particles contained in smoke. This problem is greatest when damp wood is burned, because combustion tends to be incomplete.

Biomass includes vegetation and wastes that contain significant amounts of vegetable matter. Forests are the most important source of biomass. They constitute 42% of the total available biomass of the earth. Other sources include agricultural crops, crop residues such as straw and fodder, crop-processing wastes, animal wastes, and solid wastes from cities and towns. Significant amounts of energy can be obtained from biomass sources, and these sources of energy are renewable.

Trees and shrubs that are grown for energy crops can be cultivated in dense plantings with rows of plants that are much more narrowly spaced than in forest plantings. Fast-growing varieties of trees and herbaceous plants are produced using intensive management practices such as fertilization, weed control, and increased frequency of harvesting. Such crops are well adapted to land that is not suitable for agricultural crops due to poor soils or steep slopes.

The wood from harvested biomass plants may be chipped, dried, and burned to produce steam for the purpose of generating electricity (Figure 1–24).

Figure 1–24 Biomass is a modern wood product consisting of wood chips that are dried and burned as a source of energy to generate electricity.

Electricity from this source is called **biomass power.** Other uses for biomass materials include production of paper products, construction materials, and ethanol fuels. Production of biomass as an energy crop has been practiced in woodlots for many years, but modern biomass production is much more intensive. These crops are usually harvested every three to seven years. They are known as **short rotation woody crops (SRWC)** or as **short rotation intensive culture (SRIC).**

Domestic and Wildlife Range

The **multiple use** concept of forest management provides access to forests for many uses. Examples of these uses include grazing, mining, logging, wildlife, and recreation (Figures 1–25, 1–26). Natural resources can be used in such a way that people with different interests and needs can use the same resources without damaging or depleting them. The multiple-use management system has been used successfully for many years, but this form of management depends upon responsible and prudent use of forest resources by all users.

The untimely use of a resource often becomes abuse of the resource. Riding a motorcycle across dirt trails or hillsides when they are wet leads to soil erosion. Fishing in prime spawning areas during the spawning season may significantly reduce wild fish populations. Grazing livestock in sensitive areas contributes to damaged rangelands.

Livestock grazing is a practice that allows for harvesting forages that grow in forest environments. Vast areas provide habitat for forage plants that grow on forested lands, and domestic livestock species such as cattle and sheep are able to convert these plant materials to meat. A well-managed system of livestock grazing removes vegetation before it becomes old and unpalatable, and it allows

Figure 1–25 Perhaps the best known use of the forest is the production of timber products.

Figure 1–26 The multiple-use concept of forest management makes allowances for many different types of uses of forest lands. Grazing of livestock continues to be one of the most important uses of forest lands. *(Photo courtesy of Utah Agricultural Experiment Station)*

for new plant growth. Regrowth of succulent forage on forest rangelands contributes to improved nutrition for large game animals such as deer, elk, and moose that use these areas for winter range.

The **riparian zone** is the land adjacent to the bank of a stream, river, or other waterway (Figure 1–27). A rancher who allows his or her animals to repeatedly overgraze the riparian zone may contribute to severe damage to the natural plant cover in the area. This often leads to erosion of the soil from the banks of streams and lakes, causing contamination of surface water. Tiny soil particles that become suspended in water are known as **silt.** The amount of soil that is suspended in flowing water is the **silt load.** Silt fills in lakes and reservoirs,

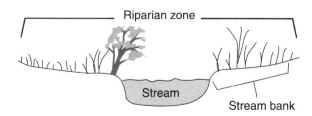

Figure 1–27 The riparian zone includes the area that is adjacent to the bank of a river or stream.

Figure 1–28 Silt consists of small soil particles that become suspended in the water of flowing streams. It is the greatest single source of water pollution.

destroys fish spawning areas, and kills young fish when water is muddy for extended periods of time (Figure 1–28).

A responsible approach to resolving abuses to forest resources is to terminate the privilege of using the resources for those who abuse our land, water, and forests. Only those who demonstrate responsible use of our natural resources should enjoy the privilege of continued use. A single abuser of a natural resource is likely to damage the trust relationship that is necessary for the multiple-use management system to work effectively.

Recreation and Wildlife

Many resources are available in our forests that have recreation value for people. Lakes and streams provide opportunities for fishing, boating, and other water sports. Many people enjoy hiking, mountain biking, horseback riding, photography, picnicking, harvesting mushrooms and wild fruits, camping, fishing, and big game hunting (Figure 1–29). All of these recreational activities draw people into our forests in large numbers. Outdoor recreation has become a huge industry, and the citizens of many communities located in or near forest lands derive much of their income from sales of food, services, or supplies related to recreational activities. Entire industries are devoted to the production of recreational products such as tents, camp stoves, boats, fishing supplies, and so on (Figure 1–30).

Forests are valuable as habitat for many kinds of wild creatures. For many people, the value of a forest lies in the wild animals that can be found there.

Figure 1–29 Forests continue to hold appeal for people who enjoy outdoor environments, and many people use forests for recreation purposes. *(Photo courtesy of Joni Conlon)*

Some people enjoy the forest as a wild place where wild animals can be observed in their natural settings. The thousands of visitors who take vacations to our national parks and monuments each year are evidence of the value that our citizens place on forests as wildlife habitat. It is difficult to measure intrinsic value, but to some people, this is the most valuable of all the forest uses. These people place great value on wild places and natural environments.

Figure 1–30 An entire industry has been developed that provides recreation equipment, clothing, and supplies for outdoor enthusiasts.

HISTORY OF THE FOREST INDUSTRY

The forest industry in North America has evolved from the harvest and sale of mast timbers to the highly mechanized and computerized wood products and paper industries of today. The production of mast timbers for the construction of ships was among the earliest commercial forest enterprises in America. The native forests of the eastern coastlines provided tall, straight timbers that were ideal for this purpose.

Early attempts to control the harvest of trees were made by Great Britain with the Broad Arrow policy in 1691. This policy reserved all trees that were 24" or more in diameter and located on public lands for the king's navy. The bark of such trees was marked with large arrows, and large fines were imposed on those who dared to cut them without the proper permits. The colonists opposed this regulation of their commerce, and it is likely that this policy contributed to their desire to become independent of British rule.

The new Congress that was established following the Revolutionary War passed laws that encouraged the sale of public land as a source of government revenue. Public lands outside the thirteen original states and Texas were surveyed, and attempts were made to sell the land. At a minimum price of $1 per acre, and a minimum purchase of 640 acres, the price was too high. The government land sale was a failure, but people went right on using the land and timber with little effort by the government to restrict them.

As the population moved west into unsettled regions, large tracts of forest land were cleared for farming, and what had seemed to be an inexhaustible timber supply began to show the effects of excessive use. Despite the laws and policies that were enacted to protect forests, there were many instances where timber and land were obtained under false pretenses. Eventually the Congress established forest reserves that it vigorously protected, and states established forest boards and commissions to set forest management policies. The concept of managing forest lands did not gain acceptance until it became evident that the forests were in danger of being destroyed.

The Division of Forestry was established by the federal government in the Department of the Interior. Gifford Pinchot became the new administrator of this agency on July 1, 1898. Pinchot had studied forestry in France prior to becoming the Chief Forester, and he believed that forests should be wisely used, but not abused. This was contrary to the dominant philosophy of the time that fostered the belief that forests should be preserved for the enjoyment of future generations. Pinchot succeeded in getting the forest reserves transferred from the custody of the Department of the Interior to the Department of Agriculture in 1905. Later that year, the Division of Forestry was changed to the Forest Service. The forest management philosophy of Gifford Pinchot still influences forest management today, after more than one hundred years.

The debate that surrounded Pinchot still rages today between those who would like to preserve the forests and those who believe that forests should be managed for timber production. The twentieth century has seen the establishment of forest reserves, wilderness, national parks and monuments, and

PROFILE ON FOREST SAFETY

Wildlife Hazards

Safety with wild animals is usually achieved by demonstrating respect for the territorial instincts of wild animals. For example, very few people experience injuries from wild animals when they are careful to keep their distance from them. This is true of most animals, but it is especially true of such animals as moose, bison, wild boar, poisonous snakes, and alligators. A person who is hiking or camping should make some noise as he or she walks through territory where these animals are likely to be encountered. This allows the wild animals to avoid surprise encounters with humans.

Many animals do not hesitate to defend their young against humans who get too close or who get between the mother and its offspring. The most common maternal animal attacks on humans involve bears, moose, alligators, and some birds. Another time when wild animals are prone to attack humans is during the mating seasons. Large male deer and some other kinds of animals sometimes view humans as competitors during these periods. They also tend to be much more aggressive during mating season than at any other time. The best way to avoid confrontations is to make enough noise that the animals will seek to avoid human contact.

It is also important that bears and other predators are not lured to campsites and tents by the smell of food. Loggers and campers should make sure that food supplies are hung above the ground at a distance from the campsite. Candy and other food should never be placed inside tents and sleeping bags or even in the immediate vicinity of the camp. On rare occasions, some wild predators seem to view humans, especially small children, as prey. Some human activities such as jogging also appear to confuse wild predators and lead to attacks. The animals most often involved in this kind of attack include large bears, alligators, and an occasional mountain lion. Many of the documented attacks of this nature have occurred when natural foods were in short supply, predator populations were excessive, or the predator had sustained an injury that interfered with its ability to catch its natural prey. The best deterrent for this kind of attack is to make plenty of noise and remain alert to avoid confrontations with wild animals. An exception to this rule would apply when small children are playing near southern waters that are known to be inhabited by alligators. Splashing and playing in the water appears to attract these large reptiles who apparently mistake the play of humans for the struggles of a prey animal.

Hunters and fishermen sometimes experience confrontations with predators (usually bears) over fish or game animals that they have bagged. The safe approach to this problem is to back away rather than to confront these wild predators. In most cases, the game will be contaminated and unfit for human use anyway after it has been claimed by a bear or other predator. It is always best to respect the territorial nature of wild animals.

CAREER OPTION

Forestry Educator

An educator who specializes in forestry is usually a person who has completed a university graduate program and successfully worked in the forest industry or a related field. A Professor of Forestry works in a college or university as a teacher and advisor of students who plan forestry careers. In addition to teaching, a forestry professor is expected to spend some of his or her time doing research in forestry or a closely related field of knowledge. Strong writing and communication skills are needed along with a strong science background.

national forests. Each of these designations imposes a different set of management practices on the forest lands. Forest recreation has become a dominant force that influences forest management practices. Forests have become highly regulated, and people are no longer allowed to cut trees without permits. Forest research has become an important tool in managing forests for sustained yields, and the concept of multiple use of forest lands has gained acceptance. Environmental considerations have a strong influence on how forests are managed today, and it is likely that this influence will continue into the next century and beyond.

LOOKING BACK

Forest management has become a political issue in North America in which social and political sciences sometimes exert as much influence in forest management decisions as the biological sciences. Healthy forests provide an abundance of forest products and perform biological functions that contribute to the comfort and health of all living creatures. They renew the atmosphere, provide clean water, and contribute to the economic stability of nations. They also provide energy sources for a variety of uses, and forest environments are popular sites for human recreation. The multiple-use concept of management allows forests to be used by many different interest groups, and it is widely practiced in the national forests. Forest policy was strongly influenced by the first administrator of the United States Forest Service, Gifford Pinchot.

QUESTIONS FOR DISCUSSION AND REVIEW

Essay Questions

1. List some important products obtained from forests that contribute to the health and comfort of people.

2. How do forest products contribute to the economy of a nation?

3. What kinds of plants make up the vegetative strata of a forest environment?

4. Name some of the life forms that contribute to the biological value of a forest.

5. Identify some naturally occurring forest functions that contribute to the biological value of a forest.

6. Describe how a watershed functions, and explain why a forested watershed is superior to a watershed that lacks forest plant cover.

7. Explain how a healthy forest environment contributes to stable populations of wild animals.

8. Distinguish between renewable resources and nonrenewable resources.

9. What are some major uses of forest products in the United States?

10. How are wood and other biomass materials converted to energy sources?

11. Explain the multiple-use concept of management for lands owned by the public.

Multiple-Choice Questions

1. What percentage of the land area in the world is forest land?
   A. 20%
   B. 30%
   C. 40%
   D. 50%

2. Broadleaf trees generally represent which of the following strata in a forest?
   A. canopy
   B. herb layer
   C. understory
   D. shrub layer

3. A measurement that takes into account the living organisms in a forest and natural forest functions is called:
   A. forensic value
   B. geological inventory
   C. economic value
   D. biological value

4. A watershed is a:
   A. region in which precipitation is readily absorbed into the soil
   B. water treatment facility in which pollutants are removed from water
   C. region where water has great difficulty penetrating into the soil
   D. municipal water storage structure

5. A forest is considered to be a natural resource that is:
   A. permanent
   B. nonrenewable
   C. renewable
   D. expendable

6. Which of the following uses consumes the greatest amount of harvested timber?
   A. plywood
   B. paper products
   C. lumber
   D. fuelwood

7. The production of lumber in the United States during the last century has:
   A. increased slightly
   B. remained constant
   C. decreased
   D. tripled

8. The sale of timber products accounts for what percentage of the gross national product (GNP) in the United States?
   A. 6%
   B. 2%
   C. 11%
   D. 21%

9. How much of the annual harvest of hardwood trees is used for fuelwood?
   A. one-tenth
   B. one-fourth
   C. two-thirds
   D. nearly half

10. Electricity that is produced using harvested plant materials as fuel to produce heat is called:
    A. hydropower
    B. nuclear power
    C. biomass power
    D. induced power

11. A forest management strategy that provides public access to forest lands for such activities as grazing, mining, logging, and recreation is called:
    A. conservation reserve program
    B. resource isolation doctrine
    C. selective resource management
    D. multiple use

LEARNING ACTIVITIES

1. Conduct an inventory of forest products that are used in your school classroom and laboratory facility. Then have each student conduct a forest products inventory in his or her home. Create a master list of all forest products that were identified by class members. Use the list to design a classroom bulletin board that focuses on forest products.

2. Visit a park or forested area near your school. Make a chart that lists "renewable resources" and "nonrenewable resources" as main topic headings. List each natural resource that was observed on the field trip under one of these headings. Call on your students to explain why a particular natural resource should be listed in the category he or she has chosen.

Chapter

2

Terms to Know

biological succession
primary succession
secondary succession
pioneer
pioneer species
softwood
hardwood
climax community
conifer
evergreen
boreal forest
deciduous
wilderness
tannin
diameter at breast high (dbh)
butt rot
coastal plain
reforestation
regeneration
silviculture
aquatic
adventitious root
alluvial fan
topography

North American Forest Regions

Objectives

After completing this chapter, you should be able to

❈ name the regional forests of North America as they are described in the text

❈ list the most important species of trees in each of the regional forests

❈ explain the principle of biological succession

❈ describe the distinguishing features between trees classed as conifers, deciduous trees, and evergreen trees

❈ identify some characteristics of the Northern Coniferous Forest region that account for the relatively low production of this forest

❈ name some important forest products in addition to wood and paper

❈ explain why it is usually important to harvest trees in a timely manner once they become mature

❈ define silviculture and give some examples of silviculture practices

❈ speculate on the process by which the geological feature known as an alluvial fan was formed

❈ identify some characteristics of the Pacific Coast Forest region that account for the relatively high production of this forest

orest boundaries do not follow strict geographic lines, and forest types seldom attain total dominance within a region. The varieties of trees that are found in a region today are often different from the varieties that existed there when European settlers first arrived in North America. Tree varieties in a region are impacted by soil characteristics, drainage, climate, and elevation. In some instances, entire forest regions have been planted to tree varieties that are not native plants. Keep these things in mind as general the characteristics and boundaries of the forest regions of North America are described in this chapter.

The forest regions of North America have been divided in scientific literature in a number of different ways. Professional foresters are quick to point out that forests are dynamic, meaning that they are always changing. Plant species change as forest environments change. The forest regions discussed in this chapter represent current thinking by leading educators and practicing professionals.

America was a land of virgin forests when the first European settlers arrived. Today, only a few virgin forests remain in North America. Most of our forests have been harvested at least two or three times since the European colonists first arrived (Figure 2–1). The land in many regions was cleared for farming as people left the Atlantic Coast settlements and moved westward. In regions where the soil was poor, farms were later abandoned and allowed to return to native species of plants. In such instances, nature has taken over, and a process called **biological succession** has begun. This process involves natural changes that occur as higher-order plants such as trees replace lower-order plants such as weeds, grasses, and shrubs.

Figure 2–1 North American forests are truly valued resources, and many forests have been harvested two or more times since America was colonized.

Biological Succession

Biological succession, sometimes called ecological succession, is an important process in forest recovery. Two forms of biological succession are known to occur. **Primary succession** occurs when organisms become established in an area where they did not exist before. **Secondary succession** occurs when an environment has been modified or damaged, and the changed environment will support only those organisms that are found naturally in an earlier stage of biological succession. Secondary succession happens when abandoned farms revert back to forests. The first plants to grow naturally in a burned or cleared area are called **pioneers** or **pioneer species.** These plants tend to be weeds, thistles, shrubs, bushes, and scrubby or inferior trees.

Pioneers are followed by two classes of trees known as **softwoods** and **hardwoods.** The most dominant species for the region eventually takes over in the process of biological succession. The plants that occupy an environment when succession of species is complete and plant populations become stable are known as a **climax community.** It is evident that different types of trees become dominant in the different regions. The dominance of one species of trees over another often can be accounted for by the elevation and climatic conditions of an area.

When the plants that populate a forest are damaged or destroyed, the environment in the region moves backward to an earlier stage of biological succession. For this reason, many of the forested areas of North America contain different types of trees than those that were found there before they were harvested. Forest managers may attempt to minimize this condition by planting the types of trees in harvested areas that represent advanced stages of biological succession (Figure 2–2).

Figure 2–2 The native cedar forests that are found in some regions of the world represent an advanced stage of biological succession. *(Photo courtesy of the College of Forestry, Wildlife and Range Sciences, University of Idaho)*

Figure 2–3 Trees known as conifers are named for the cones in which the seeds are produced.

NORTHERN CONIFEROUS FOREST

A **conifer** is a tree or shrub that produces cones containing seeds (Figure 2–3). Most of the conifers are also **evergreen** trees that bear green leaves all year. The Northern Coniferous Forest is sometimes known as the **boreal forest** because it is located in the northern zone of the continent (Figure 2–4). The region is characterized by many swamps, marshes, rivers, and lakes, and the climate is cold despite elevations that are generally low. It extends to the woodland transition zone that separates the forest lands from the tundra regions of the extreme north. The southern boundary of the boreal forest extends across southern Canada from eastern Quebec province, westward along the northern end of Lake Superior, then northwest across Alberta and British Columbia. It is the largest forest in North America.

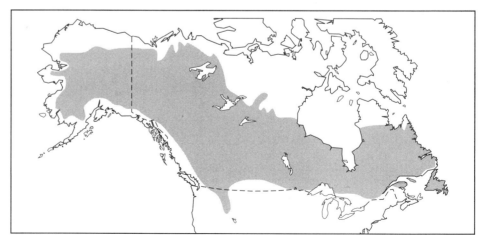

Figure 2–4 The boreal forest is located in the northern zone of the North American continent.

The most dominant types of trees of the Northern Coniferous Forest are the evergreens. In comparison with other forest regions, only a few species of forest plants are found in this region. Spruce trees are the dominant coniferous species, and broadleaf forest varieties consist mostly of birches, poplars, and willows. The Tamarack is a **deciduous** conifer because it sheds its leaves during part of each year. The Tamarack species is scattered through much of the Northern Coniferous Forest along with a few other minor species of trees.

White Spruce
(Picea glauca)

TREE PROFILE

The White Spruce is also known as the Canadian Spruce and the Skunk Spruce (Figure 2–5). It is the most important commercial tree in Canada. This tree is valuable for lumber and pulpwood production, and it produces a high-quality wood that is used to make musical instruments such as piano sounding boards, violins, and guitars. It thrives in northern climates at elevations ranging from 2,000–5,000', and it grows well in a variety of soil types.

Figure 2–5 The White Spruce is also known as the Canadian Spruce, and it is the most important commercial tree in Canada. *(Photo courtesy of Edward F. Gilman)*

Characteristics	
Height:	40–100'
Diameter:	1–2'
Needles:	evergreen, .5–.75" long, blue-green color, stiff and sharp, four angles on each needle
Cones:	1.25–2.5" long

TREE PROFILE

Jack Pine
(Pinus banksiana)

The Jack Pine is a pioneer species that populates an area following fire or logging activities. Until recently, it was considered to be a "forest weed," but it has become important as a pulpwood source for the paper industry (Figure 2–6). It is also known to provide habitat for a rare bird called the Kirtland's Warbler that is found only in the north-central region of Michigan in dense stands of young pine trees. The range of this tree extends through parts of the Northern Coniferous and the Northern Hardwoods forests.

Characteristics	
Height:	30–70'
Diameter:	up to 1'
Needles:	evergreen, two per bundle, stiff
Cones:	1.25–2"

Figure 2–6 Jack Pine is a pioneer species that has become important as a source of pulpwood. *(Photo courtesy of Edward F. Gilman)*

Three conditions that exist in the boreal forest that limit production of forest products are the short growing season, relatively poor and infertile soils, and poor drainage in surface and subsoils. Despite these factors, the vastness of this forest region contributes to a huge harvest of timber products from the Northern Coniferous Forest. Much of the land that has been designated by law as wilderness exists in this forest region.

Figure 2–7 A wilderness experience is one of the great experiences of a lifetime for people who enjoy primitive landscapes and wild country. *(Photo courtesy of John Fisher)*

Wilderness is land that is managed and protected as a wild uninhabited territory (Figure 2–7). Only limited human intervention is allowed in this environment. An abundance of wildlife ranging in size from insects to the moose is native to the region. For this reason, the early fur trade contributed to the settlement of humans in this part of North America. Migratory waterfowl in large numbers continue to use this region for nesting and summer range (Figure 2–8).

Figure 2–8 Migratory waterfowl depend upon the forest waterways for breeding grounds and resting areas.

Tree Profile

Black Spruce
(Picea mariana)

The Black Spruce is also known as Bog Spruce or Swamp Spruce (Figure 2–9). It is distributed throughout much of the boreal forest, especially in wet soils and bogs. It is smaller than the White Spruce, limiting its usefulness in lumber products; however, it is abundant in the region and it is widely used for paper production.

Figure 2–9 Black Spruce is a water-loving tree that grows in swamps and bogs. It is widely used as a raw material for paper production. *(Photo courtesy of Edward F. Gilman)*

Characteristics	
Height:	20–60'
Diameter:	4–12"
Needles:	evergreen, .25–.65" long, blue-green color, stiff and sharp, four angles on each needle
Cones:	.65–1.25" long

Tree Profile

Tamarack
(Larix laricina)

Common names for the Tamarack include Hackmatack and Eastern Larch (Figure 2–10). It is a deciduous tree that sheds its leaves each fall. The Tamarack is also a conifer. It is an important tree in the production of lumber for construction, railroad cross-ties, pulpwood, and poles. This tree is very adaptable in that it grows well in wet, swampy areas, but it also thrives on loam soils in upland regions.

Characteristics	
Height:	40–80'
Diameter:	1–2'
Needles:	deciduous, .75–1" long, soft, slender, three angles on each needle
Cones:	.5–.75" long

Figure 2–10 The Tamarack is a deciduous tree that sheds its leaves each fall. It is also a conifer that produces seeds in cones. *(Photo courtesy of Edward F. Gilman)*

NORTHERN HARDWOODS FOREST

The Northern Hardwoods Forest region reaches from southeastern Canada through New England to the northern Appalachian Mountains (Figure 2–11). It blends in with the Northern Coniferous Forest on the northern border and the

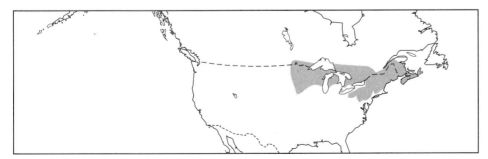

Figure 2–11 The Northern Hardwoods Forest

Central Broad-leaved Forest on the south, extending westward beyond the Great Lakes. In addition to the trees of the Northern Coniferous Forest, this forest region is now populated by a number of important hardwood species. Among these are the beech, maple, hemlock, and birch trees.

One hundred years ago, the most abundant tree species in the Northern Hardwoods Forest was the American Chestnut tree. It was used for the commercial production of chestnuts and wood products. A product called **tannin** was extracted from the wood of this tree. Tannin is also called tannic acid, and it is used in tanning hides to produce leather. It is also used in medicines. The

---

TREE PROFILE

Yellow Birch
(Betula alleghaniensis)

The Yellow Birch is a large hardwood tree that is sometimes called Gray Birch or Silver Birch (Figure 2–12). It grows best in upland regions with moist, cool climates. It is one of the most valuable of the hardwood trees. This tree should be harvested when tree diameters reach 19–20".

Characteristics	
Height:	70–100'
Diameter:	30"
Bark:	silver-gray or shiny yellow
Leaves:	3–5" long
Cones:	.75–1.25" long

Figure 2–12 Yellow Birch is also called Gray Birch or Silver Birch. It is a large tree that is important commercially. (*Photo courtesy of Edward F. Gilman*)

American Chestnut tree was once the main source of this important industrial chemical. Today, the chestnut tree is gone from many parts of this forest except for scattered tree shoots that arise from old chestnut tree roots. The parasitic fungus called the chestnut blight is still present today in this forest, and it eventually kills many of the chestnut trees of the American variety that are found there.

American Beech

(Fagus grandifolia)

The American Beech is a large shade tree that grows best in fertile upland soils where adequate moisture is available (Figure 2–13). It also thrives on well-drained lowland soils. The American Beech produces edible nuts called beechnuts that provide food for wildlife such as birds, squirrels, bears, and other mammals.

Characteristics	
Height:	60–80'
Diameter:	1–2.5'
Bark:	smooth, gray
Leaves:	two rows, spreading, 2.5–5" long, elliptical, pointed tip

Figure 2–13 The American Beech is a large tree that produces edible nuts (beechnuts). These nuts are important sources of food for birds, squirrels, bears, and other mammals. *(Photo courtesy of Edward F. Gilman)*

TREE PROFILE

Quaking Aspen
(Populus tremuloides)

The Quaking Aspen is distributed more widely in North America than any other tree (Figure 2–14). Quaking Aspen are found in abundance at high elevations from the northern to the southern Rocky Mountains and throughout the Northern Coniferous Forest. They have been used in the past as a source of fuel, but the greatest potential use of these trees may be in the pulpwood industry. These trees are propagated by both seeds and root sprouts.

Figure 2–14 The Quaking Aspen is the most widely distributed tree in North America. *(Photo courtesy of Edward F. Gilman)*

Characteristics	
Height:	40–70'
Diameter:	1–1.5'
Leaves:	1.25–3" across, nearly round, shiny above, dull beneath
Bark:	white color, thin and smooth except on old trees where it becomes thick with dark-colored ridges

ECOLOGY PROFILE

Loss of the American Chestnut Tree

One of the greatest forest disasters ever to occur in North America took place in the Northern Hardwoods Forest early in the twentieth century. It is believed that the fungus *Cryphonectria parasitica* was introduced from Asia to New York on a shipment of nursery stock. It was first observed in New York in 1904, and by the 1930s, a major epidemic was devastating the forests of the region. Forty years after it was first observed in North America, the chestnut blight had destroyed the American Chestnut tree in its native forests. Young trees continue to arise from the old stumps and root systems that remain in this forest region, but they seldom survive more than ten years before they are killed by the disease. Some recovery of this species has been reported, particularly in areas that are isolated from the disease. Populations of these trees are increasing in the western states region and other areas where the disease does not exist.

Red Pine

(Pinus resinosa)

TREE PROFILE

Red Pine, also called Norway Pine, is a large tree found in mixed forests from Southeast Manitoba to Nova Scotia, Canada, extending south to Pennsylvania and west to Minnesota (Figure 2–15). The wood is used for construction, pulpwood, and planed lumber. It is also an ornamental shade tree in parks and yards.

Figure 2–15 Red Pine is a versatile tree that is valuable for lumber and as a shade tree in parks and yards. *(Photo courtesy of Edward F. Gilman)*

Characteristics	
Height:	70–80'
Diameter:	1–3' or larger
Needles:	evergreen, 4.25–6.5" in length with two slender needles per bundle

Figure 2–16 A dbh measurement of 19–24" is a good indicator of maturity in a hardwood tree.

Hardwood trees as a group tend to mature by the time they reach 19–24" in **diameter at breast height.** This measurement is known as **dbh** (Figure 2–16). On average, trees of this size begin to decline in health due to rotting of the interior wood near the base of the tree. This condition is called **butt rot.** Trees should generally be harvested before this condition becomes widespread. When harvest of mature trees is delayed, the health of the forest rapidly declines. Eventually the weakened trees die, and they are blown down by the wind. Large amounts of combustible material on the forest floor set the stage for massive forest fires in the later stages of this cycle (Figure 2–17).

Figure 2–17 Overmature trees contribute to an abundance of fuel on the forest floor, setting the stage for massive fires in the later phases of the maturity cycle.

European Alder

(Alnus glutinosa)

The European Alder is a large tree that was introduced from Europe as a shade or ornamental tree (Figure 2–18). Young leaves and twigs are covered with gummy material, and the leaves are elliptical to round in shape with a double saw-toothed edge, a shiny green upper surface, and light green beneath with tufts of hair. This tree produces clusters of three to five small, black, gummy cones. It grows best in wet soils located in cool temperate zones with high humidity.

Figure 2–18 The European Alder is a foreign tree that was introduced as an ornamental or shade tree.

Characteristics

Height:	50–70'
Diameter:	1–2'
Bark:	smooth when young, and forming broad plates at maturity

The Northern Hardwoods Forest is heavily impacted today by people. Cities and towns abound in most of this region; however, substantial forest areas remain. Recreation is an important activity in state and national forests. Traditional management practices are often challenged in the courts by people who want the forests to remain unchanged. Unfortunately, the life cycle of a tree continues to advance, and the mature trees in a favorite campground become old and dangerous. Forest recreation areas are generally managed to maintain trees of mixed ages. This allows for unhealthy trees to be replaced by young healthy trees without destroying the scenic and recreational value of the area.

TREE
PROFILE

Sugar Maple

(Acer saccharum)

The Sugar Maple is a large tree that is sometimes called Hard Maple or Rock Maple (Figure 2–19). It grows in moist soils in the valleys and uplands of the northeastern region of the United States. The wood from this tree is used in hardwood flooring and to construct fine-quality furniture. The sap of these trees is collected in the spring and boiled down to make maple syrup. Up to 60 gallons of sap are sometimes collected from a single tree. Once the sap is boiled to remove excess water, 32 gallons of sap are concentrated to make only about one gallon of maple syrup.

Characteristics	
Height:	70–100'
Diameter:	2–3'
Leaves:	palmately lobed with five long pointed lobes, 3.5–5.5" long and wide

Figure 2–19 The Sugar Maple is important as the source of maple syrup, which is made from the sap. It also produces high-quality wood that is used for flooring and furniture. *(Photo courtesy of Edward F. Gilman)*

Eastern White Pine
(Pinus strobus)

The Eastern White Pine is also called White Pine or Northern White Pine in some of the locations where it occurs (Figure 2–20). It is the largest conifer in the northeastern region, and it used to be the most valuable tree. For many years it was selectively harvested, and it is no longer as abundant as it was at one time. It has become an important plantation tree that is popular with owners of private forest land. This tree is straight and tall, and the wood products that are obtained from it are used by the millwork, construction, and pulpwood industries.

Figure 2–20 Eastern White Pine is the largest conifer in the northeastern region of America. Selective harvesting has reduced the population of this species, but it has become important as a plantation tree. *(Photo courtesy of Edward F. Gilman)*

Characteristics

Height:	100'
Diameter:	3–4'
Needles:	evergreen, 2.5–5" long, five per bundle, blue-green
Cones:	4–8" long

Tree Profile

Eastern Hemlock

(Tsuga canadensis)

The Eastern Hemlock is an evergreen tree that is sometimes called Canada Hemlock or Hemlock Spruce (Figure 2–21). It grows best in acid soils in mountains and valleys at elevations ranging from 2000–5000'. This species is important for timber production despite the fact that the wood of this tree is somewhat brittle and weak. This limits its use and detracts from its value. The bark is a commercial source of tannin, which is used to cure hides in the production of leather.

Figure 2–21 The Eastern Hemlock is important for timber production, and its bark is a source of tannin, which is a chemical that is used to make leather from hides of animals. *(Photo courtesy of Edward F. Gilman)*

Characteristics	
Height:	60–70'
Diameter:	2–3'
Needles:	evergreen, .375–.625" long, flexible, flat
Cones:	.625–.75"

CENTRAL BROAD-LEAVED FOREST

The Central Broad-leaved Forest is an arbitrary grouping of several distinctly different forest subgroups. Much of the land in the region has been cleared for farming, and some of the most fertile and productive agricultural land in the world is located here (Figure 2–22). It is a region with few federally owned forests. Forest lands in this region are generally small and privately owned. High-quality wood from the region is used in large amounts to construct hardwood furniture. Wood of lesser quality is used by the construction industry and for building industrial pallets.

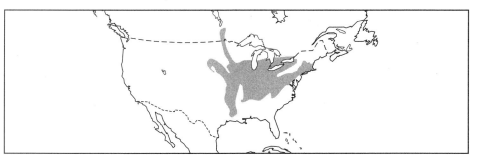

Figure 2–22 The Central Broad-leaved Forest

American Elm

(*Ulmus americana*)

TREE
PROFILE

The American Elm tree is a large shade and commercial tree that ranges across the eastern half of the United States (Figure 2–23). The wood from this tree is valuable for making furniture and wood paneling, but it is also used to make strong shipping containers. Many of these trees have been killed by a disease called "Dutch Elm disease" since the fungus that causes the disease was accidentally brought to this continent.

Figure 2–23 The American Elm tree is widely used for wood paneling and furniture construction, and it has been used as a shade tree on many city streets. *(Photo courtesy of Edward F. Gilman)*

Characteristics

Height:	100'
Diameter:	4' or larger
Leaves:	two rows, 3–6" long, 1–3" wide, elliptical in shape with double saw-toothed edges

The most abundant and valuable species of trees in the Central Broad-leaved Forest are the oak trees. Yellow-poplar has also become an important tree in this region. These trees have prospered in the area since the chestnut blight

---

TREE PROFILE

Sweetgum
(Liquidambar styraciflua)

The Sweetgum tree is a large tree with a straight trunk that is sometimes called the Redgum or the Sapgum (Figure 2–24). Next to the oak trees, it is the most important hardwood timber tree. The wood from this tree is used to make plywood and veneers, furniture, cabinets, boxes, barrels, and even pulpwood. A sweet resin can be obtained by peeling the bark on a live tree and collecting the "gum." This resin was used for medicines and chewing gum by early settlers.

Figure 2–24 Sweetgum is the second most important hardwood tree, ranking behind the oaks. The resin from this tree was used for chewing gum and medicine by early settlers. *(Photo courtesy of Edward F. Gilman)*

Characteristics	
Height:	60–100'
Diameter:	1.5–3'
Leaves:	3–6" long and wide, five or seven fingers, shiny green above, five main veins

destroyed the chestnut trees that once dominated the landscape. Many other species of trees are also found here. They include maples, hickory, black walnut, ash, sweetgum, yellow-poplar, elm, beech, and many others.

Sycamore

(Platanus occidentalis)

The Sycamore is one of the largest of the eastern hardwood trees (Figure 2–25). It ranges through much of the Central Broad-leaved Forest, extending beyond its borders. Its most productive uses include pulpwood, particleboard, and fiberboard. It is also a shade tree whose trunk grows larger than any other hardwood tree in North America.

Figure 2–25 The Sycamore is a large shade tree that is often used as pulpwood and fiberboard. *(Photo courtesy of Edward F. Gilman)*

Characteristics	
Height:	60–100'
Diameter:	2–4'
Leaves:	4–8" long and wide with three to five short-pointed lobes
Bark:	dark brown with deep furrows and wide, scaly ridges

A recent trend on privately owned land is to replace hardwood trees on the least fertile soils with pine tree plantations. Pine trees tend to do well on such soils, and the rotation time between harvests is much shorter. The growth rate of pines is also much greater than that of hardwoods.

TREE
PROFILE

Green Ash
(Fraxinus pennsylvanica)

The Green Ash is a moderate-sized tree with a dense, rounded crown (Figure 2–26). The foliage is shiny green with pinnately compound leaves in an opposite arrangement. The range of the Green Ash covers much of the eastern half of the United States, extending north into Canada. This tree has been successfully used in shelterbelts in the Great Plains region, and it makes a good shade tree.

Characteristics	
Height:	60'
Diameter:	1.5'
Leaflets:	2–5" long and 1–1.5" wide
Bark:	gray with scaly ridges

Figure 2–26 The Green Ash is an important shelterbelt tree, and it is also widely used as a shade tree. *(Photo courtesy of Edward F. Gilman)*

The Central Broad-leaved Forest is probably more impacted by people than any other forest in North America. The greatest challenge to the forests of the region undoubtedly occurred when much of the land was cleared of trees to establish farms. Most of this activity has occurred in the past one hundred and

Black Oak

(Quercus velutina)

The Black Oak is known in some regions as Yellow Oak or Quercitron Oak (Figure 2–27). It can be identified by peeling the outer bark, exposing yellow or orange inner bark. The inner bark has been used as a source of tannin for curing leather. Other products obtained from the bark are medicinal remedies and yellow dye. Favorable habitat for this tree is found on dry soils located on ridges and hillsides.

Figure 2–27 The Black Oak produces important products such as yellow dyes and medicines. It also produces tannin for making leather. *(Photo courtesy of Edward F. Gilman)*

Characteristics	
Height:	50–80'
Diameter:	1–2.5'
Leaves:	4–9" long, 3–6" wide, seven to nine lobes, shiny green above, yellowish and brown hairs beneath
Bark:	gray and smooth

TREE PROFILE

Yellow-poplar
(Liriodendron tulipifera)

The Yellow-poplar is a large tree that is also known as the Tuliptree (Figure 2–28). The most favorable habitat for this tree is on well-drained, moist soil. The range of the Yellow-poplar extends over much of the eastern region of the United States. It is an important commercial hardwood that is used for such purposes as construction of furniture, packing crates, pulpwood, and even musical instruments.

Characteristics	
Height:	80–120'
Diameter:	2–3'
Leaves:	3–6" in length and width
Bark:	thick, furrowed, dark gray

Figure 2–28 The Yellow-poplar or Tuliptree is an important commercial hardwood tree found in the eastern region of the United States. *(Photo courtesy of Edward F. Gilman)*

fifty years. A different kind of challenge from human activities exists in the forests that remain today. That challenge is the use of forests for recreation purposes (Figure 2–29). Many forest lands under both public and private ownership are impacted by recreation activities. People need outdoor recreation areas, but sometimes recreation takes precedence over management practices in our forests.

Figure 2–29 Forests are used by large numbers of people each year for recreation purposes such as hiking and sight-seeing.

Black Walnut

(Juglans nigra)

The Black Walnut is a valuable hardwood tree that has become rather scarce in native forests (Figure 2–30). Its range extends west from the Atlantic coast to the central states, with most of these trees located in the Central Broad-leaved Forest. The wood has a distinctive dark color, making it especially valuable for veneers, furniture, and gunstocks. It produces edible nuts, and a black dye is produced from the husks.

Characteristics

Height:	70–90'
Diameter:	2–4'
Leaves:	pinnately compound with leaflets 2.5–5" long
Bark:	dark brown with scaly ridges and deeply furrowed

Figure 2–30 Black Walnut is one of the most rare and valuable woods in North America. *(Photo courtesy of Edward F. Gilman)*

Forestry professionals are often criticized for their forest management practices, especially on state and federally owned lands. Sometimes they are even prevented from applying proven forest management practices that are science-based when people perceive that the practices might interfere with recreation in the area. This occurs because the people believe they have the right as citizens to use public lands. Perhaps the greatest impact of people on forests will be due to litigation in the courts of law over forest management practices.

SOUTHERN FOREST

The Southern Forest extends inland along the Atlantic coast and the Gulf of Mexico from North Carolina to Texas (Figure 2–31). A humid subtropical climate exists in this region with precipitation averaging 40–60" evenly distributed throughout the year. The combination of high temperatures, abundant moisture, and a long growing season provides ideal growing conditions for trees. Occasional droughts of several weeks duration do occur in this region, however, weakening trees and making them susceptible to insect and disease problems. Climatic conditions differ some in the region across the southwestern zone, the eastern Gulf coast, and the Atlantic coastal areas.

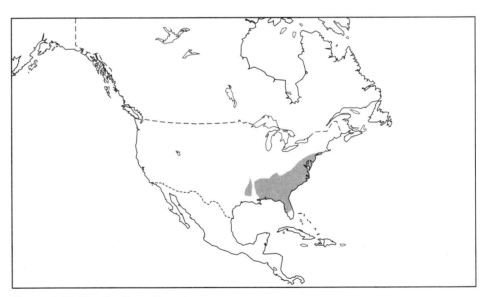

Figure 2–31 The Southern Forest

Loblolly Pine

(Pinus taeda)

The Loblolly Pine is also known as North Carolina Pine or Oldfield Pine (Figure 2–32). It is a fast-growing tree that is an important commercial pine tree in the Southern Forest. It is produced commercially for lumber and pulpwood in forest plantations. This tree grows well in wet soils (*loblolly* is another name for "mud puddle"), but it also grows well on slopes where the soils are well drained. Its range extends south from New Jersey to central Florida, and west to eastern Texas.

Figure 2–32 The Loblolly Pine is an important commercial tree in the Southern Forest for the production of lumber products and pulpwood. *(Photo courtesy of Edward F. Gilman)*

Characteristics	
Height:	80–100'
Diameter:	2–3'
Needles:	evergreen, three per bundle, 5–9" long

The **coastal plain** along the Atlantic and Gulf coasts consists of a series of terraces that generally run parallel to the coastline. Some of the lower areas tend to be poorly drained or even swampy. The soils in this area tend to be sandy, and the principal forest species include loblolly pine, slash pine, shortleaf pine, long-leaf pine, oak, and hickory. Several lesser varieties of pine, cedar, and hardwood are also found in this area.

---

*TREE
PROFILE*

Slash Pine
(Pinus elliottii)

The Slash Pine is also known as Swamp Pine or Yellow Slash Pine (Figure 2–33). It is a fast-growing southern pine that is important for lumber. It is grown near sea level in forest plantations ranging south from the coastal plain of South Carolina to southern Florida and west to southeastern Louisiana.

Characteristics	
Height:	60–100'
Diameter:	2–2.5'
Needles:	evergreen, two or three per bundle, 7–10" long

Figure 2–33 Slash Pine grows in the southern region near sea level. It is a fast-growing tree that is important for lumber production. *(Photo courtesy of Edward F. Gilman)*

Longleaf Pine

(Pinus palustris)

The Longleaf Pine is also known as Longleaf Yellow Pine or Southern Yellow Pine (Figure 2–34). This tree is a source of turpentine and resin that is obtained by tapping the tree. It is also a major source of poles, pilings, pulpwood, and lumber. Its range extends south from Virginia to Florida and west to eastern Texas at elevations generally below 600'.

Figure 2–34 Longleaf Pine is a source of turpentine and resins obtained by tapping the tree. It is also used for wood products. *(Photo courtesy of Edward F. Gilman)*

Characteristics	
Height:	80–100'
Diameter:	2–2.5'
Needles:	evergreen, three per bundle, 10–15" long

Inland from the coastal plains are the hills of the Piedmont region. Elevations range from 300–1200' above sea level. This narrow band of hill country extends from Maryland on the north to Georgia on the south. Soil erosion is a serious problem here. The United States Department of Agriculture has reported that more than 75% of the original topsoil has been lost from many

TREE PROFILE

Shortleaf Pine

(Pinus echinata)

The Shortleaf Pine is also known as Southern Yellow Pine or Shortstraw Pine (Figure 2–35). It is native to twenty-one southeastern states, and it is widely distributed in the Southern Forest. It is an important source of lumber, plywood, veneer, and pulpwood. Its range extends south from southeastern New York and New Jersey to northern Florida, west to eastern Texas, and north to Missouri at elevations up to 3,300'.

Characteristics	
Height:	70–100'
Diameter:	1.5–3'
Needles:	evergreen, two or three per bundle, 2.75–4.5" long

Figure 2–35 Shortleaf Pine is a native of southeastern and southern forests where it is important in the production of lumber, plywood, veneer, and pulpwood. *(Photo courtesy of Edward F. Gilman)*

regions of the Piedmont since the original trees were harvested. The remaining topsoils of the Piedmont tend to be somewhat acid, creating ideal conditions for the growth of pine trees. The Loblolly and Shortleaf pines along with some hardwoods are the predominant forest species in the area.

After the native trees were removed from the Southern Forest lands, the loam topsoil of the region eroded rapidly due to heavy rainfall and low

White Oak

(Quercus alba)

The White Oak is also known as Stave Oak because it is widely used to make high-quality barrels for storage of liquids (Figure 2–36). It is distributed in both upland and lowland areas through much of the eastern region of the United States in well-drained soils. It is an important white oak lumber tree, producing high-grade wood that can be used for nearly any purpose. It was widely used in colonial times for ship construction.

Figure 2–36 White Oak produces wood that is often used to make barrels for the storage of liquids. Its high-quality wood is useful for almost any purpose. *(Photo courtesy of Edward F. Gilman)*

Characteristics	
Height:	80–100'
Diameter:	3–4'
Leaves:	five to nine lobes, hairless, green above, white beneath
Bark:	light gray with scaly, loose plates or ridges

infiltration rates of water into the heavy clay subsoils. As the soils became damaged, many of the farms became unproductive, and they were abandoned or converted to commercial forest plantations. This allowed forests to grow once again in much of the region. The return of a population of forest plants to an area from which they have been previously removed is known as **reforestation** or **regeneration.** Management of forests and their environments to establish, cultivate, and promote the growth and harvest of trees for commercial purposes is called **silviculture.** Approximately 90% of the forest land in

---

TREE PROFILE

Southern Red Oak

(Quercus falcata)

The Southern Red Oak, sometimes known as Spanish Oak or Swamp Red Oak, is best adapted to loam soils in dry upland locations (Figure 2–37). Lumber from this tree along with that of Water Oak, Willow Oak, and Cherrybark Oak is marketed under the red oak label. Red oak lumber is the leading type of commercial oak because it is more abundant in our forests.

Characteristics	
Height:	50–80'
Diameter:	1–2.5'
Leaves:	4–8" long, 2–6" wide, one to three curved lobes, shiny green above
Bark:	dark gray, broad ridges or plates

Figure 2–37 The Southern Red Oak produces the leading type of commercial oak wood, known as red oak. *(Photo courtesy of Edward F. Gilman)*

the Southern Forest region is privately owned, and the practice of silviculture in this region is widespread.

Hardwood trees are abundant in the Southern Forest, and more of these are oak than any other kind of wood. Many different species of oak are known,

---

Black Cherry

(Prunus serotina)

The Black Cherry tree is widespread east of the Mississippi River, and it is a valuable native tree (Figure 2–38). It produces small, black, edible cherries, but its value is in the wood it produces. Cherry wood is used to make the bodies of musical instruments because of its beauty and durability. It is also used to make beautiful furniture and wood paneling. A well-known medicine, wild cherry cough syrup, is obtained from the bark of the tree.

Characteristics

Height:	80'
Diameter:	2'
Leaves:	elliptical, 2–5" long, 1.25–2" wide

Figure 2–38 The Black Cherry is a widely spread native tree in locations east of the Mississippi River. Its beautiful wood is used to make the bodies of musical instruments, furniture, and wood paneling. *(Photo courtesy of Edward F. Gilman)*

but they are all divided into two groups. The white oak group of trees has heartwood that is brownish or tan, and freshly cut wood does not have an unpleasant odor. The red oak group has reddish heartwood, and a distinctly unpleasant odor comes from freshly cut wood.

Two areas exist in the Southern Forest region that are known as the Interior Highlands. The Ozark Plateaus are located in the Boston Mountains north of the Arkansas River. They rise 300–600' above the lowlands that surround them on all

---

TREE PROFILE

Pignut Hickory
(Carya glabra)

The Pignut Hickory is also known as the Pignut or Smoothbark Hickory (Figure 2–39). It is widely distributed in upland hardwood forests. Its range extends south from Ontario to New England, down the Atlantic coast to central Florida, west to eastern Texas, and north to Illinois. It is an important timber resource, especially in the southern Appalachian Mountain region. The wood is in demand for tool handles and skis where toughness and flexibility are needed.

Characteristics	
Height:	60–80'
Diameter:	1–2'
Leaves:	6–10" long, pinnately compound, lance-shaped, hairless

Figure 2–39 Pignut Hickory is a tough, yet flexible wood that is often used to make tool handles and skis. *(Photo courtesy of Edward F. Gilman)*

sides with some areas elevated as high as 2,200'. The soils of this area have developed from limestone and dolomite parent materials.

The Interior Highlands also include the Ouachita Mountains. They are located near the state line between Arkansas and Oklahoma, and they range in elevation from 500–2,600'. The ridges and valleys have an east-west orientation with shallow, stony, and generally infertile soils formed from sandstone and shale. The dominant tree species in the Interior Highlands region are oak and hickory in the highest elevations along with significant populations of Loblolly and Shortleaf pines. Populations of other hardwood tree varieties are found at the lower elevations. They include black walnut, sugar maple, elm, sycamore, black cherry, white oak, and red cedar.

BOTTOMLAND HARDWOODS FOREST

The Bottomland Hardwoods Forest consists of the floodplains and swamps of the southern Mississippi Delta and the central and southern Atlantic and Gulf coastal regions (Figure 2–40). The trees in these areas are water-tolerant, meaning that they can live in environments that are flooded all or part of the time. Conifers and hardwood trees are found in the bottomlands, and they are important commercially for pulpwood, paneling, veneers, and lumber.

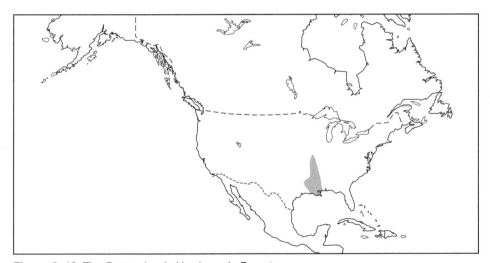

Figure 2–40 The Bottomlands Hardwoods Forest

Three conifers are found among the hardwood trees of this forest. They include Baldcypress, Atlantic White-cedar, and Pond Pine. A few of the most valuable hardwoods include the Sweetgum, Post Oak, Cherrybark Oak, Swamp Chestnut Oak, Pecan, Eastern Cottonwood, and Green Ash. Sometimes these trees grow in pure stands of a single species, but most often they are mixed.

TREE PROFILE

Baldcypress

(Taxodium distichum)

The Baldcypress tree, also known as the Cypress or Swamp-cypress, is a large **aquatic** tree, meaning that it is adapted to a water environment (Figure 2–41). This tree grows in areas where its roots are submerged in water most of the time. An unusual characteristic of this tree is the development of **adventitious roots** or prop roots. These roots grow downward from the main stem of the tree where they penetrate the soil to provide added support to the tree. The wood from this tree resists decay, and it is widely used in heavy construction.

Figure 2–41 The Baldcypress is an aquatic tree whose roots are often submerged in water. The wood of this tree is decay-resistant, making it an ideal wood for use in heavy construction. *(Photo courtesy of Edward F. Gilman)*

Characteristics	
Height:	100–120'
Diameter:	3–5'
Needles:	deciduous, .4–.75" long, single needles in two rows, flexible, dull green
Cones:	.75–1" long
Bark:	brown or gray, scaly ridges

Pecan

(Carya illinoensis)

The Pecan tree is important not only for the quality of the wood it produces, but also for the sweet-flavored nuts that it produces (Figure 2–42). It is a large tree that is best adapted to well-drained, moist soils located on river flood plains and valley floors. It is usually found in mixed hardwood forests, and the wood is used to make high-quality flooring, furniture, and veneer.

Characteristics	
Height:	100'
Diameter:	3'
Leaves:	pinnately compound, sickle-shaped leaflets, fine saw-toothed edges
Bark:	forked, scaly ridges, light brown to gray

Figure 2–42 The Pecan tree is an important source of nuts as well as high-quality wood for flooring, veneer, and furniture. *(Photo courtesy of Edward F. Gilman)*

Hardwood trees in bottomland forests are classified into three groups or forest types: (1) Cottonwood/Willow, (2) Cypress-Tupelo, and (3) Mixed Bottomland Hardwoods. Each of these forest types thrives in a particular environment, and the trees of each type are frequently found growing in close proximity to one another.

The Cottonwood/Willow forest type consists of pioneer species of trees that are among the first plants to become established on river bottom deposits of soil. Willow species tend to grow in dense stands in the wet lowland areas near the water. Cottonwoods become established above the water on the better-drained

---

TREE PROFILE

Eastern Cottonwood
(Populus deltoides)

The Eastern Cottonwood tree is also known as Carolina Poplar and Southern Cottonwood (Figure 2–43). It is one of the largest eastern hardwood trees, and its wood is used for plywood, crates, matches, and pulpwood. It grows in damp areas along streams and rivers in pure stands or mixed with willows. Its name comes from its small seeds that are attached to cottonlike fibers. The fibers blow in the wind, carrying the seeds to new locations. The Eastern Cottonwood is a fast-growing tree that averages 5' of vertical growth per year.

Figure 2–43 The Eastern Cottonwood is one of the largest eastern hardwood trees. It is used to make matches, crates, plywood, and pulpwood. *(Photo courtesy of Edward F. Gilman)*

Characteristics

Height:	100'
Diameter:	3–4'
Leaves:	3–7" long, 3–5" wide, triangular, shiny green color
Bark:	smooth, light green color

Biological Succession in Bottomland Hardwoods Forests

As biological succession progresses, willows are replaced in areas of standing water by such species as the Baldcypress and Water Tupelo. In shallow sloughs where water is present part of the time, other species such as Green Ash, Overcup Oak, and Water Hickory usually follow the willows. Cottonwoods are replaced in later stages of succession by hardwood species such as Silver Maple, Sugarberry, American Elm, American Sycamore, and others. These species are replaced in their turn by climax communities of Sweetgum and red oaks that eventually occupy the areas pioneered by the cottonwoods.

Water Tupelo

(Nyssa sylvatica)

The Water Tupelo, sometimes known as the Tupelo-gum or Cotton-gum tree, is a large aquatic tree that is often found in swamps or standing water with Baldcypress trees (Figure 2–44). Its fine-textured wood is used to make furniture, and its spongy roots can be used like cork as floats for a fishnet. It is seldom found very far from standing water or above 500' in elevation.

Figure 2–44 The Water Tupelo is an aquatic tree that produces fine-textured wood used to make furniture. *(Photo courtesy of Edward F. Gilman)*

Characteristics

Height:	100'
Diameter:	3'
Leaves:	5–8" long, 2–4" wide, tooth-shaped leaf structures
Bark:	dark brown to gray, ridges, scaly

soils. The willows and cottonwoods are generally short-lived species that grow rapidly, but they do not tolerate shade or competition from other species, and new stands seldom succeed older stands.

The Cypress-Tupelo forest type occupies habitats that are covered with water most of the year. The dominant tree species in these forests are Baldcypress, Water Tupelo, and Swamp Tupelo growing separately or in association with Redbay, Sweetbay, and Pondcypress. Stands of these trees usually require thinning to maintain optimal growing conditions. The rotation age for these trees ranges from seventy-five to one hundred and fifty years, depending on the species and growth conditions.

The Mixed Bottomland Hardwoods forest type occurs on streambeds and terraced areas consisting of old alluvial deposits of silt, sand, and gravel deposited by rivers and streams during periods of heavy spring waterflows. These areas change each time they are flooded, and over many years, they fan out to cover wide areas. For this reason, this kind of soil deposit is described as an **alluvial fan** (Figure 2–45).

As sedimentation progresses in a bottomland forest, biological succession of trees is affected. Sand, gravel, and silt deposits in streambeds eventually develop into terraces and ridges in locations where streams enter open plains and valleys. These areas are especially suited for the mixed hardwood trees that make up the intermediate and climax communities of plants in the later stages of biological succession.

The Mixed Bottomland Hardwoods forest type includes many species of trees, and management of these diverse kinds of trees consists mostly of thinning and improvement cutting. Thinning reduces the competition between trees, allowing them to grow more rapidly and to larger sizes. Competition for sunlight and nutrients can seriously stunt the growth of individual trees and reduce the production of high-grade timber for the entire forest. Improvement cutting removes problem trees and improves the health of the stand.

Figure 2–45 An alluvial fan is a soil deposit that has been created by soil components being suspended in water and settling out to the bottom when the water movement slowed down. An alluvial fan is associated with frequent flooding.

TROPICAL FOREST

The Tropical Forests of North America exist only on the southern tip of the Florida peninsula and in Mexico (Figure 2–46). Three types of tropical forests exist in the region. The characteristics that define tropical forest types are the amount and frequency of rainfall, and the **topography** or variations in elevations and surface features of the land.

The tropical rain forest type grows mostly along the Gulf coast where rainfall occurs frequently and where soil moisture is abundant. The tropical deciduous forest is located along the Pacific coast with occasional small pockets of this forest type located further east. Forests of this type are located at low elevations where a dry winter season usually prevails. The third tropical forest type is the oak and pine forest that is found at higher elevations than the other forests, and where a dry winter season is common.

Many species of trees are native to tropical forests. Scientists are anxious to protect these forests. It is widely believed that new and important medicines may still be extracted from plant species not yet studied and perhaps not even discovered.

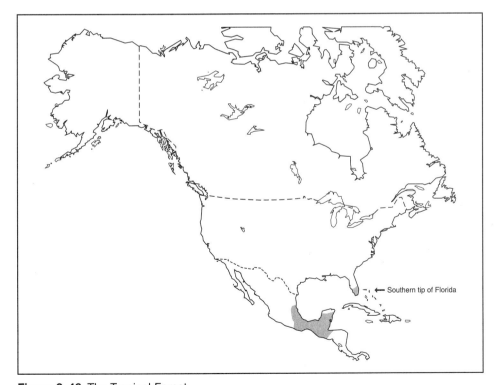

Southern tip of Florida

Figure 2–46 The Tropical Forest

TREE PROFILE

Black Mangrove

(Avicennia nitida)

The Black Mangrove tree is found in swamp forest areas from southern Florida to Brazil and Peru in South America. It is also called Blackwood or Mangle. Four mangrove species occur in North America; however, they become stunted and shrublike in their northernmost range. In tropical regions, these trees become much taller. These trees are easily killed by cold winters.

Characteristics	
Height:	40'
Diameter:	1'
Leaves:	evergreen, 2–4" long, salt crystals on surfaces
Bark:	brown or dark gray, smooth

ROCKY MOUNTAIN FOREST

The Rocky Mountain Forest stretches in a long band from British Columbia to southern Mexico (Figure 2–47). The eastern and western boundaries of this forest are dry unforested areas where natural precipitation is generally too low to sustain a forest. This forest is somewhat patchy even within its boundaries due to uneven distribution of moisture. The entire region is dry because most of the moisture that moves inland from the Pacific Ocean falls as precipitation in the coastal mountain ranges.

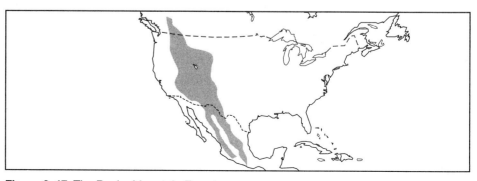

Figure 2–47 The Rocky Mountain Forest

The most numerous class of trees in the Rocky Mountain Forest is the pine in all its varieties. Pine trees are also the most important commercial trees. A strong paper industry exists in the region that depends mostly on the conifer forests. The vast lumber industry in the Rocky Mountain and Pacific Coast forests has declined in recent years due in large part to reduced and inconsistent timber harvests. The lack of a dependable supply of timber has resulted in the closing of many of the

Douglas-fir

(*Pseudotsuga menziesii*)

TREE PROFILE

The Douglas-fir is sometimes called Douglas-spruce or Oregon-pine, and it consists of two distinct varieties (Figure 2–48). Coast Douglas-fir usually grows in pure stands on well-drained soils. The Rocky Mountain Douglas-fir grows on rocky slopes in the mountains in single variety stands as well as in mixed coniferous stands. This is a very important timber tree in the United States. It ranks first in the volume of timber that is produced, total lumber production, and veneer production for plywood. The foliage and seeds are eaten by wildlife.

Characteristics	
Height:	80–200'
Diameter:	2–5'
Needles:	evergreen, two rows, .75–1.25" long, flexible, yellow-green or blue-green color
Cones:	2–3.5" long
Bark:	thick, reddish-brown, ridges

Figure 2–48 The Douglas-fir tree is the most important commercial tree in the United States, ranking first in timber volume, total lumber production, and veneer production for plywood. *(Photo courtesy of Edward F. Gilman)*

TREE PROFILE

Western Redcedar
(Thuja plicata)

The Western Redcedar is also known as Giant Arborvitae or Canoe-cedar (Figure 2–49). It is a very large tree that is resistant to rotting. It is the most important wood source for making shingles, but it is also widely used for siding, paneling, yard fences, posts, and patio construction (Figure 2–50).

Figure 2–49 Western Redcedar is a large, decay-resistant tree that is widely used to make shingles, siding, fences, and house decks. *(Photo courtesy of Edward F. Gilman)*

Characteristics	
Height:	100–175'
Diameter:	2–8'
Leaves:	evergreen, opposite in four rows, shiny dark green
Bark:	reddish-brown, fibrous
Cones:	.5" long

Figure 2–50 Rugged-looking cedar fencing materials are manufactured by splitting cedar logs to make rails and posts. *(Photo courtesy of Idaho Cedar Sales, Troy, Idaho)*

lumber mills that once operated in the area. In addition to timber harvests, other important functions of the Rocky Mountain Forest include protection of the watershed, livestock grazing, fish and wildlife habitat, and outdoor recreation.

Junipers and pinions are dominant species at the lower forest elevations. Junipers produce some wood that is used in making cedar chests, while the majority of it is used for fuel and fenceposts. Pinions produce wild nuts that are eaten by many species of birds and mammals. They are also gathered and sold commercially for human consumption. Deer depend heavily on all of these trees as they browse for winter food.

Western Larch

(Larix occidentalis)

The Western Larch, also known as the Hackmatack or Western Tamarack, is a large tree found in the hills and valleys on well-drained soil (Figure 2–51). This tree usually occurs in mixed stands with other conifers at altitudes of 2,000–6,000'. It is easily seen among the evergreen trees as the needles turn yellow and fall from the tree in the autumn. These trees are often found in burned areas either as survivors of a forest fire or as new growth. As a forest matures, this species is generally replaced by other cone-bearing trees.

Figure 2–51 Western Larch is also known as Western Tamarack or Hackmatack. It is a deciduous tree that often grows in mixed stands with evergreen trees. *(Photo courtesy of Edward F. Gilman)*

Characteristics

Height:	80–150'
Diameter:	1.5–3'
Needles:	deciduous, 1–1.5" long, grow in clusters, three angles, stiff
Cones:	1–1.5" long
Bark:	scaly, reddish-brown color, overlapping plates

Public ownership of land in the Rocky Mountain Forest is over 76%. This is a unique feature of this forest. No other forest region maintains federal and state ownership at this level. Most of the remaining private forest land is owned by farmers and commercial timber companies. A large number of state and federal parks and monuments are located in this forest region encompassing large tracts of forest lands that are reserved from commercial logging.

PACIFIC COAST FOREST

The Pacific Coast Forest is one of the most productive forests on the continent. Mixed varieties of trees abound in the region (Figure 2–52). A huge wood products industry has developed in the Pacific Northwest due to the abundance of high-quality timber. Production of lumber and paper products is one of the most important industries on the west coast. A decline in timber harvests in recent years, due mostly to litigation over environmental issues, has reduced the economic stability of the wood products industry and many of the communities that depend upon timber harvests for jobs.

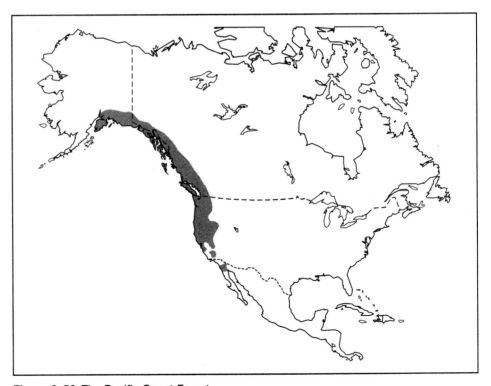

Figure 2–52 The Pacific Coast Forest

Two distinct climatic zones are found in the Pacific Coast Forest. The coastal mountains act as a barrier that captures precipitation in the form of rain or snow before it can move inland. East of the coastal mountains, the amount of annual precipitation is much lower than it is on the western slope of the mountains. This accounts for some of the differences in the two forest zones. Trees on the western slope grow taller and the forest stands are more dense than on the eastern

TREE PROFILE

Ponderosa Pine

(Pinus ponderosa)

Sometimes called the Western Yellow Pine or Blackjack Pine, the Ponderosa Pine is a very large tree that often grows in pure stands in high elevations on mountain terrain (Figure 2–53). It is the most common pine tree in North America, and it is widely distributed across several forest regions. Commercially, it is the most valuable western pine tree. The high-quality wood is relatively free from knots, and it is widely used to make door panels and window frames.

Figure 2–53 The Ponderosa Pine is a very large tree that grows in pure stands on mountainous terrain. Its wood is the most valuable of the western pine varieties because it is relatively free of knots. *(Photo courtesy of Edward F. Gilman)*

Characteristics

Height:	60–130'
Diameter:	2.5–4'
Needles:	evergreen, 4–8" long, two or three per bundle, dark green, stiff
Cones:	2–6" long
Bark:	scaly plates, yellow-brown color

slope of the Cascades. Total production of forest products is much higher in western Oregon and Washington than it is in eastern Oregon and Washington despite a much larger land area in the eastern forest zone.

Forest ownership in the Pacific Northwest differs somewhat in the different states. Private forest ownership is greater than state and federal ownership in Washington, but Oregon and California forests consist of more publicly owned forests than privately owned forests. While state and federal forest ownership is

---

TREE PROFILE

Lodgepole Pine
(Pinus contorta)

The Lodgepole Pine is also known as Tamarack Pine or Shore Pine (Figure 2–54). It is a widely distributed tree in the western regions of North America. It is adapted to high mountain environments as well as to lowland bogs. The trunk of this tree is slender in comparison with its height. The American Indians used this tree to construct their tepees, hence the name "lodgepole." It is native to the entire western region from Alaska to Mexico.

Figure 2–54 Lodgepole Pine is distributed widely in western forests. Its trunk is slender in comparison with its height, and it was used by Native American Indians to make the frames of their teepees. *(Photo courtesy of Edward F. Gilman)*

Characteristics

Height:	20–80'
Diameter:	1–3'
Needles:	evergreen, 1.25–2.75" long, two per bundle, flattened and frequently twisted
Cones:	.75–2" long
Bark:	light brown, scaly, thin

not as high in this region as it is in the Rocky Mountain Forest, public forest ownership is still very high. Public forests are often managed quite differently than privately owned forests due to the efforts of citizen groups to exert influence in the forest management process.

The most important species of trees in this forest are the conifers. The Douglas-fir is the most important commercial timber tree among the conifers. A few broad-leaved species are located in the lower valleys of Oregon and

Giant Sequoia
(Sequoiadendron giganteum)

TREE PROFILE

The Giant Sequoia tree is also known as the Sierra Redwood or Bigtree (Figure 2–55). It is a rare tree found on the western slope of the Sierra Nevada Mountains in central California. These trees are protected in public parks and in state and national forests in an effort to maintain the population of this ancient tree. The thick bark on these trees gives them resistance to fire.

Figure 2–55 The Giant Sequoia or Sierra Redwood is a very large evergreen tree found only in the Sierra Nevada Mountains of central California. The mature trees in this forest are more than one thousand years old. *(Photo courtesy of Edward F. Gilman)*

Characteristics	
Height:	150–250'
Diameter:	20'
Leaves:	evergreen, scalelike, overlapping, blue-green color
Cones:	1.75–2.75" long, maturing in two seasons
Bark:	reddish-brown, scaly, ridged, very thick.

California. Some of the most impressive trees on earth are the Giant Sequoias and the Redwoods. They are among the largest trees in the world, growing to heights of 250–325'; and they are among the oldest trees in the world with some individual trees showing annual ring growth for up to 3,200 years.

TREE PROFILE

Redwood

(Sequoia sempervirens)

The Redwood is also known as the Coast Redwood or the California Redwood (Figure 2–56). It is the tallest tree species in the world with the tallest tree measuring 368' tall. The lower trunks of these trees are much enlarged. Trees of this species mature at four to five hundred years. These trees are still logged commercially. Redwood lumber is resistant to decay, and it is used extensively to make house siding, decks, fences, and other outdoor structures.

Figure 2–56 The Redwood or California Redwood is the tallest tree species in the world, with the tallest tree measuring nearly 370' tall. The lumber is decay-resistant, and it is used for siding, decks, and fences. *(Photo courtesy of Edward F. Gilman)*

Characteristics

Height:	200–325'
Diameter:	10–15'
Leaves:	evergreen, some needlelike and some scalelike, sharp, dark-green color
Cones:	.5–1.1" long
Bark:	reddish-brown, very thick, scaly, ridged

PROFILE ON FOREST SAFETY

Dangerous and Poisonous Plants

The ability to recognize dangerous plants is important to those who work in the forests. It has been estimated that as many as one out of every hundred species of plants is poisonous to humans and animals. This is because they contain substances that cause harmful reactions when they are eaten or even when they are touched.

Among the substances that are most likely to injure or poison people are alkaloids, glycosides, oxalates, saponins, resinoids, and others. The poison may be distributed throughout the plant or concentrated in a particular part of the plant. For example, poison hemlock has poison distributed throughout the plant, while water hemlock accumulates poison in the roots. Some of the most common forest plants that are dangerous to people include Poison Ivy, Poison Oak, and Poison Sumac. These plants produce an allergic reaction when they are touched that can cause serious injuries to the skin.

Poison Oak and Poison Ivy have distinctive leaf patterns in which three leaflets are grouped together in each leaf (Figure 2–57). They also develop clusters of small white or yellowish berries in late summer or fall. One or the other of these two plants is found throughout southern Canada and the United States. These plants are often found in forests in the form of vines climbing up tree trunks or trailing over the surface of the forest floor. Poison Sumac is a tall shrub that is found in swampy areas in the eastern region of North America. The plant has pinnately compound leaves with several pairs of leaflets arranged opposite each other and a single leaflet at the tip (Figure 2–58).

Figure 2–57 Poison Oak and Poison Ivy are poisonous plants with a vinelike growth pattern. They cause serious skin irritation. *(Photo courtesy of Edward F. Gilman)*

PROFILE ON FOREST SAFETY

continued on page 76

PROFILE ON FOREST SAFETY — *continued*

Figure 2–58 Poison Sumac is a tall shrub with pinnately compound leaves arranged opposite each other and a single leaflet at the tip. *(Photo courtesy of Edward F. Gilman)*

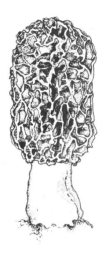

Figure 2–59 Morel mushrooms are edible, but they are sometimes confused with poisonous varieties. *(Illustration by Laurette Richin)*

The best way to prevent injuries from these plants is to recognize them on sight and to avoid touching them. The plant resin can be spread to humans from contaminated clothing, pets, or even tools. Once the poisonous resin has touched the skin, it should be immediately scrubbed off with soap and water. Symptoms of poisoning from these plants include reddening of the skin along with severe itching. This is followed by the emergence of water blisters, and the skin can become discolored. Calamine lotion or boric acid seems to give some relief from itching, but severe cases should be taken to a doctor for treatment. Nettles are also known to cause skin irritations, but they are not nearly as serious as the three plants that have been discussed here.

Of the many other forest plants that can cause poisoning, wild mushrooms may be among the most dangerous. This is because people sometimes confuse the poisonous and edible varieties or eat mushrooms without making a positive identification of the variety. The wild Morel species of mushrooms is hunted by many people in the spring season (Figure 2–59). They are excellent edible mushrooms that are highly prized. On the other hand, wild mushrooms such as *Amanitopsis vaginata, Amanita muscaria,* and *Marasmius urens* are poisonous. Each year, people lose their lives due to poison from mushrooms. It is never safe to eat a wild mushroom without first making certain that it is an edible variety. To be safe, a person must be able to identify edible mushrooms in each stage of maturity because the appearance of most mushrooms changes rapidly as they mature.

CAREER OPTION

Regional Forester

A forester is a person who is educated in the sciences related to the propagation, growth, management, and harvesting of trees (Figure 2–60). He or she is responsible for administering all of the activities that occur in the forest. This includes maintaining forest health, planning and managing timber sales, preparing environmental impact studies, and ensuring that forest activities are compatible with the wildlife that live in the forest environment. A degree in forestry with education in the related sciences is needed along with broad experience in dealing with public issues and resource management.

Figure 2–60 A forester is responsible for administering all of the activities that occur in a public forest. *(Photo courtesy of Cliff Coles, Photographic Services, Oregon)*

LOOKING BACK

America was a land of virgin forests when the first colonists arrived from Europe. Today, only a few virgin forests remain. The major forest regions of North America are the Northern coniferous, Northern hardwoods, central broad-leaved, Southern, bottomland hardwoods, tropical, Rocky Mountain, and Pacific coast forests. Most of the forests have been harvested two or three times, and large tracts of forest have been cleared for farming. Serious soil erosion has occurred, particularly in the Piedmont and other areas in the Southern forest region. Some of this land was later abandoned and allowed to revert to forests, although in many cases, the dominant species of trees are different today than they were in the original forests. Many of the hardwood forests have been replaced by pines. Forest management, especially on western public lands, is heavily impacted by litigation and citizen initiatives.

QUESTIONS FOR DISCUSSION AND REVIEW

Essay Questions

1. Name the regional forests of North America, and locate them on a map.

2. What are the most important species of trees in each regional forest?

3. Explain the concept of biological succession, and distinguish between primary succession and secondary succession.

4. What are the characteristics that distinguish conifer, deciduous, and evergreen trees from one another?

5. Identify some characteristics of the Northern Coniferous Forest that may account for its low level of production of forest products in comparison with other North American forests.

6. List some important products besides wood and paper that are obtained from forests.

7. Why is it important to harvest trees in a timely manner once they have reached maturity?

8. Define the term *silviculture*, and list some examples of important silviculture practices.

9. How is the geological formation known as an alluvial fan formed?

10. What characteristics of the Pacific Coast Forest account for its high production of wood products in comparison with other forests?

Multiple-Choice Questions

1. A natural process that occurs as higher-order plants replace lower-order plants in an environment is called:
   A. primary succession
   B. secondary succession
   C. biological succession
   D. succession sequence

2. The population of plants that occupies an environment when the succession of plant species is complete and the plant population has stabilized is known as a:
   A. climax community
   B. boreal forest
   C. pioneer species
   D. terminal forest

3. A conifer is a tree that:
   A. is called an evergreen
   B. produces seeds in cones
   C. sheds its leaves or needles every year
   D. produces acorns

4. Which of the following important commercial hardwood trees was eliminated from commercial production by a parasitic fungus?
   A. yellow birch
   B. black spruce
   C. American beech
   D. American chestnut

5. The most abundant and valuable trees in the Central Broad-leaved Forest are the:
   A. oaks
   B. walnuts
   C. pines
   D. hickories

6. Which of the following destructive forces became a major problem on native forest lands in the southern and eastern regions of the United States soon after the land was cleared for farming?
   A. hurricanes
   B. soil erosion
   C. climate changes
   D. flooding

7. Management of forests and their environments for the commercial production and harvest of trees for lumber and other wood products is called:
   A. silviculture
   B. reforestation
   C. horticulture
   D. forest regeneration

8. Which of the following trees found in the Bottomland Hardwoods Forest is *not* a conifer?
   A. sweetgum
   B. Atlantic white-cedar
   C. baldcypress
   D. pond pine
9. Which of the following forest types consists of pioneer species of trees?
   A. Cypress-Tupelo
   B. Oak/Hardwoods
   C. Cottonwood/Willow
   D. Hickory-Walnut
10. Cypress and Tupelo trees are adapted to environments that are covered with water during much of the year. Which of the following terms best describes these trees?
    A. aquatic
    B. evergreen
    C. pioneer
    D. pulpwood

LEARNING ACTIVITIES

1. Obtain slides or video materials from a biological supply house or other vendor that can be used by students to learn to identify common trees. Teach students to use the key in a field guide to identify unknown trees.
2. Visit an arboretum, either as individual students or as a class. Obtain permission to gather samples of leaves that can be pressed and dry-mounted by the students. Other leaf samples might be gathered from the trees located in a local community. Prepare, properly label, and display the leaf collections for a class grade or for extra credit.0

DENDROLOGY
The Scientific Study of Trees

Chapter

3

Classification and Anatomy of Trees

Terms to Know

dendrology
anatomy
physiology
needle-leaf
scale-leaf
resin
lobed simple leaf
pinnately compound
bipinnately compound
palmately compound
cell
cell wall
cell membrane
permeable
nucleus
gene
protoplasm
cytoplasm
nucleoplasm
vacuole
chloroplast
tissue
parenchyma
petiole
blade
midrib
spine
vein
margin
collenchyma
sclerenchyma
xylem
tracheid

Terms to Know continued

Objectives

After completing this chapter, you should be able to

❋ list and describe the most common tree groups in North America on the basis of leaf structure

❋ distinguish between the anatomy and the physiology of a tree

❋ name the basic structures of a plant cell

❋ describe the different tissue systems of a tree

❋ identify the external parts of a tree leaf

❋ explain the importance of xylem tissue in a tree

❋ compare the methods of seed production between angiosperms and gymnosperms

❋ describe the importance of meristem tissue as it relates to growth of trees

❋ illustrate the basic structure of a tree root

❋ name the basic parts of a flower

vessel element	pith	secondary tissue
vessel	heartwood	vascular cambium
phloem	sapwood	cork cambium
sieve element	organ	secondary growth
sieve tube	vegetative organ	pistal
gymnosperm	reproductive organ	stigma
angiosperm	root cap	ovary
epidermis	primary growth	ovule
cuticle	primary tissue	style
cork	vascular cylinder	stamen
meristem	root hair	anther
apical meristem	cortex	receptacle
cambium	endodermis	filament
annual ring	semipermeable	petal
vascular ray	pericycle	sepal

The scientific study of trees is known as **dendrology.** Under this broad heading is the more narrowly defined study of **anatomy** that examines the structure of an organism. This includes the arrangement and relationships of individual organs to each of the other organs. Another aspect of dendrology is **physiology,** the branch of biology that deals with the life functions and processes of living organisms. Studied together, the sciences of anatomy and physiology deal with the relationships between the parts of a tree and the biological processes that nourish it and allow it to grow and reproduce itself.

TREE CLASSIFICATION

As you study the different kinds of trees, it is often convenient to organize them into groups that have similar characteristics. Many different groupings are possible, and a particular species of trees may move from one group to another, depending on the characteristics that are used. For example, in the first two chapters of this textbook, the trees studied have been described as softwoods and hardwoods or as deciduous and evergreen.

For convenience in identification, the trees of North America are grouped in this chapter according to their leaf structure. This system is used by the National Audubon Society in its field guides to North American trees.

Needle-leaf Conifers

Conifers have leaves of two basic types, although there are many distinctly different leaf arrangements. The **needle-leaf** is unique in that the leaves are narrow in width, and quite long in length. These leaves are called needles, and each leaf tends to resemble a sewing needle in its proportion of length to width.

The leaf length differs greatly among tree species and even on an individual tree. The Black Spruce, for example, has needles that are as short as ¼–⅝", and the Longleaf Pine has needles up to 15" long. Some trees, such as the redwoods, have needles that extend along the smaller branches as well as on the tips. In contrast, the needles of the larch trees are grouped, and single needles are not evident along the stem (Figure 3–1).

Needle shape is another distinguishing feature in tree identification. Seen in cross-section, some needles are relatively flat; others are three-angled or four-angled (Figure 3–2). Some needles have rounded points while others are sharp or blunt. Some needles are flexible and others are stiff; some are thick and others

Needle-leaf Conifers

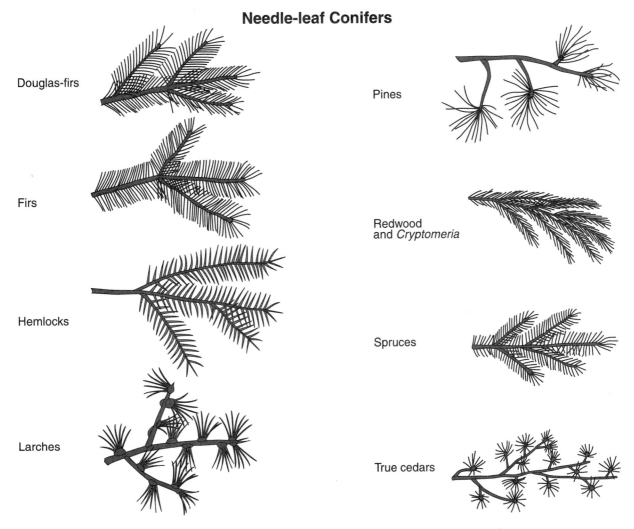

Douglas-firs

Firs

Hemlocks

Larches

Pines

Redwood and *Cryptomeria*

Spruces

True cedars

Figure 3–1 The arrangement of needles differs considerably from one species of tree to another.

Needle-leaf Shapes

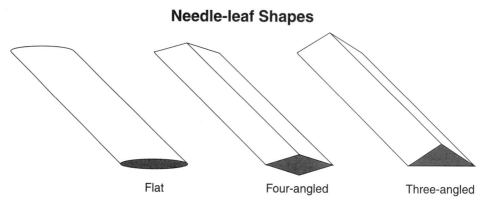

Flat   Four-angled   Three-angled

Figure 3–2 Needle-leaves have three distinctly different shapes.

are slender; some are wide and others are narrow. Some needles emerge from the stem in bundles; others are single.

Needle-leaf conifers include the Douglas-firs, firs, hemlocks, larches, pines, Redwood/*Cryptomeria*, spruces, and true cedars.

Scale-leaf Conifers

Three kinds of trees are included in the **scale-leaf** conifer group. They are the cedars, cypresses, and junipers (Figure 3–3). In each instance, the leaves are shaped like tiny overlapping scales (Figure 3–4). From a distance, these leaves

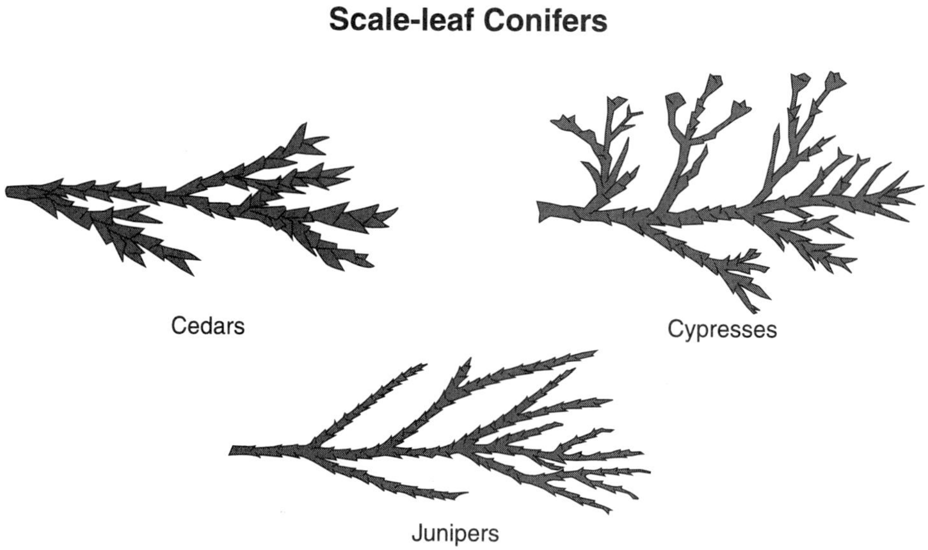

Scale-leaf Conifers

Cedars   Cypresses

Junipers

Figure 3–3 The scale-leaf structure is different both in shape and arrangement for these conifers.

Scale-leaf Structure

Figure 3–4 Scale-leaf structure consists of a series of small scaly leaves that overlap. They are attached to branching stems.

appear to be quite dry, but under magnification, they appear to be quite succulent. Leaves that make up the new growth of a tree frequently ooze small amounts of **resin** or sticky sap on the leaf surfaces.

The cedars, junipers, and some cypresses are evergreen trees that grow well in environments where water is so scarce that most other trees cannot survive. The small scale-leaf surfaces are generally smooth and shiny like the surface of a cactus plant. This characteristic is an indicator that this kind of leaf structure is capable of conserving water.

Untoothed Simple Leaves

A simple leaf is one that has only one set of leaf parts. It has a single blade and only one petiole. An untoothed leaf has a smooth leaf margin. Several North American trees display untoothed simple leaves. Among them are the catalpas, dogwoods, eucalyptus, magnolias, oaks, redbuds, sumacs, and willows (Figure 3–5).

Toothed Simple Leaves

Many broadleaf trees have simple leaves with teeth on the edges. The leaf margin may have teeth that are uniform and point forward. These leaves are described as saw-toothed. Doubly saw-toothed leaves have alternating large and small teeth. Many variations occur among the toothed simple leaves, but all of them have some kind of serrated leaf margins (Figure 3–6). Trees in this group include alders, birches, cercocarpuses, cherries, cottonwoods (also aspens and poplars), crabapples (also apples), elms, hawthorns, hophornbeams, mulberries, oaks, and willows.

Untoothed Simple Leaves

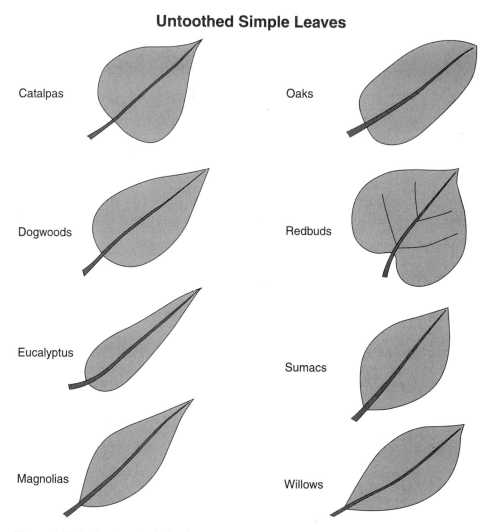

Catalpas

Oaks

Dogwoods

Redbuds

Eucalyptus

Sumacs

Magnolias

Willows

Figure 3–5 Each untoothed simple leaf consists of a single lobe with a smooth leaf edge or margin.

Lobed Simple Leaves

The leaves of many of the broadleaf trees have the appearance of rounded divisions along the margins of their leaves. Some of them appear to be shallow, while others are deeply cut. Some leaf lobes are quite broad; others are narrow. Despite these differences, each of these is known as a **lobed simple leaf.** Among the species of trees having lobed simple leaves are cliffrose, California fremontia, ginkgo, hawthorns, maples, mulberries, oaks, poplar, sassafras, sweetgums, sycamores, and yellow-poplar (see Figure 3–7 on page 90).

Toothed Simple Leaves

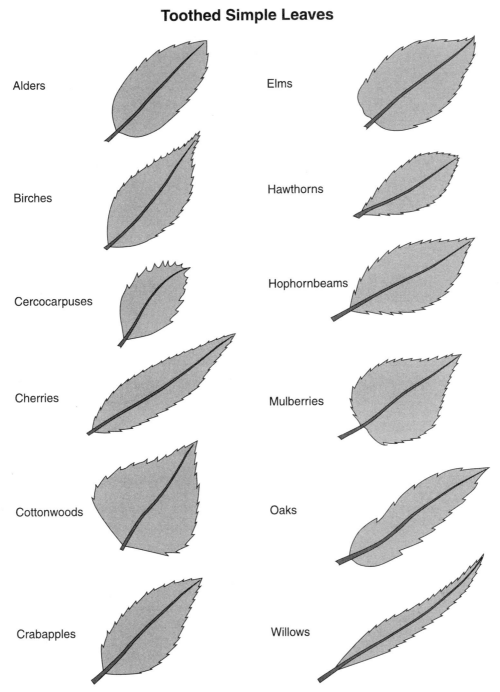

Alders

Birches

Cercocarpuses

Cherries

Cottonwoods

Crabapples

Elms

Hawthorns

Hophornbeams

Mulberries

Oaks

Willows

Figure 3–6 Each toothed simple leaf consists of a single lobe with a serrated leaf margin.

Lobed Simple Leaves

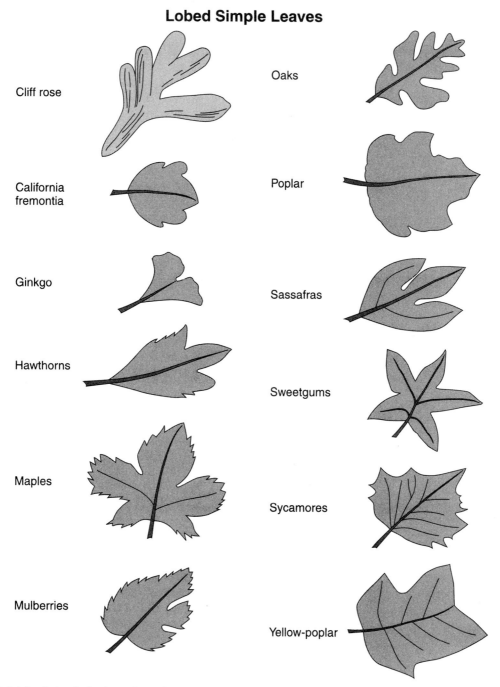

Cliff rose

California fremontia

Ginkgo

Hawthorns

Maples

Mulberries

Oaks

Poplar

Sassafras

Sweetgums

Sycamores

Yellow-poplar

Figure 3–7 A lobed simple leaf consists of two or more leaf lobes with a smooth or toothed leaf margin.

Compound Leaves

A compound leaf consists of three or more small leaflets. The leaflets may be arranged on the leafstalk in a number of different configurations. Leaflets that are arranged along a central leafstalk are considered to be **pinnately compound.** Trees having leaves of this type include ashes, elders, hickories, locusts, pecans, sumacs, and walnuts (Figure 3–8). Leaflets that are attached to side-branches arranged along a central leafstalk are **bipinnately compound.** This group of trees includes acacias, Kentucky Coffeetree, mesquites, and paloverdes.

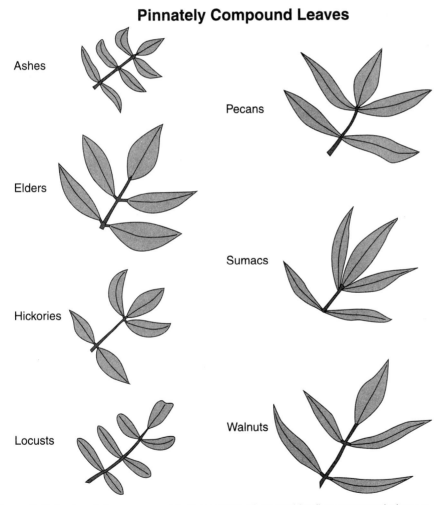

Pinnately Compound Leaves

Ashes

Pecans

Elders

Hickories

Sumacs

Locusts

Walnuts

Figure 3–8 A pinnately compound leaf consists of several leaflets arranged along a central leafstalk.

Bipinnately Compound   **Palmately Compound**

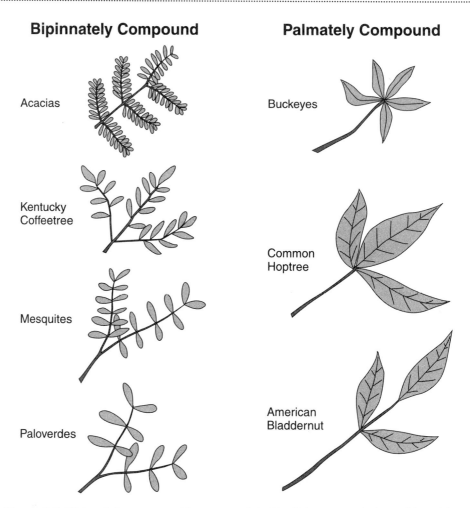

Acacias

Kentucky
Coffeetree

Mesquites

Paloverdes

Buckeyes

Common
Hoptree

American
Bladdernut

Figure 3–9 Bipinnately compound leaves consist of leaflets arranged along either side of the side-branches attached to a central leafstalk. A palmately compound leaf consists of leaflets that are attached at the end of the leafstalk.

When the attachment of the leaflets is at the end of the leafstalk, the leaf is **palmately compound.** These trees include the buckeyes, Common Hoptree, and the American Bladdernut (Figure 3–9).

PLANT STRUCTURES

Most plants have physical characteristics that make them somewhat similar to other plants. In spite of their similarities, however, each of the species of plants and trees is distinctly different from all of the others. Examples of these differences that are easily observed include size, shape, and color. There are also some very subtle differences between closely related plants that can be distinguished only through careful observation. The importance of plant classification takes

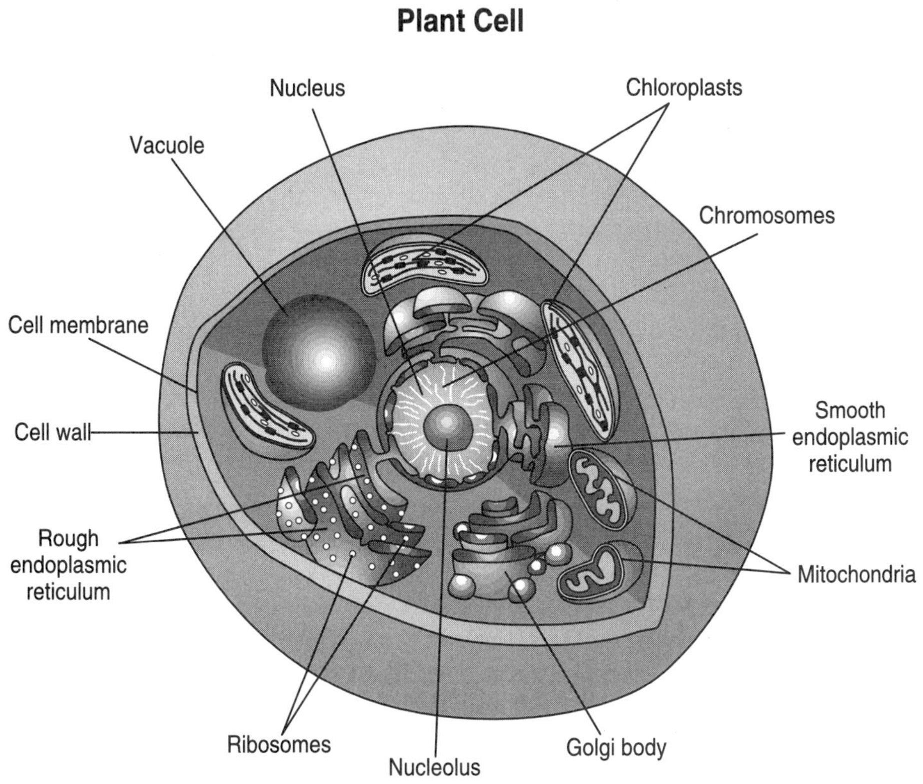

Plant Cell

Figure 3–10 The cell is the basic unit of life, and specialized structures in the cell perform life-sustaining functions.

on meaning and value when we realize that plants with similar structures tend to respond in similar ways to the environments in which they live.

Cellular Biology

The most basic unit of life is the **cell.** Each living cell contains specialized structures that are used to perform life-sustaining functions (Figure 3–10). The **cell wall** is a rigid outer covering of the cell that contains high amounts of fiber. The cell wall makes it possible for plants to maintain their shapes. A **cell membrane** is found inside the cell wall. It is **permeable,** meaning that body and plant fluids can pass through it to deliver nutrients and to remove waste materials.

Each cell has an oval-shaped **nucleus.** This structure contains the hereditary material through which a living organism passes its traits to its offspring. The chromosome is a structure found within the nucleus (Figure 3–11). It is a long strand of material on which many smaller structures called **genes** are located. Chromosomes are found in pairs, and each male and female parent

Figure 3–11 A chromosome can be described as a long spiral similar to a twisted ladder. The double-helix structure is held together by the attraction of hydrogen bonds. This model illustrates the structure of a chromosome. *(Photo courtesy of Utah Agricultural Experiment Station)*

furnishes one of the chromosomes of each pair that is present in the offspring. The genes contain the information that determines the characteristics an organism inherits from the parents.

Protoplasm is made up of all of the structures and substances located within the cell. It consists of a thick liquid composed of salts, water, proteins, fats, and carbohydrates. Two kinds of protoplasm are found in cells. **Cytoplasm** includes all of the cell contents except for the nucleus. The other kind of protoplasm includes the material found in the nucleus of a cell. This material is called **nucleoplasm.** The **vacuole** is a round structure that collects excess water and wastes within the cell and discharges them through the cell wall. The **chloroplast** is a cell structure that contains materials capable of capturing energy from the sun as the plant produces sugars and starches.

Plant Tissue Systems

Groups of plant cells that contribute to a single function are called plant **tissues.** The anatomy of a tree is very much like the anatomy of other seed plants (Figure 3–12). Plant tissue systems perform a number of specialized functions in a tree. The different types of plant tissue systems are the ground tissue system, the vascular tissue system, and the dermal tissue system (Figure 3–13). They strengthen the stem or trunk, they provide structures through which water and

Anatomy of a Tree

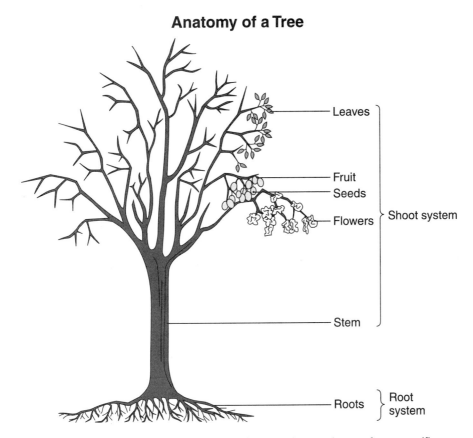

Figure 3–12 Plant cells become specialized, forming tissues that perform specific functions such as transporting cell nutrients or absorbing water.

nutrients can flow within the tree, and they protect the internal tissues and prevent water loss from the plant.

The ground tissue system consists of the following kinds of specialized cells: parenchyma, collenchyma, and sclerenchyma. **Parenchyma** cells are thin-walled living cells that are loosely packed together to form spongy tissues with air spaces interspersed between the cells. These cells make up much of the material in plant leaves, roots, stems, and fruits. They perform functions in photosynthesis and other plant processes.

The **collenchyma** cells have extra thick cell walls that add strength to plant stems and stalks. Another kind of plant cells called **sclerenchyma** cells strengthen tissue by adding fibers. Examples of material added to plant tissues by these cells include the long, stringy material found in the bark of some kinds of trees. These cells also provide the hard material for the shells of nuts (Figure 3–14).

The vascular tissue system is responsible for movement of nutrients and plant food between locations in the plant (Figure 3–15). This is accomplished by

Plant Tissue Systems

Tissue System	Specialized Cell/Tissue Types	Major Functions
Ground tissue	Parenchyma Collenchyma Sclerenchyma	Photosythesis Plant food storage Strength Protection
Vascular tissue	Tracheids Vessel elements Vessels	Transport of nutrients and water
	Phloem Sieve elements Sieve tubes	Transport of sugars
Dermal tissue	Epidermis Cork	Protection Prevention of H_2O loss
Meristem tissue	Apical meristem Cambium	Plant growth

Figure 3–13 Groups of specialized cells combine to form plant tissues that perform life-sustaining functions.

the **xylem,** a water-conducting woody tissue through which dissolved nutrients are carried from the roots to stems and leaves. Specialized plant cells called tracheids and vessel elements become important components of xylem after they have died and become hollow. **Tracheids** are long, tapered cells that have pits in their cell walls through which water is able to flow from one cell to another. Many tracheids growing side by side in the stem of a plant create passages

Figure 3–14 The hard shells of nuts are composed of collenchyma cells, which have thick cell walls that add strength, and sclerenchyma cells that add fiber to the shell.

Plant Vascular Tissue

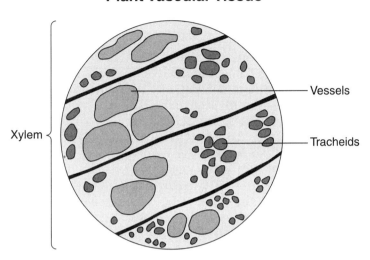

Figure 3–15 The vascular tissue of an angiosperm such as an oak, maple, or elm tree contains both tracheids and vessels.

through which water and dissolved nutrients can move to other locations in the plant (Figure 3–16).

A **vessel element** functions in a similar manner to a tracheid cell. After it

Tracheid Cells

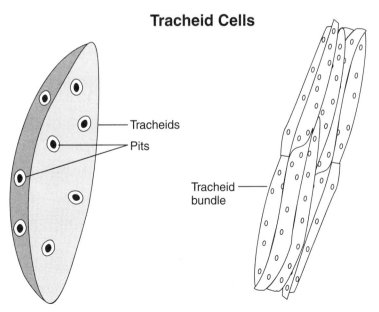

Figure 3–16 The hollowed-out interiors of dead tracheid cells become passages for transporting dissolved nutrients in living plants.

ANATOMY OF A TREE

External Leaf Structure

The external parts of a leaf include several specialized structures organized into two main parts, the blade and the petiole (Figure 3–17). The **petiole** or leafstalk is the point of attachment to the tree. The petiole consists of vascular tissues for transporting water, minerals, and nutrients to the leaf cells. These tissues transport sugars from the leaf cells to the stem and roots of the plant. In addition to vascular tissues, the petiole contains collenchyma cells that strengthen the leafstalk. The flat part of the leaf is usually green or red in color. This part of the leaf is the **blade.** The remaining leaf structures are found within the blade. The **midrib** is composed of the same materials as the petiole. It gives shape to the leaf and distributes dissolved nutrients within the leaf. The **spines** of a leaf function in the same manner as the midrib. Leaf **veins** connect to the vascular tissues in the leaf spines, and they distribute materials to and from the cells of a leaf. The leaf **margin** differs greatly among tree species. It is useful for tree identification purposes.

External Leaf Structure

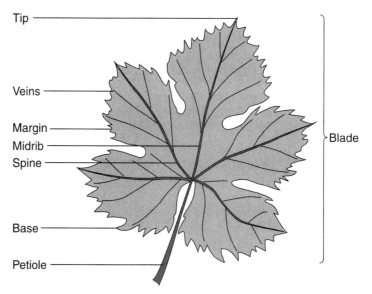

Figure 3–17 The external parts of a leaf consist of specialized tissues that perform specialized functions.

has died, it becomes a hollow tube through which dissolved nutrients can pass. Several vessel elements that have grown together end-to-end form long tubes inside stems, roots, and leaves. These are called **vessels,** and they are capable of conducting large volumes of plant fluids. They function much more efficiently than tracheids.

Plant food consisting of sugars manufactured in the leaves is carried to stems and roots through a vascular tissue called **phloem.** This is a specialized tissue consisting mostly of **sieve elements** that join together to form long **sieve tubes.** These structures are located in the bark and outer protective coverings of stems and roots. Sugar solutions flow both directions in the phloem from areas of high concentration to areas of low concentration.

Trees that bear seeds in cones as the conifers do are called **gymnosperms.** Some of the most common trees classed as gymnosperms include the pines, spruces, and cedars. Trees that produce seeds inside an ovary or fruit are called **angiosperms.** The vascular tissue in the xylem of gymnosperms, such as the pines, spruces, and cedars, contains only tracheid cells. Movement of dissolved plant materials is quite slow in conifers. Vascular tissue in angiosperms contains both tracheid cells and vessel elements, making it possible for plant materials to be transported rapidly within these trees.

The dermal tissue system in plants performs some important functions. One of these is protection of the soft material that is found inside plant cells from losing the plant fluids. Another function is to keep harmful microorganisms out of the cells. The leaves, stems, flowers, seeds, and roots of plants are protected by an outer layer of cells known as the **epidermis** and a covering of waxy material called the **cuticle.** Woody stems and roots are covered with a different kind of cells. A layer of these cells containing waxy material in their cell walls is deposited over the surfaces of root and stem tissues. These cells mature and die, and the material that remains is called **cork.** This cork covering prevents water loss from the plants. Some plant structures such as leaves and roots consist of several plant tissues that function in a coordinated manner (Figure 3–18).

Another group of specialized plant tissues is the meristems. A **meristem** is a rapidly dividing mass of cells that causes plants to grow. When it is located on the ends of branches, twigs, or roots, it is known as an **apical meristem** (Figure 3–19). When these cells divide, roots and twigs grow longer, and trees get taller. Meristem tissue is also located between the phloem and xylem layers of roots and stems. In this location, it is part of a layer of cells known as the **cambium** layer.

When a cell in the cambium layer divides, one cell remains as a cambial cell until it divides again. The other cell becomes part of either the xylem or the phloem tissue. Each cell division in the cambium layer increases the diameter of the stem or root. This accounts for the increased thickness that occurs each growing season in the trunk of a live tree. Each growing season deposits a new layer of xylem in the woody portion of a tree trunk or stem. Each new layer, which is distinct from the one before, is called an **annual ring** (Figure 3–20). This difference occurs because the vessels that form in the spring season are larger and the tracheids have thinner walls than those formed during the dryer summer season. When water is less plentiful to the plant, the cell walls of the tracheids become thicker, and the vessels that form are smaller.

The cambium also forms rows of parenchyma cells that radiate to the center of the stem (Figure 3–21). They become the **vascular rays,** and they transport dissolved materials across the woody section of a stem between the phloem and

Internal Structure of a Leaf

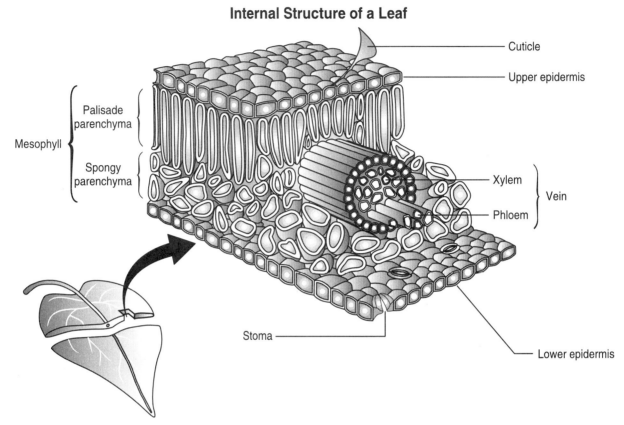

Figure 3–18 A leaf contains several examples of plant tissues consisting of specialized plant cells.

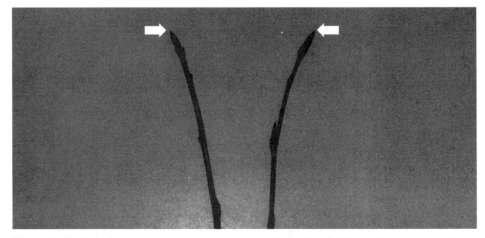

Figure 3–19 Apical meristem consists of a mass of rapidly dividing cells located on the tips of branches and stems that cause branches to grow longer and stems to grow taller.

Figure 3–20 Each annual ring consists of a new layer of woody xylem tissue that is deposited throughout the year. The age of a woody plant can be determined by counting the annual rings.

Tree Cross-Section

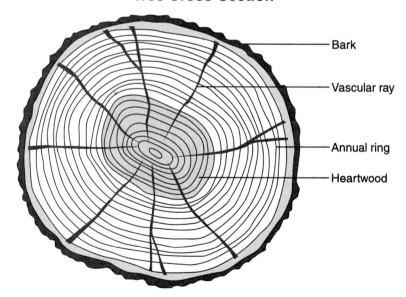

Bark

Vascular ray

Annual ring

Heartwood

Figure 3–21 Vascular rays are composed of rows of parenchyma cells that penetrate from the cambium layer to the center of a stem. They transport dissolved materials across the woody section of a stem.

the **pith** that is located at the center of the stem. The pith consists of parenchyma cells that have the primary function of storing plant food. As a plant matures, the pith is surrounded by xylem. When a tree matures, the xylem of the tree becomes old, and it begins to fill with such materials as tannins, gums, and resins. When this happens, it becomes darker in color, and it no longer conducts water. This wood is now known as **heartwood** (Figure 3–22). Once the heartwood has formed, the pith can no longer function. The lighter-colored wood through which water still moves is called **sapwood.** In a very old tree, the heartwood tends to rot away, causing the tree to become hollow. It can continue to live, however, as long as enough sapwood remains to transport plant nutrients.

An **organ** is a group of several tissues that function together as a single unit. Roots, stems, and leaves are the **vegetative organs** of a plant, and the flowers are the **reproductive organs.**

The phloem area in the root merges with the phloem region in the stem, and it performs the same functions in both locations. This tissue is deposited throughout the length of the root and the stem to facilitate the flow of plant nutrients. Plant nutrients are transported through the phloem tissue to nourish the cells of the stem and roots. In the same manner, the xylem located in the plant root merges with the xylem tissue in the stem to facilitate the flow of minerals, nutrients, and dissolved gases from the roots to the cells of the leaves and stems. Xylem and phloem are located in the **vascular cylinder** in the innermost part of the root.

Figure 3–22 Heartwood is old wood located in the center of a stem that loses its ability to conduct water. This change occurs as the xylem tissue fills with tannins, gums, and resins.

The Root Tip

Plant roots are specialized organs that anchor the trees in the soil and transport water and plant nutrients into the plant from the soil. There are several different types of roots, but the anatomy of each type is similar (Figure 3–23). A growing root contains meristem tissue located near its tip. As it divides, some of the cells develop toward the root tip. These cells will become the **root cap,** consisting of a group of specialized cells that deposit slimy material in the soil to help the root pass through as it grows. The cells of the root cap protect the tender root tissue. As the root grows in length, the cells of the root cap wear off. They are continually being replaced by new cells.

As meristem cells divide, some of the new cells become phloem, but most of them develop into xylem. This part of the root is called the region of cell division. The portion of the root in which the new cells are deposited is known as the region of elongation. These cells double or triple in length and become a little wider during this period of development. This elongation of root cells accounts for the lengthwise growth of the root and is called **primary growth.**

When root cells have stopped growing, they mature. During this period they develop or differentiate into specialized tissues such as root hairs or cortex (Figure 3–24). This part of the young root is called the region of maturation. The first specialized root tissues of a new plant that form in the area of maturation are the epidermis, cortex, and vascular cylinder. These three tissues compose the **primary tissue** of the root.

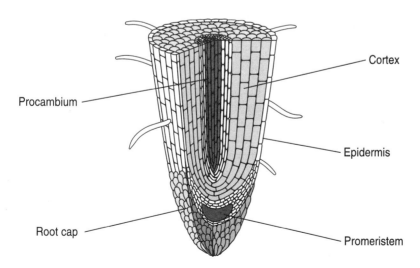

Figure 3–23 There are several different types of roots, but all of them have similar characteristics and functions.

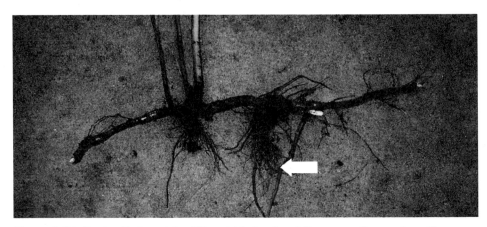

Figure 3–24 Root cells form into different kinds of root tissues as they mature. For example, some root cells develop into root hairs that perform the important function of absorbing water and dissolved nutrients into the plant.

The anatomy of a root is very similar in many ways to that of a stem. The outer layer of cells is called the epidermis (Figure 3–25). This tissue protects the root and absorbs water and dissolved plant nutrients. Epidermal cells develop long, thin, threadlike projections called **root hairs** that extend out into the soil. Root hairs increase the amount of surface area of the root and make it possible for the root to absorb large amounts of water and minerals. Root hairs only live for a short time before they die and are replaced by new root hairs. The cells of the epidermis and the root cap also slough off the root, and they are replaced with new cells.

Young roots contain a large amount of tissue called **cortex.** This tissue is composed of parenchyma cells that are organized in a loose arrangement in the interior of the root. Plant foods such as sugars and starches are stored in the parenchyma cells of the cortex. Water and dissolved materials move easily through this area until they get to the inner boundary of the cortex. Movement of fluids is stopped there by the endodermis. The **endodermis** is a single layer of cells that is surrounded by a waxy waterproof material. The wax prevents water from flowing through the spaces between cells. The membranes of the endodermal cells are **semipermeable,** allowing only certain materials to pass through. In this way, plant roots have some control over the kinds of materials that enter plants.

Just inside the row of cells that form the endodermis is another row of parenchyma cells called the **pericycle.** The pericycle is the outer layer of the vascular cylinder. Lateral roots develop from the pericycle, growing out through the cortex and epidermis of the primary root. The pericycle becomes an extension of the tissue of the primary root, increasing the ability of the root system to absorb water and nutrients from the surrounding soil (Figure 3–26).

Root Cross-Section

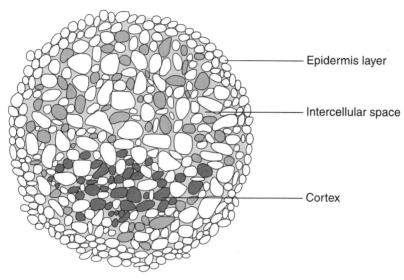

Epidermis layer

Intercellular space

Cortex

Figure 3–25 Roots and stems are quite similar in structure. Both have outer layers of epidermal cells that provide protection. In the root, the epidermal cells also absorb water and nutrients.

Once the primary tissues of the root have formed (epidermis, cortex, and vascular cylinder), all of the root tissues that are produced are called **secondary tissues.** All of the secondary tissues develop from the meristem tissue or **vascular cambium** that forms in a continuous ring between the xylem and

Figure 3–26 Most trees have complex branching root systems that reach out into the soil surrounding the tree. The roots serve a dual function of anchoring and stabilizing the tree, as well as absorbing nutrients from the soil.

ANATOMY OF A TREE

The Flower

The female flower parts include the stigma, style, and ovary. These three parts constitute the **pistil** of the flower (Figure 3–27). The **stigma** is an organ located on the female structure of a flower where it functions as a pollen receptor. The **ovary** is the flower part that produces the egg cell. A small chamber inside the ovary known as the **ovule** is the site where the seed eventually forms. The **style** is the structure that connects the stigma to the ovary.

The male flower parts consist of the anther and the filament. These two parts together compose the **stamen** of the flower. The **anther** is the organ in which pollen grains develop and mature. It is supported and connected to the **receptacle,** or base of the flower, by the **filament.** Several anthers are usually present in a flower, and each anther contains four pollen sacs in which pollen grains develop. Other flower parts are the brightly colored **petals** whose color attracts pollinators, and the leaflike **sepals** that close over the flower to protect it.

Parts of a Flower

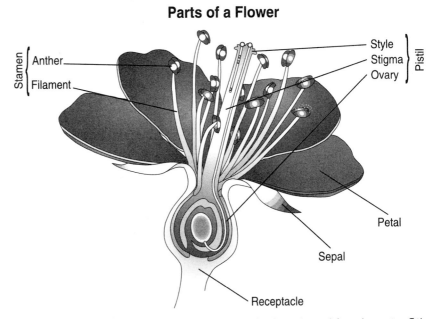

Figure 3–27 Some trees have flowers that contain both male and female parts. Other tree species have separate male and female flowers on the same tree. Some species of trees bear male flowers and female flowers on separate trees.

CAREER OPTION

Forest Botanist

A botanist with a specialty in forestry must be highly educated in the anatomy and physiology of trees. A strong science background is required, and a person in this career will use microscopes and other scientific instruments to study the internal and external structures of trees. He or she will study the effects of temperature, precipitation, climate, soil, elevation, and other factors on trees and other forest plants. This career will require a college or university degree in botany or a related science.

phloem layers of the vascular cylinder. Secondary xylem and secondary phloem are formed from the vascular cambium. They add to the thickness of the root, causing pressure to develop between the epidermis tissue and the soil in which the root is growing.

As a root grows larger, the epidermis sloughs off, and it is replaced by cork. This material is waterproof. It does not absorb materials from the soil, but the root continues to perform all other root functions. Cork tissue is produced from meristem tissue called **cork cambium.** Growth in the diameter of a root is **secondary growth.**

Reproduction occurs in trees in much the same manner as it does in other plants. Seeds are produced from flowers, which are specialized organs.

LOOKING BACK

Tree classifications based on leaf structure include needle-leaf conifers, scale-leaf conifers, toothed simple leaves, lobed simple leaves, and compound leaves. Anatomy is the study of an organism's structure, and physiology is the study of its life functions and processes. In the study of forestry, these two sciences deal with relationships between the parts of the tree and the biological processes that nourish it and allow it to grow and reproduce. Plants are organized with cells being the most basic structures, followed by tissues, tissue systems, organs, and the complete organism. Examples of tissues are epidermis, cortex, and vascular bundles. Tissue systems in plants include the ground, vascular, and dermal tissue systems. Organs include structures such as roots, stems, and leaves.

QUESTIONS FOR DISCUSSION AND REVIEW

Essay Questions

1. What are the names and distinguishing characteristics of the tree groups found in North America?

2. What is the difference between the anatomy and the physiology of a tree?

3. What are the basic structures of a plant cell?

4. Describe the different tissue systems of a tree.

5. What are the external structures of a leaf, and what are their purposes?

6. How is xylem tissue adapted to its function of transporting dissolved materials within a tree?

7. In what ways is seed production different in angiosperms than in gymnosperms?

8. What role does meristem tissue play in the growth of a tree?

9. What are the basic structures of a tree root?

10. Name and describe the basic functions of the parts of male and female flowers.

Multiple-Choice Questions

1. A study of the structure of an organism is called:
   A. physiology
   B. permeability
   C. anatomy
   D. taxonomy

2. A permeable structure found in plant cells that restricts the kind of materials that can enter a cell is the:
   A. cell membrane
   B. nucleoplasm
   C. vacuole
   D. cell wall

3. Which of the following is *not* one of the basic tissue systems of a plant?
   A. vascular tissue
   B. nucleoplasm
   C. ground tissue
   D. dermal tissue

4. A plant cell that has thick cell walls that add strength to plant stalks and stems is called:
   A. parenchyma
   B. nucleoplasm
   C. collenchyma
   D. sclerenchyma

5. The petiole is a plant structure found in a:
   A. flower
   B. leaf
   C. stem
   D. root

6. Phloem is a conductive tissue that includes which of the following types of structures?
   A. sieve tube
   B. tracheid
   C. vessel element
   D. sclerenchyma

7. A plant root tissue that stores starches is:
   A. cortex
   B. cambium
   C. endodermis
   D. epidermis

8. A plant structure that transports dissolved materials across the woody section of a stem is called:
   A. apical meristem
   B. pith
   C. sieve tube
   D. vascular ray

9. A male flower part in which pollen grains develop and mature is the:
   A. filament
   B. anther
   C. stigma
   D. sepal

10. A female flower part in which the seed forms is the:
    A. ovule
    B. stamen
    C. receptacle
    D. style

LEARNING ACTIVITIES

1. Examine the scale-leaf structure of a juniper, cedar, or cypress tree (ornamental shrub). Assign the students to draw what they see. Speculate with the class about how this leaf structure is able to conserve water. Name some other life forms found in desert environments that have scales, and consider whether they function in a similar way.

2. Obtain the flowers of several different types of woody plants and examine them under magnification. Sketch the flowers, and identify the male and female structures of each. Assign a class member to describe the role of each of these structures.

Chapter

4

Terms to Know

photosynthesis
chlorophyll
light reaction
adenosine triphosphate (ATP)
hydrogen ion
NADPH
Calvin cycle
respiration
starch
dehydration synthesis
cellulose
lipid
mitosis
interphase
prophase
chromatid
centromere
metaphase
centriole
spindle
anaphase
telophase
propagation
sexual reproduction
gamete
meiosis
homologous chromosome
homologue
spore
haploid
diploid
pollen grain
fertilization
microspore mother cell
tetrad

Terms to Know continued

Physiology of Trees

Objectives

After completing this chapter, you should be able to

❈ explain the importance of photosynthesis in sustaining plant and animal life

❈ describe two kinds of high-energy molecules that are produced by the light reactions during photosynthesis

❈ identify some products that are formed in plants from the simple sugars that are formed during photosynthesis

❈ explain the importance of dehydration synthesis in the formation of starch

❈ distinguish between the structures of starch and cellulose molecules

❈ name the type of cell division that accounts for most of the growth in trees

❈ describe the process of meiosis in the production of male and female gametes

❈ distinguish between sexual and asexual propagation of plants

❈ explain the difference between a high forest and a low forest

❈ compare the different forms of vegetative reproduction

❈ explain how tissue culture technology is used to propagate plants

microspore
megaspore mother cell
megaspore
micropyle
polar nuclei

embryo sac
asexual reproduction
vegetative reproduction
low forest
high forest

coppice method
sprout method
layering
tissue culture
callus tissue

Physiology is the branch of science that deals with life processes. This chapter deals with the life processes of trees. These processes include the production and storage of plant foods. The process by which energy from sunlight is captured and plant foods are produced is the most basic of all life processes. The survival of all plant and animal life depends upon it. Other processes that sustain plant life are plant growth and plant reproduction. Each of these processes plays a vital role in forestry.

LIFE FUNCTIONS OF PLANTS

Photosynthesis

All living organisms require energy to sustain life. Some of them, such as animals, fungi, and some parasites, obtain energy by eating or extracting energy from other organisms. Plants are unique among organisms because they are able to convert energy obtained from sunlight into food. Trees and other plants capture the energy found in sunlight through a process called **photosynthesis.** This is a chemical reaction in plant leaves that captures energy from the sun and combines it with carbon dioxide from the atmosphere to form plant tissues (Figure 4–1).

Photosynthesis produces two of the most basic needs of animals: food and oxygen. Animals use the plant tissues that are produced during plant growth as sources of food. Oxygen is a waste product from the photosynthesis process, but it is a life-sustaining element for many other forms of living things. Oxygen is used in animal cells to release energy from nutrients obtained from plant sources. Even those animals that eat flesh of other animals obtain their nutrition indirectly from plants.

Chlorophyll is the green substance found in plants that is capable of capturing energy from sunlight. Plants use this energy to make sugar from carbon dioxide and water. Chlorophyll is a substance that is located in a plant cell structure called a chloroplast. Chlorophyll is more abundant in the leaves than in any other part of the plant. It is required in two important chemical reactions that convert light energy to chemical energy. These two reactions are called **light reactions** (Figure 4–2).

The first light reaction occurs when light is absorbed by chlorophyll molecules. This is followed by several steps that lead to the formation of a high-energy molecule called **adenosine triphosphate (ATP).** This molecule stores

Photosynthesis

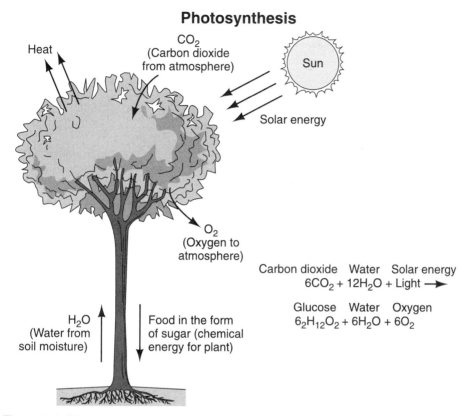

Figure 4–1 Photosynthesis is the process by which carbon dioxide is combined with water to store energy obtained from sunlight. Chlorophyll supports this reaction in which sugar and oxygen are produced.

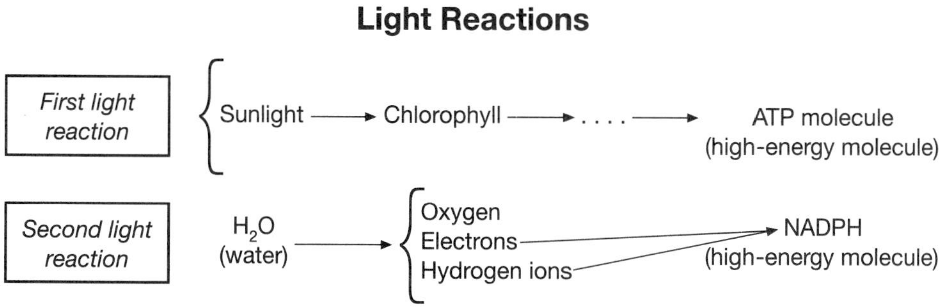

Light Reactions

First light reaction	Sunlight $\longrightarrow$ Chlorophyll $\longrightarrow$ . . . . $\longrightarrow$	ATP molecule (high-energy molecule)
Second light reaction	H_2O (water) $\longrightarrow$ Oxygen, Electrons, Hydrogen ions	NADPH (high-energy molecule)

Figure 4–2 Two different kinds of light reactions occur during the process called photosynthesis.

Glucose and Related Molecules

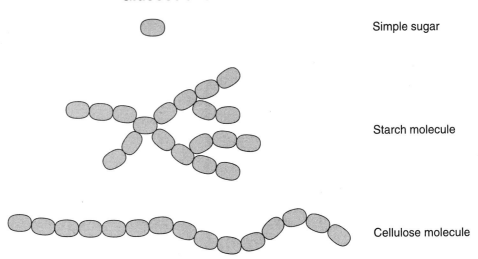

Simple sugar

Starch molecule

Cellulose molecule

Figure 4–3 Glucose is a simple sugar molecule that is converted to large complex molecules when many glucose molecules bond together in long chains.

electrons in the form of chemical energy in the electrical bonds that attract and hold its phosphate molecules together.

The second light reaction occurs at the same time that light energy is converted to chemical energy during the formation of ATP. Some of the water molecules located in the leaf are split to create oxygen, electrons, and positively charged hydrogen particles called **hydrogen ions.** The hydrogen ions use energy obtained from electrons to bond with a molecule called nicotinamide adenine dinucleotide phosphate (NADP), forming another kind of high-energy molecule called **NADPH.**

The next phase of photosynthesis is called the **Calvin cycle** during which carbon dioxide from the environment reacts with ATP and NADPH obtained from the light reactions to form simple sugars. These sugars may be used as energy sources for the plant, or they may be converted to more complex sugars, starches, oils, and proteins for storage (Figure 4–3).

During hours of darkness, stored plant materials tend to decrease. When the photosynthesis process is interrupted, the enzyme reacts with oxygen and reverses the process of photosynthesis. This process is called **respiration** (Figure 4–4).

Energy Storage

The storage of energy reserves is an active process in plants. It involves the transport of the sugars that were formed during photosynthesis from the leaves of the plant to its roots, stems, fruits, and seeds. In these locations, the sugars are converted to starches or cellulose. Starches are usually converted back to sugars

Respiration

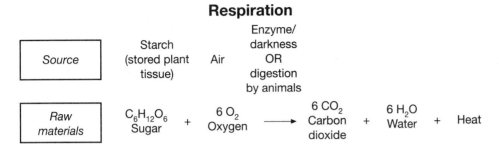

Source	Starch (stored plant tissue)		Air	Enzyme/ darkness OR digestion by animals				
Raw materials	$C_6H_{12}O_6$ Sugar	+	$6\ O_2$ Oxygen	⟶	$6\ CO_2$ Carbon dioxide	+	$6\ H_2O$ Water	+ Heat

Figure 4–4 Respiration is the process by which plant tissues are broken down to produce heat, water, and carbon dioxide.

as plant nutrients during seasons when photosynthesis does not occur rapidly enough to meet the needs of the plant.

Transport of sugars within a plant occurs through the system of tubes located in the phloem tissue. Sugar molecules that are dissolved in water are capable of moving through this system to the different plant tissues. A tree usually converts some sugar molecules to starch molecules as part of its food reserves that are stored in the cortex tissue of roots and in the seeds. A **starch** molecule is formed when large numbers of glucose molecules bond together in long branching chains. As each glucose molecule becomes attached to a starch molecule, a molecule of water is removed. For this reason, the chemical reaction that forms starch from sugars is referred to as **dehydration synthesis** (Figure 4–5).

Large amounts of the glucose manufactured by a tree are also used to make the **cellulose** that eventually becomes wood fibers. This is a very important biological process to the forest industry. It is this process that forms the xylem tissue in a tree that eventually becomes lumber products. A single cellulose molecule consists of an unbranched chain in which up to three thousand glucose molecules are bonded together.

Dehydration Synthesis

$C_6H_{12}O_6$ (Glucose)	+	$C_6H_{12}O_6$ (Glucose)	Dehydration ⟶ (loss of water)	$C_{12}H_{22}O_{11}$ (Maltose, disaccharide)	+	H_2O Water
Glucose	+	Starch	Dehydration ⟶	Starch	+	Water
			OR			
Glucose	+	Cellulose	Dehydration ⟶	Cellulose	+	Water

Figure 4–5 Each time that a glucose molecule is bonded to another similar molecule such as glucose, starch, or cellulose, a molecule of water is removed. Starch and cellulose molecules grow larger each time this reaction occurs.

Figure 4–6 Pine cones and other reproductive structures form nuts or seeds from which the next generation of trees is produced. The nuts that are produced by trees contain highly concentrated energy deposits in the form of fats and oils.

Plants store some energy in compounds known as **lipids.** Examples of these compounds include fats, oils, and waxes. Most of the oils found in trees or other plants are found as deposits in the seeds. For example, the nuts formed by the conifers and many of the hardwoods are high in plant oils (Figure 4–6). Energy deposited in the form of fats and oils is a highly concentrated source of stored energy. Birds and mammals of many kinds depend on seeds and nuts as sources of energy in their diets (Figures 4–7, 4–8).

Figure 4–7 Many wild birds and mammals depend on nuts for their food. *(Photo courtesy of Texas Parks and Wildlife Department)*

Figure 4–8 Squirrels are known to harvest nuts and pine cones, and they hide them for later use when they are less plentiful. *(Photo courtesy of Texas Parks and Wildlife Department)*

Growth

Growth occurs in plants as plant tissues become longer or thicker. This is accomplished through cell division and, in some cases, through cell elongation. Cells divide to form new cells throughout the life span of an organism. The type of cell division that occurs to cause growth in a tree or other forest plant is called **mitosis** (Figure 4–9). Several steps occur during mitosis. Cells exist for most of their life spans in a resting or nonreproductive stage. This is called **interphase.** The first stage of active cell reproduction is known as **prophase.** During prophase, the membrane around the nucleus disappears, and the chromosomes appear. Each chromosome has been replicated, and each half of the doubled chromosome is known as a **chromatid.** The point where the chromatids are attached is called the **centromere** (Figure 4–10).

The next step in cell reproduction is **metaphase.** The chromosomes are pulled to the center of the cell by fibers attached to cell structures called **centrioles** that have migrated to opposite sides of the cell. These fibers or **spindles**

Mitosis

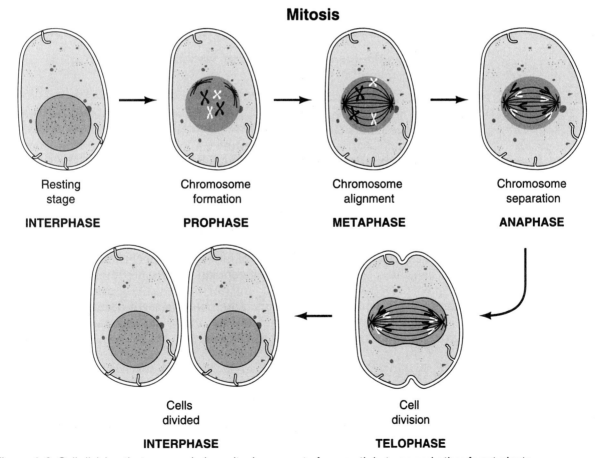

| Resting stage | Chromosome formation | Chromosome alignment | Chromosome separation |
| INTERPHASE | PROPHASE | METAPHASE | ANAPHASE |

| Cells divided | Cell division |
| INTERPHASE | TELOPHASE |

Figure 4–9 Cell division that occurs during mitosis accounts for growth in trees and other forest plants.

Cell Structures During Cell Division

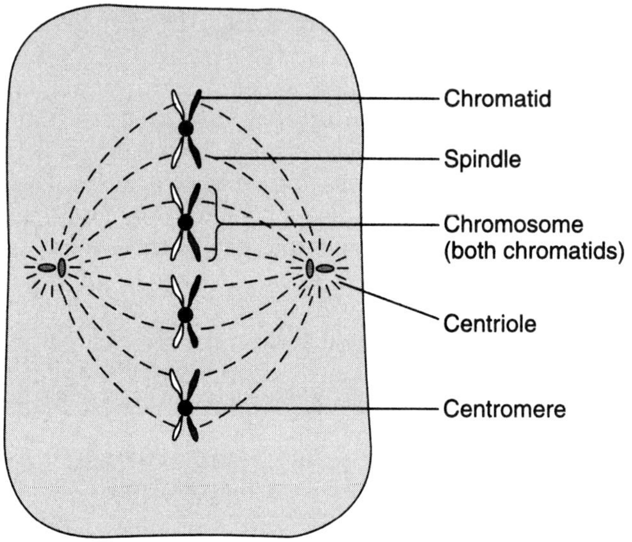

Figure 4-10 Some cell structures are evident only while the cell is dividing.

are attached to the centromeres that connect the pairs of chromosomes together. The chromatids are pulled apart by the spindles as the cell elongates. This step of cell division is known as **anaphase.** A full set of chromosomes becomes evident on opposite sides of the cell during anaphase.

Telophase is the last phase of mitosis. The cell becomes constricted with the cytoplasm shared equally by the two new cells that are forming. A full set of chromosomes eventually becomes separated into each new cell. The membrane around each cell nucleus forms once again, and in plant cells, a cell wall begins to develop between the two new cells.

Cells divide through the process of mitosis to form clusters of cells. Cell clusters become specialized to form different kinds of tissues in the organism, such as phloem or xylem. Mitosis accounts for most of the growth that occurs in plants, trees, and other organisms. Some plant growth occurs in the zone of elongation found in young roots and the tips of growing stems. This type of growth was discussed earlier in this chapter.

Reproduction

A term that is associated with reproduction in plants is **propagation.** Propagation may occur in nature, or it may be controlled and manipulated by forest workers. In either situation, it occurs in plants in two distinct ways, sexual reproduction and asexual or vegetative reproduction. The most common form of plant reproduction in trees found in forest environments is **sexual**

reproduction. This involves the production of seeds by mature trees. Natural reproduction of seeds occurs in most flowering species of trees.

Cell division that occurs to produce seeds is different from the cell division that occurs as plants grow. Part of the seed production process is the formation of reproductive cells. These cells are known as **gametes.** This form of cell division utilizes a process known as **meiosis** (Figure 4–11). The first step in meiosis occurs when chromosomes are duplicated and become aligned in the middle of the cell with their matching chromosomes. These matching chromosomes are known as **homologous chromosomes** or **homologues.**

Once the chromosomes have become aligned, meiotic divisions begin. In the first division, one homologue from each pair of duplicated chromosomes is separated into a different cell mass. The division of cytoplasm is equal in the formation of male gametes known as **spores.** In female gamete production, this division of cytoplasm is unequal. One of the cell masses is larger than the other.

Each half of a duplicated chromosome is called a chromatid. The second meiotic division results in the separation of the paired chromatids in each cell mass, resulting in four cell masses from the two that existed after the first meiotic cell division. Once again, the division of cytoplasm between cell masses is equal in the formation of male gametes (spores), but unequal in the formation of female gametes (ova).

The new gamete that is formed through meiosis consists of one chromatid from each original chromosome pair. It is a **haploid** cell because it contains only half of the genetic material of the cell from which it was formed. The parent cell is a **diploid** cell, meaning that it contains both homologues of each chromosome.

Sexual Reproduction

Most trees and other plants reproduce by sexual reproduction through the formation of flowers. A fertile seed is produced when a **pollen grain** (male gamete) from a male flower part called the stamen merges its genetic material with that of the female gamete or ovule (Figure 4–12). The process by which this occurs is called **fertilization.** After the ovule has been fertilized, it matures into a seed that is capable of growing into a new plant.

The stamen consists of the anther and the filament or stalk. The stamen is the male portion of a flower. The anther is the organ in which pollen grains develop and mature. It is supported and connected to the receptacle, or base of the flower, by the filament. Several anthers are usually present in a flower, and each anther contains four pollen sacs in which pollen grains develop.

Pollen formation begins with the production of **microspore mother cells** inside the pollen sacs. These are diploid cells that contain chromosome pairs. As they begin pollen formation, each of these cells divides through the process of meiosis that was discussed earlier. A **tetrad** consisting of a cluster of four haploid cells is formed. Later, these four cells pull apart, forming four cells called **microspores.** The nucleus of each microspore divides one more time, forming a pollen grain consisting of two cells. One cell contains the generative nucleus

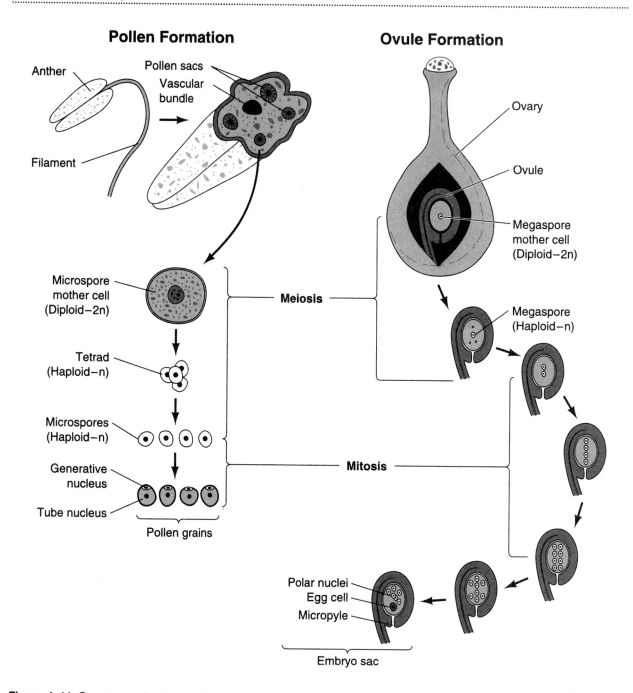

Figure 4–11 Gamete production in plants

Fertilization

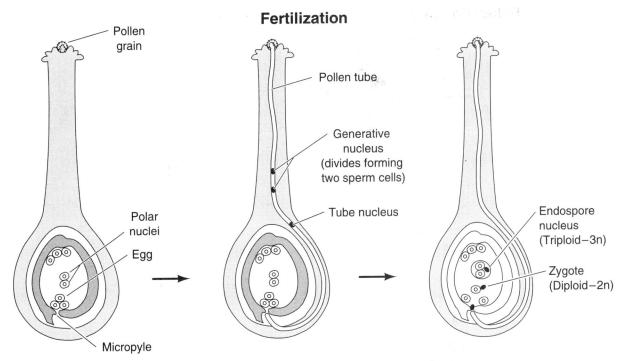

Figure 4–12 Fertilization in plants occurs when a pollen grain unites with the ovule.

from which two sperm cells will develop during fertilization of the ovules. The other cell contains the tube nucleus, which is destroyed once it enters the ovule.

Ovule formation begins when a **megaspore mother cell** is produced inside the ovule. It is a diploid cell. This cell divides during meiosis, forming four haploid cells called **megaspores,** of which three die. The remaining megaspore grows in size, and its nucleus divides to form two nuclei. Two more divisions occur, and eight nuclei are produced. Each of these nuclei is haploid.

Next the eight nuclei move to different locations in the ovule. Three of the nuclei gather near a small opening in the ovule called the **micropyle.** One of these nuclei enlarges and becomes the egg cell. Three other nuclei migrate to the opposite end of the ovule. They will eventually develop cell membranes along with two of the nuclei near the egg cell. The other two nuclei are called **polar nuclei,** and they migrate to the center of the chamber where they form a single cell. They will produce food for the tiny plant embryo.

A total of seven cells is formed, encompassing the eight original nuclei. These seven cells together comprise the female gamete. This structure is also called the **embryo sac.** At this stage of reproduction, the ovule is mature. Once the pollen grains and the ovules have matured, pollination and fertilization can occur.

Figure 4–13 Vegetative or asexual reproduction in plants occurs when new plant growth develops from the root, stem, or leaf tissues of plants.

Asexual Reproduction

One characteristic of plants that is different from many other living organisms is the capability of a plant to replace missing parts by growing new parts. This process is called regeneration. Some kinds of trees can be propagated from leaf, root, or stem tissue. This is a form of **asexual reproduction,** also known as **vegetative reproduction** (Figure 4–13).

A forest that has been regenerated from the roots, stumps, or branches of other trees is known as a **low forest.** A forest that has been propagated from seeds is called a **high forest** (Figure 4–14). Both of these types of forests are

Figure 4–14 The production of tree seedlings from seeds is an important forest management practice.

common in intensively managed plantation forests. Vegetative reproduction has been widely practiced in the forests of Europe since ancient times. Several different vegetative reproduction methods are used to regenerate North American forests.

The **coppice method** or **sprout method** of regenerating trees is practiced by cutting all of the trees in a particular stand at the same time. This is necessary because live trees are believed to release hormones into the soil that inhibit sprouting. They also compete with young tree sprouts for nutrition. When all of the trees of a particular species are cut, a new generation of trees is frequently induced to grow from the stumps of the old trees. The best example of the coppice method as an effective management practice is in the aspen forests of the Great Lakes region. Entire stands of aspen shoots arise from root suckers when care is taken to cut all of the trees.

The stump-sprout method of forest regeneration is the most commonly practiced form of vegetative reproduction in silviculture (Figure 4–15). With the exception of the Redwood, vegetative reproduction is ineffective in conifers, but for many other species of trees, regeneration from sprouts is both common and effective. For the best results, all of the trees should be cut during the dormant season.

Stump-sprouting is an effective reforestation method, but it does have some problems associated with it. Sprouts that arise too far above the ground seldom develop into useful trees. Sprouts that arise from stumps where stump rot was present in the harvested trees are sometimes at risk of developing rot in the shoots of the young trees. This risk appears to be different among different species of trees. The risk of stump rot can be reduced by using seedling sprouts

Figure 4–15 The growth of new plants from the stumps of harvested trees is an example of the stump-sprout method of forest regeneration.

instead of stump sprouts. The difference in these two forest regeneration methods is that stumps of trees that are less than two inches in diameter appear to consist mostly of sapwood, and little or no heartwood is present from which the stump rot problem arises.

Seedling roots appear to provide more vigorous root systems for saplings, resulting in faster growth rates in comparison with the stump-sprout method of propagation. In general, timber rotations should be shorter for stump-sprout stands than for trees produced from seed. This is because the root systems are more likely to die before the trees become large enough for sawlogs. The system works fine, however, for such uses as pulpwood or fuelwood.

Another form of vegetative reproduction that is effective in forest regeneration is called **layering.** This occurs when live branches are buried in the debris on the forest floor. When adequate moisture and sunlight are available, roots are often generated, and live stems arise from the buried plant material (Figure 4–16). It has been observed that the shoots that arise from the plant material of older trees are more vigorous than those produced from young trees. It is also recommended that cuttings be harvested during the period when a tree is dormant. This practice results in more vigorous sprouting than that which is observed when late spring or summer cuttings are used.

Tissue Culture

A relatively recent technology that will impact forestry is **tissue culture.** This is a method of propagating plants asexually by reproducing entire plants from a single plant cell. This is not unlike the sprouting method of plant propagation, but it is much more sophisticated. Sprouts arise from **callus tissue,** which is

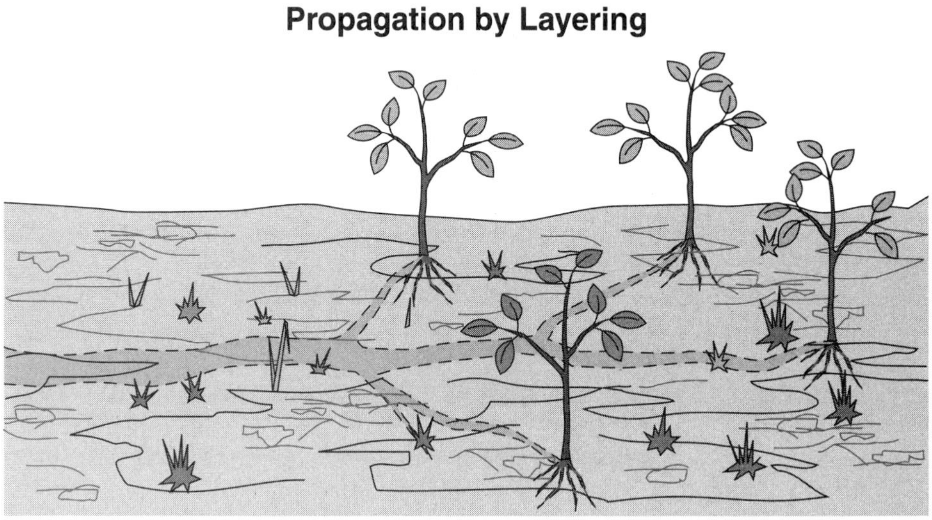

Propagation by Layering

Figure 4–16 Live branches that are buried under the debris on the forest floor will often develop roots and stems that grow into trees.

undifferentiated plant tissue that has lost its identity as root, stem, or leaf tissue (Figure 4–17). Callus tissue can be manipulated in the laboratory to become either root or stem tissue, depending on the plant hormones that are introduced into its environment.

In combination with genetic engineering techniques, genes that control the expression of desirable traits in specific trees can be introduced into the chromosomes of unrelated trees. This raises the possibility for the fast growth rates of poplar trees to be expressed in the more valuable hardwood trees. We can certainly expect the development of genetically engineered trees that will be resistant to many of the insects and diseases that now devastate some of our forests.

Figure 4–17 Individual plants that are produced by tissue culture methods must be carefully protected while they develop roots, stems, and leaves. *(Photo courtesy of Utah Agricultural Experiment Station)*

CAREER OPTION

Genetic Engineer

A career in genetic engineering involves changing the genetics of living organisms by isolating desirable genes such as those that provide resistance to diseases. Such genes, along with their resistant qualities, are removed from their original chromosomes and inserted in the chromosomes of other plants that lack resistance to the troublesome disease. A college degree in the combined fields of engineering and the biological sciences or biotechnology is required for entry into this emerging profession. It is anticipated that people engaged in this career will develop new technologies that lead to widespread changes in the kinds of trees that are produced in our future forests.

LOOKING BACK

Physiology is the branch of science that deals with life processes. Among the processes discussed in this chapter are photosynthesis, including the production of high-energy ATP and NADPH from the two light reactions, and the production of simple sugars during the Calvin cycle. Energy storage is a process that occurs as simple sugars are converted to more complex molecules such as starch and lipids (fats, oils, and waxes). Plant growth occurs as plant cells divide through the process of mitosis. A second form of cell division called meiosis occurs in the division of cells to form gametes. Sexual reproduction occurs when male and female gametes join to produce a new plant. Asexual reproduction, also known as regeneration, occurs when new plants are formed from vegetative plant parts. A new method of propagating plants asexually is known as tissue culture. This high-tech process generates new plants from single plant cells.

QUESTIONS FOR DISCUSSION AND REVIEW

Essay Questions

1. Explain the importance of the process of photosynthesis in sustaining plant and animal life.
2. What are the two high-energy molecules that are formed during the light reactions as part of the photosynthesis process?
3. Name some products that are formed in plants from the simple sugars that are produced during photosynthesis.
4. Describe the formation of starch from simple sugars through the process of dehydration synthesis.
5. Name the type of cell division that accounts for most of the growth in trees, and list the steps in the process.
6. Describe the process of meiosis during the formation of male and female gametes.
7. What are the differences between sexual and asexual propagation of plants?
8. How is a high forest different from a low forest?
9. Compare the different forms of vegetative reproduction.
10. What steps are involved in the propagation of plants using tissue culture technologies?

Multiple-Choice Questions

1. Energy from sunlight is captured and stored in plant tissues through the process of:
   A. meiosis
   B. photosynthesis
   C. dehydration synthesis
   D. mitosis
2. The green substance found in plant cells that plants use to capture energy from sunlight is called:
   A. chloroplast
   B. NADPH
   C. chlorophyll
   D. cellulose
3. A high-energy molecule that is formed during the first light reaction is:
   A. ATP
   B. NADPH
   C. cellulose
   D. chlorophyll
4. The phase of photosynthesis during which carbon dioxide reacts with ATP and NADPH forming simple sugars is:
   A. respiration
   B. dehydration synthesis
   C. light reactions
   D. Calvin cycle
5. Dehydration synthesis is a plant process that is responsible for the formation of:
   A. starch
   B. glucose
   C. cellulose
   D. chloroplast
6. The fats, oils, and waxes that are found in plants are found mostly in:
   A. cellulose
   B. starch
   C. seeds
   D. lignin
7. Which of the following terms is *not* a stage of mitosis?
   A. prophase
   B. comatose
   C. telophase
   D. metaphase
8. Meiosis is a form of cell division in which:
   A. gametes are produced
   B. asexual reproduction occurs
   C. plant growth occurs
   D. callus tissue is formed

9. When sexual reproduction occurs in plants, the name of the male gamete that fertilizes the egg cell is:
   A. anther
   B. pollen
   C. coppice
   D. chromatid

10. Which of the following terms is *not* associated with propagation of plants from live plant parts?
    A. regeneration
    B. asexual reproduction
    C. vegetative reproduction
    D. megaspore mother cell

11. In plant tissue culture, a term that describes the plant tissue from which new plants develop is called:
    A. sprouts
    B. filament
    C. embryo sac
    D. callus

LEARNING ACTIVITIES

1. Prepare some planting pots and obtain some fresh cuttings from the suckers of a poplar tree or other similar tree species. Make sure there is a well-developed bud on each cutting. Have the students "plant" their cuttings beneath a layer of soil and care for them for several weeks. The soil should be kept damp, and the pots should be placed in warm locations for best results. Observe the plantings daily, and record the results.

2. Obtain several different kinds of medium to large flowers from a greenhouse or floral shop. Divide the class into small groups, and assign each group to identify and label the parts of their flower using straight pins with labels attached. Rotate the groups around the room until they have inspected the flowers at each station, checking to be sure that the flower parts have been correctly identified. Have each student draw and label the parts of a flower.

Chapter

5

Terms to Know

ecology
industrial waste
surface water
groundwater
conservation of matter
acid precipitation
elemental cycle
nitrogen fixation
nitrogen-fixing bacteria
denitrification
nitrogen cycle
soil texture
humus
horizon
soil profile
eluviation
illuviation
parent material
Alfisols
Spodosols
Ultisols
erosion
soil conservation
water cycle
smog
growth impact
terminal growth
radial growth
habitat
adaptive behavior
herbivore

Terms to Know continued

Forest Ecology

Objectives

After completing this chapter, you should be able to

* explain the relationship between soil erosion and pollution of surface water

* understand the basic law of physics known as the conservation of matter

* describe how natural cycles function to prevent pollution and to renew the environment

* discuss the importance of the element *carbon* to living organisms

* suggest some reasons why soil is considered to be one of our most important natural resources

* distinguish among the three soil orders that are of significance to forestry in North America

* explain the major functions and significance of watersheds

* illustrate the different events that occur in the water cycle

* suggest some effects of air pollution on forests

* describe the similarities and differences among food chains, food webs, and food pyramids

predator
carnivore
omnivore

food chain
producer
primary consumer

secondary consumer
food web
food pyramid

A forest is much more than a population of trees. Within its boundaries are many other living organisms and natural resources. The study of relationships between living organisms and the environments in which they live is the science of **ecology.** This branch of the biological sciences seeks to understand the needs of all of the living things that are found in an environment. Ecology also includes the study of effects that a particular living organism has on other living organisms and natural resources.

PRINCIPLES OF ECOLOGY

The organisms that inhabit forest environments have some common needs. Plants, animals, and insects all require nutrition to meet their need for energy (Figure 5–1). Sunlight, soil, and water are our most basic resources. Plants depend upon nutrients and moisture from the soil along with sunlight to supply their energy needs. All animals and insects derive their energy from the food they eat. Some of them eat plants in their diets, but even those that eat diets of meat derive their energy from the plants that were eaten by their prey. All living organisms are dependent for survival upon sunlight, water, and nutrients from the soil.

One of the great tragedies of our time has been the loss of significant amounts of the topsoil from the land (Figure 5–2). Most of this soil has been carried to lakes

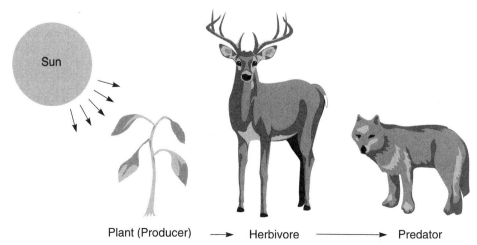

Plant (Producer) ⟶ Herbivore ⟶ Predator

Figure 5–1 The energy flow in an ecosystem proceeds from sunlight to plants to herbivores, and last of all, to carnivores.

Figure 5–2 Most of the soil erosion that occurs in a forest takes place immediately following intensive logging operations or forest fires. It is important that ground cover is quickly reestablished to protect the soil.

and oceans where it lies in great deposits of silt beneath the water. Silt contamination of surface water is the number one cause of surface water contamination, and serious damage is also sustained by the soil. For many years, it was a common practice to clear the forests from the land, and to produce crops until the soil was no longer fertile or until erosion of topsoil made farming unprofitable. Many farms were then abandoned, particularly in the Atlantic Coast and Southern regions of North America. Eventually, many of these lands were reclaimed by forest vegetation, but the damaged soils remain.

A serious problem that affects most and perhaps all of the environments in the world is pollution. Among the restrictions that humans introduce into natural environments are the substances that we make. Some of these materials are not easily degraded when they are no longer useful. On the other hand, waste materials generated by plants and animals seldom become serious problems in nature because they cycle back to the soil to be used again.

Every segment of society must become accountable for the waste materials that it creates. The industry of agriculture must assume full responsibility for the agricultural chemicals and other toxic materials that are by-products of agricultural production. The industry must find ways to produce adequate supplies of food and fiber without abusing the soil and water resources. Industries must assume responsibility for all of the waste materials that they produce.

Industrial waste has been a serious environmental problem for many years (Figure 5–3). Waste materials include a variety of harmful chemicals, poisonous metallic compounds, acids, and other caustic materials that are by-products of mining or industrial processes and manufacturing activities.

Much of our **surface water** is polluted by industrial wastes. In some instances, the fish and other aquatic animals and plants are harmed or even killed by water environments that have been poisoned. Most of the pollutants that affect surface water are also problems in **groundwater.** This is the water that is located beneath the soil surface. It is stored in large underground reservoirs where it occupies the space between soil particles such as sand, gravel, and rocks (Figure 5–4).

A basic law of physics is the **conservation of matter.** Nothing is wasted in nature. Matter may change from one form to another, but it cannot be created or destroyed by natural physical or chemical processes. This law of nature applies to all living and nonliving resources. The law of conservation of matter applies as much to waste materials as it does to anything else. We may change the form of waste materials to make them more compatible with the environment, but we cannot destroy them. With this in mind, we must properly dispose of all waste materials to prevent serious environmental problems.

In addition to man-made pollutants, a serious pollutant in a forest environment is the silt that comes from soil erosion following fires. Inadequate planning of logging activities such as massive unprotected clear-cuts and poorly

Figure 5–3 Mine tailings and industrial wastes are responsible for some serious pollution problems that harm natural resources.

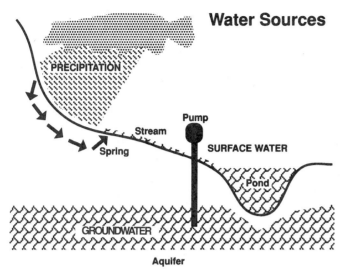

Water Sources

Figure 5–4 Water moves in a continuous cycle, and it can be polluted at any stage of the cycle. A forested watershed acts as a biological filter system to remove pollutants from the water, affecting all of the water sources in the system.

constructed roads can also contribute to serious erosion of the soil (Figure 5–5). Other pollutants that cause serious damage to forest environments include **acid precipitation** and chemicals from mine tailings.

Figure 5–5 Poor construction or improper use of forest roads can cause serious silt pollution in the area's streams and rivers during seasons of heavy rainfall.

Much of the air pollution that is found in North America comes from exhaust fumes of automobiles and from the by-products of burning. The burning process releases large amounts of carbon monoxide and sulfur dioxide into the air along with other gases and ash. Each of these materials can be harmful to the health of trees, animals, and people. Entire forests are at risk due to acid rain, and life no longer exists in some rivers and lakes that have been polluted by waste materials from mining or industry. Pollutants become especially dangerous when they get in our water supply. They create serious threats to the health and lives of plants, animals, and humans. New technologies have been developed to prevent pollution, but there is still much that must be done.

Natural Cycles

The most abundant elements in the tissues of trees and other living organisms are carbon, hydrogen, oxygen, and nitrogen. These four elements account for 96% of the material that is found in living organisms. Over thirty other elements are known to make up the other 4% of living tissue.

These thirty elements are so important in the formation of living tissue that they are used over again and again. An atom of carbon may exist as a starch molecule in a pine nut that is later eaten by a squirrel. It may then pass from the tissue of the squirrel to that of an owl when the owl makes a meal of the squirrel. The owl may eventually die of old age and decompose on the forest floor. Finally, the carbon atom is deposited in the soil to be taken up as a nutrient by another plant (Figure 5–6). In this manner, elements cycle from living organisms to nonliving materials and back, again and again. This circular flow of elements from living organisms to nonliving matter is known as an **elemental cycle.** Cycles exist for all of the elements that make up living tissue.

Carbon is the most abundant element found in living organisms. It makes up the framework of the molecules that are found in living tissue. In the absence of water, nearly half of the dry matter found in trees and other plants consists of carbon. Carbon moves readily between living organisms, the atmosphere, the ocean, and the soil (Figure 5–7). The respiration process of both plants and animals releases carbon dioxide (CO_2) into the atmosphere. Using light in the process of photosynthesis, plants take carbon dioxide from the atmosphere and turn it into sugars that they use to make new tissue such as roots, stems, and leaves. When plant tissue decays, carbon dioxide is released back to the atmosphere as a gas or converted over a long period of time to fossil fuels such as natural gas, crude oil, or coal. Sometimes plant materials are eaten by animals. When animals die, their bodies decompose, releasing carbon in the same manner as that described for decayed plant materials.

People mine or extract fossil fuels from the surface of the earth for use as fuels and other purposes. When these materials are burned, the combustion process releases carbon dioxide to the atmosphere. The oceans absorb large amounts of carbon dioxide from the atmosphere when the carbon content of the atmosphere is high, and they release carbon dioxide to the atmosphere when atmospheric carbon dioxide decreases. Until recent years, the carbon

Carbon Cycle

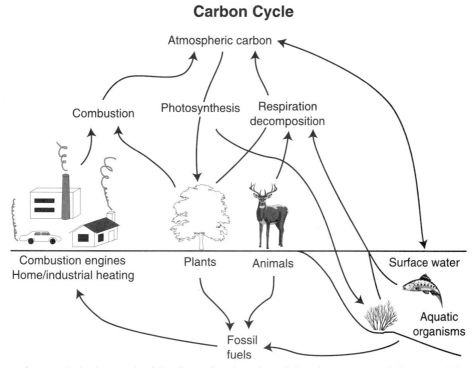

Figure 5–6 The carbon cycle is the result of the flow of carbon from living tissues to nonliving materials and back again in a continuous flow.

Figure 5–7 The oceans of the earth act as storage reservoirs to stabilize the amount of carbon in the atmosphere. This is accomplished by absorbing carbon compounds from the atmosphere when levels become too high, or releasing carbon compounds to the atmosphere when levels become too low.

content of the atmosphere remained nearly constant due to this action by the oceans.

During the past one hundred years, the large amounts of fossil fuels that have been burned have increased the levels of carbon dioxide in the atmosphere. We are burning these fuels faster than the oceans can absorb the extra atmospheric carbon. We do not know the effect of our massive CO_2 inputs to the atmosphere, but one effect is thought to be global warming. Long-term climatic change could have drastic consequences on our forests and all other living things.

Nitrogen is the most abundant element in the atmosphere. It makes up about 80% of the air supply. In its elemental form (N_2), it is a colorless, odorless gas that cannot be used by plants or animals. It must be combined with oxygen or other elements before it is available as a nutrient for living organisms. Plants and animals use nitrogen compounds to form protein and other important molecules.

Nitrogen fixation is a process by which nitrogen gas is converted to nitrates. This occurs in several different ways. **Nitrogen-fixing bacteria** can convert nitrogen gas to nitrates. Some forms of these bacteria live in the soil. Others live in nodules on the roots of legume plants. Some types of blue-green algae and fungi are also capable of nitrogen fixation.

Nitrogen fixation takes place naturally in the atmosphere when lightning strikes occur. The electrical current that passes through the atmospheric nitrogen converts some of the nitrogen gas to nitrogen compounds that can be used by plants. Nitrates are also released from animal wastes, and from plants and animals that die and decay.

At the same time that nitrates are being produced from nitrogen gas, other nitrates are breaking down to release nitrogen gas back to the atmosphere. This process is called **denitrification.** It occurs when some forms of bacteria come into contact with nitrates. A similar process occurs when nitrates are carried by runoff water into surface water which constantly exchanges nitrogen with the atmosphere.

The circular flow of nitrogen from free nitrogen gas in the atmosphere to nitrates in the soil and back to atmospheric nitrogen is known as the **nitrogen cycle** (Figure 5–8).

FOREST SOILS

Soil is one of the world's most important natural resources. It is from the soil that nutrients are made available to all living organisms. Of equal importance with soil is clean water. When both soil and water are present, plant and animal life is possible. Without water, great barren deserts dominate the landscape, and nothing grows.

Soil is the surface layer of the earth's crust consisting of particles of different sizes and originating from a variety of materials. **Soil texture** is a measurement of the proportion of mineral particles of different sizes that are found in a sample of soil (Figure 5–9). The smallest of the mineral particles is clay (less than 0.002 millimeter in diameter). Silt includes mineral particles between

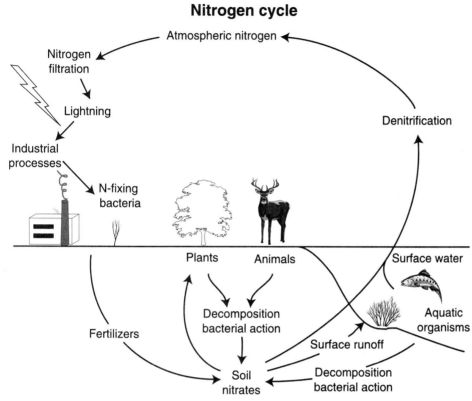

Figure 5–8 The nitrogen cycle is the result of a flow of nitrogen from living tissues to nonliving materials and back again in a continuous flow.

0.002 and 0.05 millimeter in diameter. The largest soil particles are sand ranging between 0.05 and 2 millimeters in diameter.

Soil color is influenced by the kind of minerals that make up the soil particles. Other factors that contribute to soil color include the amount of organic matter contained in the soil, and the drainage of the soil by which excess water is removed. Soil colors range from the red iron oxide soils of the southern and southwestern United States to the black organic soils of the corn belt region.

Organic matter or **humus** is important in soils because it provides food to soil organisms, releases nutrients to plant roots, increases the water-holding capacity in the soils, and increases the movement of air through the soil. It also tends to insulate the soil, causing soil temperatures to be more constant. Forest soils tend to build up deposits of litter on the forest floor (Figure 5–10). As the litter decomposes, it creates humus in the soil. Humus is also added to soil by the decomposition of small roots.

Soils tend to have differences from one region to another, and even within a localized area, major differences can be observed. One of the ways that scientists have developed to describe the soil in an area is to describe the texture and

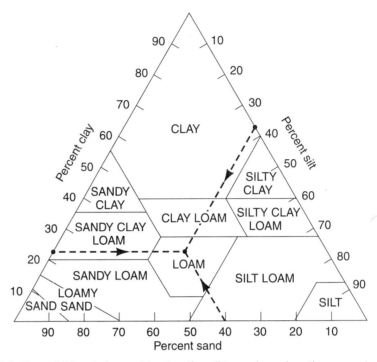

Figure 5–9 The soil triangle is used to classify soil types based on the percentages of sand, silt, and clay that make up a particular sample of soil.

Figure 5–10 Forest soils tend to be high in organic matter from the decay of large humus deposits on the forest floor.

content of the different **horizons** or layers in the soil. This description of a soil is called a **soil profile.**

Each of the horizons in a soil profile is represented by a letter. For example, the top horizon in a soil is the O horizon. This soil layer consists of dead and decomposing plant vegetation that is deposited on the soil surface. The O horizon is much more evident in many forest soils than it is in most agricultural soils. This is because the amount of vegetation that is deposited on the forest floor is quite high in comparison with many other soils. It is in the O horizon that most of the organic matter in the form of humus is added to the soil profile.

A complete soil profile includes the following horizons: O, A, E, B, C, and R. The A horizon is composed of mineral soil formed at the surface or beneath the O horizon in areas where it exists. The A horizon tends to be composed of mixtures of humus and mineral soil. It is the topsoil, and it is the most productive soil layer. The roots of many shallow-rooted plant varieties become concentrated in this layer of soil.

The E horizon is composed of mineral particles such as silt and sand. Most of the clay particles, humus, and other water-soluble materials have been leached out by the downward movement of water through the soil profile. This loss of soil components is called **eluviation.** The E soil horizon tends to be light in color in comparison to other horizons.

Materials that have leached out of the A and E soil horizons tend to accumulate in the B horizon. This buildup of translocated soil components is known as **illuviation.** This horizon is also called subsoil, and much of the clay in the soil profile becomes concentrated here. Subsoils that consist of high concentrations of clay are often poorly drained. This problem can seriously restrict the kinds of plants that can grow in the area. This is especially true of plants with deep root systems.

The C horizon contains the **parent material** (usually rocks), not including the bedrock material. Little or no evidence of soil formation can be seen in this soil horizon. The R horizon of a soil profile consists of deposits of granite, limestone, sandstone, and so on that form the bedrock layer.

The United States Department of Agriculture has identified ten soil orders and the types of forest cover that are usually associated with them. This soil classification method is based somewhat on color, texture, and the mineral content of a particular soil. Of the ten soil orders, three are of real significance to forestry in North America (Figure 5–11). These three soil orders are the Alfisols, Spodosols, and Ultisols. The remaining soil orders are restricted mostly to agricultural uses.

The **Alfisols** are high in calcium, magnesium, sodium, and potassium. The B horizon tends to be high in clay content. The Alfisol soils are slightly acid. They are best suited to such types of trees as the oaks, hickories, aspens, birches, Ponderosa pine, and Lodgepole pine.

The order of soils known as **Spodosols** are products of cold, damp climates and coarse silica parent material. The subsoil is illuvial, meaning that it is composed of materials (humus, aluminum, and iron) that have leached into the B

Soil Orders

Alfisols	Spodosols	Ultisols
Oak	Spruce	Loblolly Pine
Hickory	Fir	Shortleaf Pine
Northern hardwoods	Eastern White Pine	Oak
(New York)	Northern hardwoods	Hickory
Aspen	Aspen	
Birch	Birch	
Ponderosa Pine	Western Hemlock	
Lodgepole Pine	Sitka Spruce	
	Longleaf Pine	
	Slash Pine	

Figure 5–11 Of the ten soil orders, only the alfisols, spodosols, and ultisols are of real significance to forestry.

horizon from the A and E horizons. These are acid soils that tend to be light in color. They support the growth of forest species like spruce, fir, white pine, northern hardwoods, aspens, birches, Western Hemlock, Sitka Spruce, longleaf pines, and slash pines.

The third soil order that is of importance to the forest industry is the **Ultisols.** These soils are found in warm, humid climates, and they generally show evidence of heavy weather action as evidenced by illuvial deposits in the subsoil. The B horizon is composed of clay and reddish-colored iron deposits and silicate-based clays, and the E horizon is yellow in appearance. These soils support mostly Loblolly and Shortleaf pines along with oaks and hickories.

Many of the soils upon which forests are found have lost large amounts of top-soil as a result of soil erosion. **Erosion** is a destructive process that occurs when land is unprotected against the forces of flowing water or strong winds. Soil erosion is the number one source of water pollution in North America (Figure 5–12). It damages wildlife populations by polluting water supplies, killing young fish and aquatic animals, and filling reservoirs and lakes with deposits of soil.

Forest lands are most susceptible to soil erosion following forest fires that have destroyed the plant cover on the soil. Fires also kill the roots of many plants on the forest floor, and they damage the topsoil by breaking down the structure of the soil granules. Soil particles that have endured extreme heat no longer adhere to other soil particles, and the pores in the soil tend to become clogged with small soil particles, reducing the permeability and absorption of water by the soil. These damaged soils become like powder that is easily washed away by heavy rainfalls. Heavy flows of water from rapidly melting snow also contribute to severe erosion.

Harvest practices that disturb the soil surface over large tracts of land can also contribute to soil erosion. This is especially true when all of the trees are harvested from a large area using the clear-cut method, or when soil surfaces have been disturbed to construct loading sites or roads. Special practices must

Figure 5–12 Soil erosion is the number one source of water pollution in North America.

be used in damaged areas to cause precipitation to be absorbed by the soil and to prevent it from flowing overland across soil surfaces. One of the methods that is used frequently to prevent erosion is to create an uneven surface that will trap and hold excess water until it is absorbed by the soil (Figure 5–13).

Figure 5–13 Soil erosion from logging areas can be significantly reduced or eliminated by gouging the forest floor to create an uneven surface. This causes water to be absorbed instead of flowing across the soil surface.

Figure 5–14 Many forest managers have developed "silt traps" for the purpose of removing soil from flowing water by slowing down the rate of flow, allowing the silt particles to settle out.

Soil erosion is also prevented in disturbed areas by using straw bales, plastic sheeting, and other types of silt traps to slow the flow of water, allowing heavy silt particles to settle to the bottom of the flow (Figure 5–14). These structures are also used to disperse the flow of water over a larger area. This improves the prospects that the water will be absorbed into the soil instead of flowing across the soil surface. **Soil conservation** is the practice of protecting soil from erosion caused by strong winds or flowing water. Any action or method that contributes to the protection of soil surfaces is considered to be a soil conservation practice.

WATER AND WATERSHEDS

Water is one of the most important resources in the environment (Figure 5–15). It provides a living environment to many species of organisms, and supports the growth of plant life. Water is a required nutrient for living things. It makes up approximately seventy percent of the weight of living plants and animals (Figure 5–16). Water is used to control the temperature of organisms. Evaporation of water cools the surfaces of leaves and plants. Water also performs many other functions related to maintaining life.

Figure 5–15 Pure water is one of the most valuable and important resources in the environment.

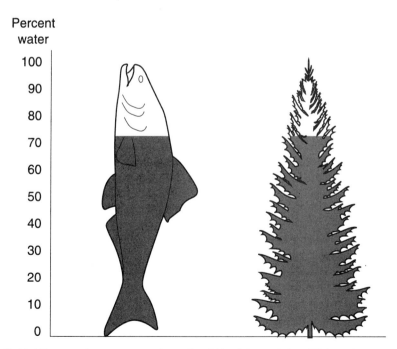

Figure 5–16 Water makes up approximately 70% of the weight of living plants and animals.

Figure 5–17 Marshes and swamps act as natural water treatment facilities by removing a variety of contaminants from surface water. *(Photo courtesy of USDA)*

Some of the most valuable resources in nature are marshes and swamps (Figure 5–17). These areas are natural water treatment facilities, and they remove a variety of contaminants from surface waters. They require no supplemental energy sources, and they function without human interference. The cleansing agents are plants, bacteria, and other aquatic organisms.

Vast expanses of forests and other vegetation are needed to regulate a uniform flow of water in rivers and streams. Precipitation on land surfaces can cause severe soil erosion unless plants are available to slow the flow of water over the land. Slow-flowing water infiltrates into the soil and comes out of the earth purified from most contaminants. An area where precipitation is absorbed in the soil to form groundwater is called a watershed (Figure 5–18). A watershed is an area in which water from rain and melting snow is absorbed to emerge as springs of water or artesian wells at lower elevations. Each watershed is separated from other watersheds by natural divisions or geological formations, and each is drained to a particular stream or body of water. Watersheds are valuable because they act like huge sponges, soaking up water from precipitation and melting snow and releasing it slowly. Forested areas provide ideal conditions for controlling the flow of water as it enters and leaves a watershed.

Water is also cleansed by plants as they take in contaminated water and release clean water vapor into the atmosphere through the process called transpiration (Figure 5–19). Plants give up large amounts of water to the atmosphere through this process. Transpiration is a controlled evaporation process by which plants lose water through pores in their leaf surfaces. These pores open wide

Figure 5–18 A watershed is an area in which rainwater and melting snow are absorbed to emerge as springs of water or artesian wells at lower elevations. *(Photo courtesy of Utah Agricultural Experiment Station)*

when leaf surfaces are hot, and they are closed when leaf surfaces are cool. Transpiration creates a negative pressure that draws water and nutrients into the plant from the soil.

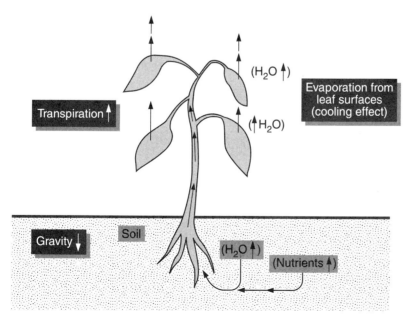

Figure 5–19 Transpiration occurs when water is released to the atmosphere from the surfaces of plant leaves.

Figure 5–20 Water environments are used by the beaver and some other animals as protection from predators.

Water in living tissues is part of the water cycle. It dissolves nutrients and transports them to the tissues that need them. It stores heat and helps maintain a more constant temperature in the environment. Water action helps to create soil by breaking down rock into small particles. It also cleans the environment by diluting contaminants and flushing them away from vulnerable areas. Water even provides protection to some species of animals living in forest environments (Figure 5–20). Living things require large amounts of water to survive.

The **water cycle** occurs when water moves from the oceans, to the atmosphere, to a land mass (in the form of rain or snow), then flows back to streams, rivers, and finally, to the oceans (Figure 5–21). The energy that drives this cycle comes from two sources: solar energy and the force of gravity. Water is constantly recycled. A molecule of water can be used over again and again as it moves through the water cycle. Solar energy that is trapped by the ocean is a source of heat that causes water to evaporate. Additional water enters the atmosphere by evaporating from soil and plant surfaces, especially in areas of hot temperatures and high precipitation.

Moisture from all sources builds up in the atmosphere forming clouds. Clouds release stored water to the earth surface in the form of rain or snow. Gravity draws the water back into the earth, and causes it to flow from high elevations to low elevations. Large amounts of the earth's water supply are stored for long periods of time. Storage occurs in aquifers beneath the surface of the earth, in glaciers and polar ice caps, in the atmosphere, and in deep lakes and

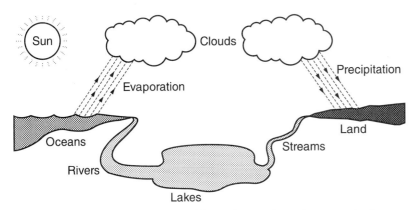

Figure 5–21 The water cycle is a natural flow of water from the ocean to the atmosphere, to a land mass, to rivers and streams, and back to the ocean.

oceans. Sometimes this water is stored for thousands of years before it completes a single cycle.

AIR AND CLIMATE EFFECTS

Air is a natural resource upon which plants and animals depend for survival. Among the greatest hazards to a forest and the wild creatures that live there are the effects of acid precipitation on the surfaces of plants and on the surface waters. The weak acids that are formed when rain combines with pollutants in the atmosphere are capable of killing forests and destroying living organisms in streams and lakes (Figure 5–22). These problems are evident in North America and in many other industrialized nations.

The exhaust gases from cars and trucks are the greatest source of atmospheric pollution, but factories and electrical power plants that burn large amounts of coal and petroleum products also emit large amounts of polluted gases into the atmosphere. Ultraviolet light from the sun reacts with atmospheric pollutants, adding to the atmospheric haze. The result of this pollution to the atmosphere is a great cloud of polluted air called **smog.** Part of the problem has eased since the development of vehicles that burn unleaded gasoline, but exhaust gases continue to pose a threat to the forests and the environments near large population centers because people drive more vehicles and travel longer distances every year.

One serious aspect of air pollution is that pollutants are carried by the wind to other areas. This creates damage to forests and wild environments over much larger areas, including those that are somewhat distant from major cities and towns. A considerable amount of damage has been sustained from acid precipitation by the forests and wildlife habitats of large regions of North America, especially in the vicinities of the Atlantic and Pacific coastal regions.

Air is cleansed as the water cycle operates; contaminants are trapped by falling rain or snow, and they are carried to the ground. The best solution we

Acid Rain

2 SO$_2$		O$_2$		2 H$_2$O	NO, NO$_2$	2 H$_2$SO$_4$
Sulfur	+	Atmospheric	+	Water	⟶	Sulfuric
dioxide		oxygen			Nitrogen oxides (catalysts)	acid

OR

2 NO		O$_2$			2 NO$_2$	
Nitric	+	Atmospheric	⟶		Nitrogen	
oxide		oxygen			dioxide	

3 NO$_2$		H$_2$O		2 HNO$_3$		NO
Nitrogen	+	Water	⇄	Nitric	+	Nitric
dioxide				acid		oxide

Figure 5–22 Acid precipitation occurs when acids are formed from pollutants in the atmosphere, reducing the pH of the rain or snow until it is at or below pH 5.

have to this problem, however, is to remove as much of the pollutants from the emission gases of cars and trucks as we can. It would also be wise to cut back on the amount of gases that are produced and released to the atmosphere. The best opportunity that we have for doing this is to create effective mass transit systems in our cities that reduce dependence on personal vehicles. We should also research new industrial processes that require less electricity or generate the electrical power that is needed using cleaner production methods that are less damaging to the atmosphere.

As people begin to experience greater health problems due to the effects of pollution to the atmosphere, there is likely to be greater motivation to solve the problems that are created by harmful pollution of the air that people, animals, and plants depend upon for survival. It is unlikely that humans will do very much to improve the quality of the air for the forests and wildlife, but the forests and wild animals will benefit when humans improve the air quality for themselves.

INTERACTIONS WITH INSECTS

Insect interactions with trees include both good and bad effects. Some insects that are found in forest environments are pollinators. They help ensure that fertile seed will be produced by the trees and other plants in the forest. Some insects are predators. They eat other insects, and in some cases, they play important roles in controlling harmful insects. Some insects are parasites that infect and control populations of harmful insects. A delicate balance sometimes exists between predatory and parasitic insects and the insects that damage the forest. A similar balance exists between insect populations and the diversity of the species of trees that make up a forest. A forest composed of a mix of different

Figure 5–23 Insect damage to forest resources causes greater annual losses to trees than all of the forest fires.

kinds of trees is less vulnerable to insects than a forest composed of a single variety of tree.

Insect damage to forest resources is indisputable, and it causes greater loss of timber than all of the forest fires (Figure 5–23). Insect damage is measured in terms of **growth impact,** which calculates timber losses due to reduced growth rates and death of the trees. It has been estimated by forestry professionals that as much as 123 million cubic meters of timber have been lost in the past fifty years due to insect damage.

Insect damage to every forest ecosystem is an ongoing destructive force that continues every day. Damage occurs to terminal buds, causing trees to lose **terminal growth** (vertical growth). It also occurs beneath the bark and inside the woody tissues of trees, interfering with **radial growth** or growth resulting in increased diameter of a tree. Massive insect infestations sometimes occur in forests, but the greatest insect losses are due to the day-to-day reductions in the growth of trees. Chapter 6 includes a description of the most important insects that impact the forest industry.

Figure 5–24 A forest habitat provides food, shelter, and protection from natural enemies for many kinds of wild animals.

WILDLIFE RELATIONSHIPS

Forests provide many of the essential elements for survival of wild birds and animals. Many wild animals obtain their food from plants and other animals that are found in the forest. The other basic need of an animal is a **habitat** or place to live that provides shelter from the effects of weather and climate, and that provides protection from its natural enemies (Figure 5–24).

Habitat

The need for a favorable living environment is greatest during extreme conditions in the weather that occurs. For example, during the winter months when an animal is weakened by an inadequate supply of food, it depends on shelter to conserve its energy. If the animal is able to get out of the wind, it will be able to maintain its body temperature with less food than when it is subjected to chilling conditions. The other period of greatest need for shelter occurs during and immediately following the birth of the young. Plenty of cover is needed to hide and shelter baby animals that are born into wild environments. Adequate habitat provides a place of shelter from weather and climatic conditions, and a place to hide or escape from predators.

Animals learn to use their environment to improve their chances of survival. For instance, a deer may develop the habit of bedding down during the hot part of the day on a hillside where it is not easily seen, but from which it can see anything that approaches its position. Once it has been detected, the same deer might follow a predictable route up the hill and away from the area to escape its enemy. These are learned or **adaptive behaviors** that improve the chances of the deer to avoid predators and hunters.

Food Webs

An animal that eats food obtained directly from plants is a **herbivore.** Herbivores may eat the foliage, seeds, fruits, or even the roots of plants. In some cases they may eat only specific plant parts, such as the seeds of a plant. For example, the Douglas Squirrel lives in pine forests where it gathers and caches pine cones from which it obtains the seeds for its food. The wild turkey has been returned to much of its original range, and it has been introduced into many new areas. This bird eats a varied diet, but when it has the opportunity, it consumes a diet that consists mostly of seeds such as nuts and acorns.

An animal that eats other animals is known as a **predator** or **carnivore** because it kills other creatures and eats meat. The Canada Lynx is a good example of a predator that lives in a forest environment. Most of these animals live in the Northern Coniferous Forest region. The favorite food animals of this predator are wild hares, but they also eat birds, rodents, foxes, and even small deer. Another example of a well-known predator is the Great Horned Owl. It ranges across most of the North American continent, and it eats rabbits, hares, small rodents, and birds.

Some animals and birds eat both plants and animals. These animals are called **omnivores.** The bear and the racoon are examples of omnivores. Such animals are usually successful in adapting to new environments and are in little danger of becoming extinct because they do not depend on a narrow range of foods. Some birds and animals depend on a single source of food, and when it becomes scarce, such creatures cannot survive.

A **food chain** is made up of a sequence of living organisms that eat and are eaten by other organisms that live in the community of animals. Each member of the chain feeds upon lower-ranking members of the food chain. The general organization of a food chain progresses from organisms known as **producers** (usually considered to be food plants) to herbivores. These plant-eating creatures are also called **primary consumers.** Carnivores eat the primary consumers, and they are known as **secondary consumers.**

A typical food chain begins with a plant as a food source, and it ends with a large predator (Figure 5–25). For example, a food chain may begin with a small

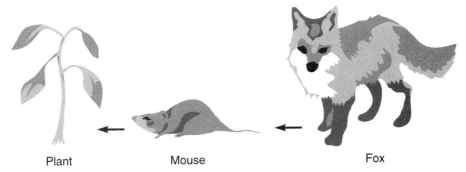

Plant   Mouse   Fox

Figure 5–25 The organization of a food chain progresses from food plants (producers), to herbivores (plant-eating animals), to carnivores (flesh-eating animals).

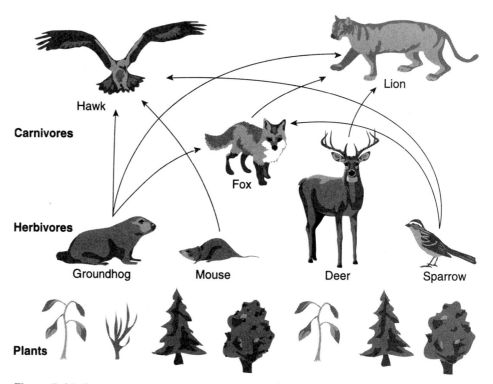

Figure 5–26 Food chains are interwoven with other food chains to form a food web.

forest shrub (producer) from which berries are obtained as food for a pine vole (primary consumer). This small rodent lives in colonies in underground tunnels beneath the forest floor. A shrew (secondary consumer) enters a burrow where it kills and eats a pine vole. Later, the shrew is killed and eaten by a great horned owl (secondary consumer).

Most food chains become quite complicated because many predators will eat nearly any animal that they can catch and kill. Each food chain is interwoven with other food chains to create a **food web** (Figure 5–26). A **food pyramid** arranges creatures in a ranking order according to their dominance in a food web.

The most versatile predators occupy the highest rank in a food web. These mammals or birds usually have few natural enemies, and they are capable of preying on a large variety of other species of birds and animals. They maintain their positions at the top of the food pyramid unless a stronger predator migrates into the area that is capable of competing more strongly for the existing food supply. Sometimes a new predatory species even preys upon the predatory species that formerly occupied the highest rank in the food pyramid.

When changes occur in the kinds of organisms that occupy an ecosystem, they affect nearly every other species in the ecosystem. The movement of humans into a new area has the effect of displacing the predators that occupy the top ranks in a food pyramid. Humans often assume these top positions by

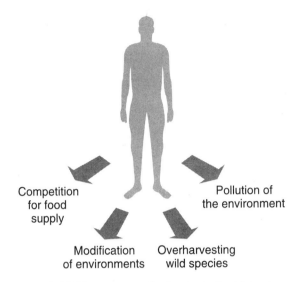

Competition
for food
supply

Pollution of
the environment

Modification
of environments

Overharvesting
wild species

Figure 5–27 Humans usually dominate the highest ranking in the food web.

preying on the herbivores in competition with the predators. They also control the size of the predator population by killing these animals or driving them out of the area.

HUMAN RELATIONSHIPS

The dominance of humans in forest environments and in the food web has contributed to the controversy that surrounds the movement to maintain and restore natural environments (Figure 5–27). It is clearly evident that humans are a very competitive species that tends to exploit natural resources to benefit themselves. It is equally evident that using natural resources beyond their capacity for renewal will eventually destroy them. Extreme positions have been taken on both sides of these issues. Some people contend that human dominance of other species of organisms is a natural process that has evolved since the beginning of human existence, and as the dominant species, humans have the right to exploit all other living organisms. Other people recognize humans as the only one among the living organisms that is capable of changing its behaviors to preserve other species of organisms. They believe that humans have the moral obligation to protect all other forms of life.

These extreme positions regarding acceptable uses of forests and forest environments by humans have resulted in political debate and controversy. The forest management plan that is most likely to succeed will need to be positioned somewhere between these two extremes. The human population must accept sustainable levels of use for natural resources, and human activities that exploit natural resources must end.

CAREER OPTION

Soil Specialist

A person who enters a forestry career as a soil specialist will work in the field classifying soils according to soil types. This data is then used to develop soil maps. The maps are used by forest managers as they make forest management decisions. The soil specialist is involved in solving problems related to soil damage due to erosion, fire, or other causes. A specialization in soils along with a university degree in forestry or a related field of study is a common approach to preparing for this career.

LOOKING BACK

The study of relationships between living organisms and the environments in which they live is the science of ecology. Conservation of matter is a principle of physics based on the evidence that matter can be changed from one form to another, but it is not destroyed. Pollutants that are created by modern society have negative effects on the environment that must be addressed. Natural cycles of water and elements have a cleansing effect on the environment. Soil and water are resources that make it possible for plants and animals to live and grow. A watershed collects water from precipitation and releases it slowly from springs located at lower elevations. The water cycle operates as water moves in a predictable pattern from oceans to clouds, falling as precipitation on land areas. It emerges and flows from springs to streams and rivers as it returns to the ocean. Clean air is important to forest health. Insects also play a role in the ability of a forest to grow and produce. Forests contribute shelter and food to many living creatures, and food relationships exist among plants, insects, and animals as food chains, food webs, and food pyramids. Humans are the most dominant members of food pyramids because they are able to adapt to changing environments.

QUESTIONS FOR DISCUSSION AND REVIEW

Essay Questions

1. How does soil erosion contribute to the pollution of surface water?

2. Explain the basic principle of physics known as the law of conservation of matter.

3. What are the key elements of the carbon and nitrogen cycles, and how do such cycles prevent pollution of the environment?

4. What makes carbon so important to living organisms?

5. Why is soil considered to be one of our most important natural resources?

6. What are the three most important classes of forest soils, and what characteristics separate them into distinct classes?

7. How does a watershed function, and why are healthy watersheds important?

8. Describe each of the events that is known to occur as the water cycle functions.

9. In what ways does polluted air affect forests?

10. How are food chains, food webs, and food pyramids similar? What are the differences among them?

Multiple-Choice Questions

1. The source of all energy used by plants and animals is:
   A. water
   B. sunlight
   C. soil
   D. air

2. The greatest single pollutant of surface water is:
   A. silt
   B. air pollution
   C. industrial waste
   D. mine tailings

3. Acid precipitation occurs when raindrops are converted to weak acids by absorbing:
   A. industrial waste
   B. animal waste
   C. sulfur dioxide
   D silt

4. Which of the following natural cycles is not an elemental cycle?
   A. carbon cycle
   B. nitrogen cycle
   C. water cycle
   D. oxygen cycle

5. A process by which bacteria are able to convert atmospheric nitrogen into nitrates that are useful to plants is called:
   A. nitrogen fixation
   B. nitrogenization
   C. denitrification
   D. metabolism

6. A process by which a buildup of translocated soil components occurs is known as:
   A. eluviation
   B. particle concentration
   C. erosion
   D. illuviation

7. A soil order that is high in concentrations of calcium, magnesium, sodium, and potassium is:
   A. alfisol
   B. ultisol
   C. spodosol
   D. humus

8. A destructive process that sometimes occurs in soils that are not protected against forces of flowing water or strong winds is:
   A. denitrification
   B. conservation of matter
   C. erosion
   D. soil conservation

9. A common description of a watershed is:
   A. an indoor toilet
   B. a small building that protects a well equipped with a water pump
   C. an area in which rain and melting snow are absorbed to emerge as springs of water at lower elevations
   D. a swampy lowland area

10. Rain or snow that is contaminated with smoke from engine exhausts or industrial plants forms a destructive product called:
    A. acid precipitation
    B. water cycle
    C. smog
    D. transpiration

11. Another name for a predatory animal that eats other animals is:
    A. secondary consumer
    B. producer
    C. herbivore
    D. primary consumer

12. The dominant position in a food pyramid is usually occupied by:
    A. an omnivore
    B. a predator
    C. a primary consumer
    D. a herbivore

LEARNING ACTIVITIES

1. Following a discussion about food chains, food webs, and food pyramids, arrange for the class to visit an outdoor site near the school. Assign class members to carefully observe and record the plants, insects, and animals that live there. After returning to the classroom, have each student prepare an illustration of a food web that they observed. Display the illustrations in the classroom, and use them in a discussion to determine the order in which the organisms should be arranged to create the food pyramid that most correctly describes the area.

2. Invite a person to visit the class who has expertise in the identification of insects. Extension educators, foresters, and urban foresters in many areas are competent to lead discussions about the ways that insects affect trees and forests. Ask the visitor to use pictures, slides, or actual insect collections to raise the interest levels of the students. Assign pairs of students to bring a picture or illustration of a particular insect to class. Make a collage on the classroom bulletin board with the pictures and illustrations that the students contribute.

Chapter

6

Diseases and Pests of Trees

Terms to Know

disease
biotic disease
abiotic disease
fungi
parasite
saprophyte
white rots
brown rots
heart rot
conk
fruiting body
rotation age
cutting cycle
rhizomorphs
canker
rust
wilt
root graft
vector
parthenogenesis
mechanical control
biological control
chemical control
insecticide
integrated pest management (IPM)
recombinant DNA technology
genetic engineering
rodents
depredation hunt

Objectives

After completing this chapter, you should be able to

* distinguish between biotic and abiotic diseases

* describe the symptoms of the heart-rot diseases and explain the significance of these diseases to the forest products industry

* explain the roles that fungi play in causing diseases of trees

* describe the symptoms of the canker diseases and identify some control methods that are used in the forest industry

* describe the symptoms of the rust diseases and identify some control methods that are used in the forest industry

* identify the symptoms of vascular wilt infections in trees and recommend a method for controlling these diseases

* name and describe some forms of abiotic diseases in trees

* name the most important classes of destructive forest insects, and describe the nature of the damage that they inflict upon trees

* distinguish between biological control and chemical control of insects, and list some examples of each control method

* discuss the merits of integrated pest management (IPM) and genetic engineering as insect control methods for the future

* identify other pests that damage or kill trees, and suggest methods for controlling them

157

Insects, pests, and diseases have always been part of the forest environment. There was no real effort to manage these disruptive influences in the forests, however, until long after the European settlers arrived in North America. During this period in history, the forests of this continent seemed much too vast to raise concerns over shortages. Even the damage to the forests that was caused by insects, pests, and diseases seemed insignificant. Today we see these forests as finite resources that may be depleted if careful oversight and management practices are not used to protect and maintain them.

DISEASES

All living organisms are affected by destructive disorders that interfere with health or cause illness. When a particular disorder occurs that can be traced to a specific cause with consistent symptoms, it is called a **disease** (Figure 6–1). A disease has predictable symptoms, and the causes of the disease are the same in each occurrence. Diseases that are caused by living agents of infection such as bacteria, fungi, viruses, micoplasmas, parasites, or nematodes are described as **biotic diseases.** A disease that is caused by a nonliving factor or condition is an **abiotic disease.**

The most destructive agents of disease in the forest are the **fungi.** These organisms are threadlike plants that lack chlorophyll. Since they cannot make their own food, they obtain their nourishment from other organic materials. Fungi that obtain nourishment from living organisms are called **parasites.** These are organisms that live in the tree and obtain their nourishment from the tree, but they do not contribute to the good of the tree. Fungi that obtain their nutrients from dead organic materials are called **saprophytes.** These organisms perform the useful function of converting dead vegetation to humus, but they also invade the heartwood and other mature tissues of living trees. Nearly all of the fungi that affect trees are saprophytes.

Biotic Diseases

Wood rots are diseases that occur when fungi cause decay in trees. Two types of wood rot are known to occur, and each type utilizes a different decay

Tree Disease Agents

Biotic Diseases	Abiotic Diseases
Bacteria	Drought conditions
Fungi	Temperature extremes
Viruses	Poisonous effects
Micoplasmas	Nutrient deficiencies
Parasites	Acid precipitation
Nematodes	

Figure 6–1 Tree diseases are caused by both living (biotic) and nonliving (abiotic) factors.

process. The least damaging of these are the **white rots.** The fungi that cause white rots break down both the cellulose and lignin components of cell walls. The **brown rots** occur when decay fungi break down the cellulose in the cell walls. It is estimated that over 70% of all timber losses due to forest diseases is caused by heart rot, which is a form of brown rot.

The presence of a fungus infection in a tree is sometimes difficult to detect on the outside. There are some external indicators that indicate that decay is present

Heart Rot

FOREST DISEASE

Fungi are responsible for **heart rot,** which is the decay of the core of deadwood that accumulates at the center of mature trees (Figure 6–2). This disease occurs when fungi invade the heartwood. The sapwood of young trees is usually resistant to wood rot fungi; however, the heartwood no longer sustains the flow of sap, and it becomes vulnerable to infection. As this disease progresses, the trunks of affected trees become hollow and weak. Many of these trees are blown down by wind in the later stages of heart rot. Timber that is harvested in the early stages of heart rot requires trimming to eliminate diseased wood, but in the late stages of the disease, entire sections of badly infected logs must be discarded because the diseased wood is of little or no value. This disease causes greater losses in North American forests than any other.

Figure 6–2 Heart rot occurs in a tree when the core of woody material at the center of the tree begins to decay.

Figure 6–3 A fruiting body or conk is a growth that is seen on the outside of a tree that extends inside the tree to an area that is infested with a fungus. The fruiting body releases fungal spores into the forest, causing the infection to spread.

in a live tree. One of these is the presence of a specialized growth on the trunk of a tree. A **conk** is a growth that arises from an infection caused by fungi inside the tree. It is a fruiting body that is filled with the reproductive spores from a growth of fungus (Figure 6–3).

The most effective way to deal with decay in timber is to try to prevent infections that cause decay. Fire prevention is important in preventing decay and rotting of trees. This is because the scars that are sustained around the bases of trees during a fire become entry points for fungus infections. Other silviculture practices for dealing with rot in trees include reducing the **rotation age** or maturity stage at which a stand of trees is harvested. For example, if a serious problem exists with heart-rot in a stand of timber that is harvested

SCIENCE PROFILE

Fungi

A fungus is an organism consisting of specialized cells, but it lacks chlorophyll for making its own food, and it has no way to eat the nutrients it needs to sustain life. All of the fungi are organisms that derive their nutrients by absorbing them from other living things or from dead organic material. These organisms release their digestive juices on the materials in their environment, and these food sources are digested outside the cells of the fungus. When digestion is complete, the fungus absorbs the nutrients that were released by this process. Mushrooms and molds are examples of common fungus forms. A fungus reproduces by forming a **fruiting body,** which is a reproductive structure that releases spores into the environment. A spore is capable of growing into a complete fungus.

at one hundred years of age, a reduction to a rotation age of seventy-five to eighty years in the next generation of trees may solve the problem.

Another strategy to reduce rot in a forest is to increase the frequency of harvesting in forests of mixed ages. Reducing the **cutting cycle** from twenty-five years to twenty years can result in immature problem trees being removed from the stand during the regular harvest cycle along with mature timber. This improves the chances for forest managers to prevent the spread of fungus infections in the forest.

Root Rot

FOREST DISEASE

The most important disease agents associated with root rot are fungi. Root rots are hard to control because they are difficult to detect. Some fungi attack the succulent roots of young trees, while other fungi destroy the woody tissues of mature roots (Figure 6–4). Destruction of any part of the root system of a tree will cause the tree to decline in vigor and health. When a large part of the root system of a tree is damaged or destroyed, the tree is likely to be killed. The shoestring root rot fungus is a common forest disease in many forests of the world. This fungus becomes established in a stump or in the roots of a tree, and then using the stump or the plant roots as a food source, the fungus sends out thin strands of tissue called **rhizomorphs.** These strands enter into the surfaces of any tree roots that they encounter where they establish new colonies of fungi. It is also possible that new trees become infected when their roots touch the roots of infected trees.

Figure 6–4 Fungi are often associated with root rot such as the destruction of the woody tissue of this mature root. *(Photo courtesy of Boise National Forest)*

Butt rot and root rot are caused by fungi of many kinds. Infection spreads when fungal spores become wind-borne and land on freshly cut tree stumps. When conditions are favorable, the stumps become colonized by the spores, and a new cycle of infection begins. Each individual kind of fungus causes specific symptoms, and in some instances a particular fungus is more harmful to one kind of tree than it is to another. Among the methods of control that are in use is the treatment of the freshly cut stumps with applications of borax or other similar materials. Some attempts have been made to reduce fungal infections by removing stumps from the area. Some fungi do not appear to damage trees. These organisms are sometimes intentionally introduced to colonize stumps in competition with more harmful fungi.

Some fungi cause the wood to become stained and discolored (Figure 6–5). This usually happens when moisture is present and temperatures are warm. This is a problem with logs that have been cut and stored to await processing. Logs that are stored in ponds of water or that are kept wet by sprinkler systems are less affected by these fungi than logs that are allowed to dry quickly (Figure 6–6). This is because less oxygen is available to the fungi when a log is wet. Logs that are wet also tend to produce lumber with fewer cracks due to the center of the log drying out at a similar rate to the outer layers. Cracks occur in the logs as the outer tissues shrink faster than the inner core. Cracks in the log surface provide points of entry into the log for fungi.

Some kinds of fungi attack soft tissues of the tree such as the cambium and the bark, causing them to die. An infection of this type that kills patches of tissue on the trunk or branches of a tree is called a **canker.** The cankers affect trees of both the conifer and hardwood varieties.

Figure 6–5 Wood sometimes becomes stained and discolored due to fungal infections. *(Photo courtesy of Boise National Forest)*

Figure 6–6 Stored logs that are kept wet by spraying them with water are less susceptible to fungal damage than dry logs. They are also less likely to crack due to uneven drying of the interior and the exterior of the log.

Canker

Two general types of cankers, annual and perennial, are known to infect trees. The fungi that cause most cankers enter trees through wounds in the bark. Injuries to the tree trunk or branches from any source make the tree susceptible to canker infections. Annual cankers move rapidly through the bark and cambium of a tree. When the tree is girdled by the infection, it soon dies.

Perennial cankers persist in the tree for several years. They cause lesions along the edges of the infected areas that produce callus tissue. This tissue causes damaging growths to develop that extend into the woody tissue, causing it to decay and become weakened. These growths also result in large lesions located on the outer surfaces of the tree. The best control method for most cankers is to remove infected trees from the stand, eliminating the source of infection.

A class of plant diseases that results in spotted red or brown discoloration of the stems and leaves is known as **rust.** The discoloration comes from the spores of the fungi that cause this disease. These diseases attack trees in different ways. For example, rust infections may invade the cones, needles, and stems of some conifer trees, but they may infect only the leaves of certain hardwoods.

All of the fungi that cause the rust diseases spend part of their life cycles on two different unrelated host plants. In addition to tree hosts, the fungi that cause white pine blister rust also infect currant and gooseberry bushes (Figure 6–7). An

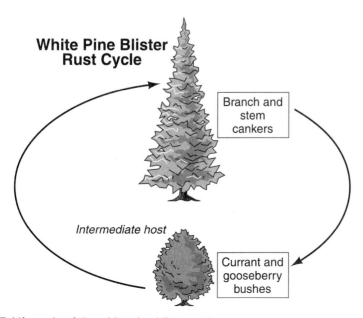

White Pine Blister Rust Cycle

Branch and stem cankers

Intermediate host

Currant and gooseberry bushes

Figure 6–7 Life cycle of the white pine blister rust

FOREST DISEASE

Rust

White Pine Blister Rust

The fungus organism that causes this disease produces up to five different spore stages. Some of these spore stages infect trees, and the others infect currants and gooseberries, the secondary plant hosts. This disease, which is of Asian origin, was introduced in the United States nearly one hundred years ago on white pine planting stock imported from Europe. It now infects trees in many regions of the world. The North American white pine tree varieties proved to be very susceptible to this disease (Figure 6–8). The best hope for controlling this disease is to use resistant strains of plants to regenerate white pine forests.

Fusiform Rust

The fusiform rust disease organisms infect southern pines such as the loblolly and slash pines. The fungus that causes this disease also infects oak trees. When the original pine forests containing fusiform rust–resistant populations of longleaf pine were cut, the oak trees increased in abundance, and the loblolly and slash pines were planted to speed forest recovery. These management choices favored the increase of the disease. Current efforts to control this disease include the use of fungicides in forest nurseries, and restoring forests with fungus-resistant planting stock.

Figure 6–8 White pine blister rust causes the formation of pustules that contain the spores by which the disease spreads from infected trees to healthy trees. *(Photo courtesy of Boise National Forest)*

intense effort was made over a period of many years to eliminate fungal infections by destroying gooseberry and currant bushes throughout the forests where this fungus is a problem. This proved to be impossible, and the effort was later abandoned.

Fungi of some types invade and grow in the vessels of the xylem tissue where they block the flow of water in the trunk and branches of the tree. This kind of disease is called a **wilt.** The Dutch Elm disease that destroyed many of the American elm trees in North America is this kind of disease. Oak wilt is similar to the Dutch Elm disease; however, it is a native disease to which oak trees are vulnerable.

Some of the wilt infections are diseases that are spread easily from one tree to another. It is believed that beetles and other insects that eat tree sap play a role in spreading this disease. The disease also spreads through root connections that occur naturally between trees when roots grow together. Such a connection between roots is called a **root graft** (Figure 6–9).

Other disease organisms that infect the trees in North American forests include bacteria, nematodes, fungi, and viruses. Some insects and disease organisms even work together to invade trees. For example, some wood wasps are known to carry fungal spores in their egg sacs that are deposited in the tissues of trees along with the eggs of the wasp. In this manner, the fungus invades the tunnels created inside the tree by the wasp larvae. Still another disease agent consists of the seeds of the dwarf mistletoe, a parasitic plant.

Figure 6–9 A root graft occurs when the roots of two different trees come in contact with one another and grow together.

FOREST DISEASE

Vascular Wilt

The class of diseases known as wilts is of greatest importance in angiosperms (flowering trees). The leaves of affected trees tend to show symptoms of water loss such as limpness. This is due to the blockage of the vessels that carry water and nutrients to the foliage of the tree. The most important of the wilts are Verticillium wilt, Dutch Elm disease, and oak wilt.

Control methods for vascular wilts include the use of caution to avoid injuries to trees that allow fungi to infect them. Other control efforts include the removal of affected trees, and trenching or fumigating with chemicals between trees to break root connections. Some success has been observed when trees are injected with fungicides, but the best long-term solution is probably the development of disease-resistant trees (Figure 6–10).

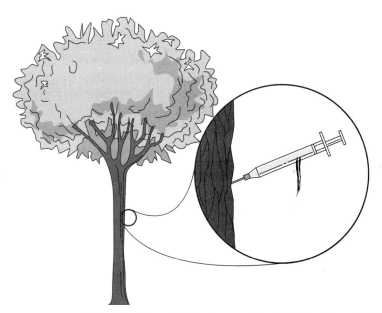

Figure 6–10 When small, localized fungal infections are identified in a forest, individual trees are sometimes injected with fungicide to control the disease.

Abiotic Diseases

Abiotic diseases include damage caused to trees by conditions such as drought, heat, cold, poisons, air pollution, nutrient deficiencies, mechanical damage, and weather-related factors. These conditions are not contagious, but they may affect large numbers of trees in an area.

Dwarf Mistletoe

Dwarf mistletoe is a plant that grows out of the branches of conifer trees (Figure 6–11). Despite the ability of the mistletoe plant to engage in photosynthesis, a tree that is a host to mistletoe is seriously stunted in its growth and sustains high losses in timber production. This disease ranks second behind heart rot in the damage that it causes to conifer forests in the western United States. It also affects the black spruce in eastern forests and in the Great Lakes region. Dwarf mistletoe spreads by seeds. These plants are parasites that grow directly into the tissues of the tree branches and stems from which they draw nourishment. Control of this forest disease is done mostly by physically removing or burning affected trees or their parts.

Figure 6–11 Mistletoe is a parasitic plant that grows into the branches of conifer trees. It obtains its nutrition from the plant fluids obtained from the host tree.
(Photo courtesy of Boise National Forest)

Sunscald is a condition in which the bark of trees is damaged by direct sunlight. This condition may follow logging or thinning operations that allow direct sunlight to penetrate below the foliage of the forest canopy. Damage can be severe enough to interfere with the function of the bark. Similar damage can occur due to freezing temperatures following a few warm days in the late winter or spring seasons.

Long periods of drought are known to cause declines in deposits of hardwood in the annual rings of a tree. When a drought condition becomes so severe that it causes the loss or death of plant tissue, it is called blight or dieback. Symptoms include wilting of foliage that may be severe enough to cause branches or even the entire tree to die.

Air pollution is another source of abiotic disease. Sulfur and nitrogen compounds are often given off as exhaust fumes by motor vehicles and industrial factories. When these materials combine with raindrops, weak acids are formed (Figure 6–12). These acids are destructive to all kinds of plants, especially trees. The foliage of affected plants is sometimes damaged so badly that a decline in photosynthesis appears likely.

Trees require nutrients of the proper kinds and in the right amounts. When nutrients such as phosphorous are missing or too little is present in the soil, trees suffer deficiency symptoms, and they are unable to sustain normal growth and development. Too much of a particular nutrient can poison the tree. Both of these conditions are nutritional diseases. Herbicides are chemicals that were developed to kill weeds. Unfortunately, some herbicides also kill or damage trees, and they must be used with great caution. Care should be taken to make sure that these chemicals do not get on trees or in the soil near tree roots. Using a contaminated sprayer to apply insecticide to a tree might kill the tree if

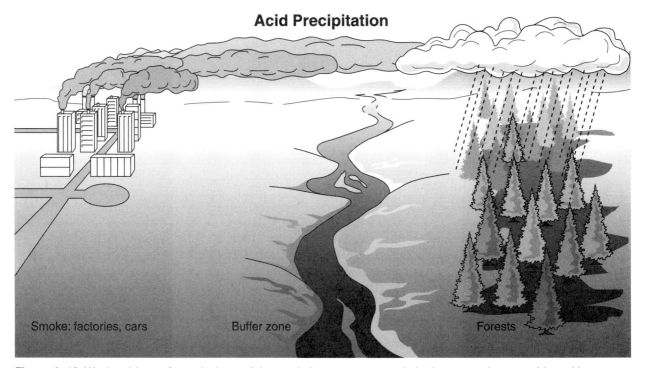

Acid Precipitation

Smoke: factories, cars   Buffer zone   Forests

Figure 6–12 Weak acids are formed when sulphur and nitrogen compounds in the atmosphere combine with moisture such as snow or raindrops. The resulting acids are capable of defoliating and killing a forest.

herbicide residues remain in the cannister. Sprayers should be thoroughly cleaned immediately after they are used to avoid chemical contamination.

One of the serious effects of abiotic diseases is that the trees they infect become weak. In this weakened condition, trees are unable to resist more serious diseases, and biotic disease agents are often able to invade and become established in them.

INSECTS

The ecology of insects and forests was discussed in Chapter 5 of this text. Relationships between insects and the forest environments were the topic of that discussion. The discussion in this chapter will deal with insect types and management of forests, including methods for controlling the insect pests that are found there.

Bark Beetles

The mode of action of the bark beetles is to bore an entrance hole through the bark of a tree. Tunnels are cut through the area between the bark and the woody part of the tree trunk (Figure 6–13). After mating, the females lay their eggs inside the entrance tunnels or in a series of tunnels that have been excavated beneath the bark to form galleries. The larvae feed upon the phloem and xylem tissues of the tree (Figure 6–14). Sometimes they completely girdle the tree beneath the bark, interrupting the flow of nutrients through the tree. These trees eventually die.

Figure 6–13 Bark beetles excavate galleries beneath the bark of susceptible trees in which they lay their eggs. After the eggs hatch, the larvae feed on the tissues of the tree.

Figure 6–14 The larvae of bark beetles and some other insects cause considerable damage to the phloem and xylem tissues of trees. Sometimes, they completely girdle and kill the trees on which they feed. *(Photo courtesy of Boise National Forest)*

Figure 6–15 Mature adult bark beetles cause serious damage to many different kinds of live trees, and several similar species exist in North American forests. *(Photo courtesy of Boise National Forest)*

Trees that are healthy and vigorous have a defense mechanism against beetles (Figure 6–15). Resins and sap flood the tunnels of the adult beetles, causing them to drown (Figure 6–16). Trees that are overcrowded or past maturity may not have enough sapwood to combat these pests. They are more vulnerable to the bark beetles than are the young trees.

Figure 6–16 The flow of sap from tree tissues damaged by insects is a defense mechanism of the tree. Insects drown when they become trapped in the flowing sap. *(Photo courtesy of Boise National Forest)*

Some beetles carry disease organisms in their bodies, and they distribute them as they move between trees. An insect or other organism that spreads disease organisms in this manner is a **vector.** One example of the spread of a deadly disease by a vector is the Dutch Elm disease that has been spread by bark beetles. This disease has killed nearly all of the American elm trees in many regions of North America.

Defoliators

Some insects do damage to trees by feeding on the leaves and needles (Figure 6–17). A large population of defoliators such as caterpillars, webworms, sawflies, or tussock moths can completely remove the foliage from a tree (Figure 6–18). Sometimes many of the trees in an area are defoliated by insects.

The loss of leaves or needles affects different trees in different ways. Complete defoliation of some species of pines with needles that persist from year to year may kill the trees. This is partly due to their dependence for plant food on foliage that has been accumulated for several years. Some broad-leaved trees, such as the Yellow Birch, Black Maple, and Sugar Maple, may also be killed in a single season if they are defoliated during midsummer. Most other broad-leaved trees and deciduous needle-bearing trees are able to withstand defoliation for several seasons before they are killed.

Some defoliating insects eat only the tissue inside leaves. Some of these insects damage leaves by chewing tunnels through the internal leaf tissue. Other insects, such as the skeletonizers, eat all of the soft tissue inside the leaves. All that remains of these leaves are the leaf veins.

Figure 6–17 Some insects injure or kill trees by feeding on the leaves and needles. Such insects are called defoliators. *(Photo courtesy of Boise National Forest)*

Figure 6–18 Defoliating insect populations are capable of removing all of the leaves from a sapling. *(Photo courtesy of Boise National Forest)*

Defoliators often weaken trees enough that they can no longer resist other types of deadly diseases and insects. Sometimes the secondary invasion of the tree by disease organisms or insects is fatal (Figures 6–19, 6–20). In many instances, terminal and radial growth of trees is seriously reduced. When such factors as temperature, light, moisture, and wind are favorable to defoliating insects, the insects are capable of rapid and massive population growths. When this happens, large areas of forest land may be defoliated by these insects.

Root-feeding Insects

The insects that feed on tree roots cause death to trees, or stunt their growth by causing damage to their root systems. These insects consist mostly of wire-worms and white grubs, both of which are beetle larvae. The wireworms are immature forms of a group of insects known as click beetles, and the white grubs are larval forms of May beetles and June beetles. These insects feed on the roots, causing young plantings of trees to be stunted or killed.

Another type of root insect is the weevil (Figure 6–21). Mature Pale weevils lay eggs in the stumps of freshly harvested pine trees. When the adult weevils emerge from the stumps, they eat the bark off the roots and stems of young seedling trees. Entire plantations of young trees have been killed in this manner.

The pine root collar weevil is an insect that infests the trunks of some pine species in the area of the root collar where roots are attached to the tree trunks. It feeds upon this tissue, causing it to become weak. Damaged trees sometimes break off level with the ground during windstorms, and others are often severely stunted in their growth. All of the root insects are capable of killing trees.

Figure 6–19 Lodgepole Pine killed by insects

Figure 6–20 Insect infestations sometimes become so serious that large tracts of trees are killed.

Figure 6–21 Weevils nibble plant tissues with jaws that are located at the end of the snout. They also use their snouts to drill holes in plant parts in which they lay eggs. *(Photo courtesy of Boise National Forest)*

Terminal-feeding Insects

The terminal-feeding insects cause serious deformities to trees by killing the meristem tissue located on the tips of growing branches and the tips of the central leaders on the main tree trunks (Figure 6–22). A tree that sustains this kind of damage usually responds by sending up a lateral shoot to take the place of the damaged central leader. This results in tree trunks that are crooked, and the quality and value of the lumber that these trees produce is poor in comparison with lumber produced from straight tree trunks. Terminal-feeding insects seldom kill a tree, but they sometimes cause the main trunk of a tree to fork, producing two leaders or trunks (Figure 6–23).

Insects that are responsible for damage to shoots and buds include the pine tip moths and the white pine weevil. They cause their worst damage in stands of trees that are about the same age. They are especially destructive in freshly planted and immature tree plantations. This is probably because an abundant food supply is available to terminal-feeding insects when all of the trees are in the early stages of growth.

Sucking Insects

The sucking insects that infect trees include the aphids, mites, leafhoppers, plant lice, scales, cicadas, and spittlebugs. All of these insects are equipped with mouth parts that penetrate young immature shoots, twigs, and foliage to feed upon the resin and sap of the tree. Large concentrations of these insects can cause the growth rates of trees to be reduced due to the continual loss of plant nutrients. The balsam wooly aphid is an exception. It frequently kills the balsam fir tree.

Figure 6–22 Some insects cause terminal damage by eating the meristem tissue. Other insects cause terminal damage by destroying the phloem and xylem of the immature branch tissue. *(Photo courtesy of Boise National Forest)*

Figure 6–23 Death of the main terminal bud of a tree often results in two lateral twigs becoming dominant. This results in a forked tree that is of low quality and value.

Aphids and plant lice tend to feed upon the softer plant tissues (Figures 6–24, 6–25). They eat sap, and they secrete a sticky, sweet liquid called honeydew. The honeydew falls down beneath the feeding area, attracting ants,

Figure 6–24 Sucking insects such as aphids and plant lice feed on tree sap that they obtain from the leaves. *(Photo courtesy of Boise National Forest)*

Figure 6–25 Trees can become seriously weakened by sucking insects, making them susceptible to other insects and diseases. *(Photo courtesy of Boise National Forest)*

SCIENCE PROFILE

Parthenogenesis

Aphids are a type of sucking insect with an unusual mode of reproduction. Adult males and females mate in the fall, and the females lay their shiny black eggs on the individual needles of white pine trees. The eggs hatch in the spring, and all of the offspring are females that have no wings. These aphids give birth to live young throughout the spring and summer seasons. Many generations of aphids are produced with no males present. This kind of reproduction in which females do not mate with males to produce offspring is called **parthenogenesis** (Figure 6–26). As the fall season progresses, both male and female aphids are produced. These insects mate and the females lay eggs to preserve the species until the next growing season.

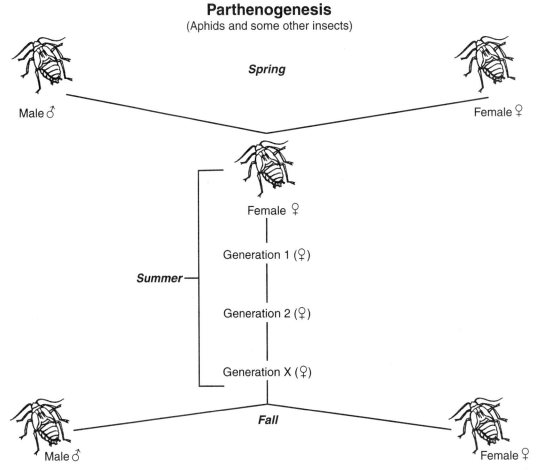

Parthenogenesis
(Aphids and some other insects)

Spring

Male ♂   Female ♀

Female ♀

Generation 1 (♀)

Summer—

Generation 2 (♀)

Generation X (♀)

Fall

Male ♂   Female ♀

Figure 6–26 Aphids are capable of reproducing several generations of offspring without mating. This reproductive method is called parthenogenesis.

wasps, and flies to the area. The honeydew also provides nutrients for a fungus that grows rapidly and produces black spores. The black-colored spores soon cover the surface of the tree, interfering with light absorption by the leaves. In the case of the white pine and some other trees, many of the leaves eventually shrivel, die, and fall off.

The scale insects are unusual in that the female secretes a large, round scale made of wax. The insect lives beneath this scale where she eats and reproduces live offspring called crawlers. Some of these crawlers are carried to new trees by strong winds while others remain on the tree where they were born. These insects feed by inserting their mouthparts into a tree branch, and drawing nutrients from the tree through a long hollow tube. Female scales lose their legs when they molt, and they are permanent residents of the tree where they live (Figure 6–27). They also excrete honeydew, and the fungus that feeds on the honeydew eventually causes leaves to die and fall from the tree in a similar manner as was observed with the aphids.

Mites infest leaf surfaces where they suck the cell contents out of leaf cells. The loss of chlorophyll eventually causes the leaves to become brown in color. Some kinds of mites spend part of their lives inside tree leaves where they eat leaf tissue. These mites are sometimes put in a distinct grouping of their own. In most instances, trees sustain more damage from fungus infections and loss of chlorophyll than they do from the loss of fluids.

Figure 6–27 Female scales deposit cottony egg masses on twigs and branches such as those deposited by the cottony maple scale on this maple tree.

Wood Borers

Wood-boring insects attack the heartwood and sapwood of trees that are weakened by stress or trees that are overly mature. Sometimes they even destroy wood in harvested trees that have not yet been processed. They are the most damaging insect pests in North American forests (Figure 6–28). They damage trees by tunneling through the mature wood as they eat. Most of these insects hatch from eggs that were deposited in cuts in the bark of the tree. After they hatch, they burrow into the phloem tissue. Eventually they work their way through the woody tissue to the center of the tree. Their tunnels become quite large, and the wood of affected trees is seriously damaged (Figure 6–29).

Among the wood-boring insects are two types: the long-horned, round-headed borers and the short-horned, flat-headed borers. The antennae of the long-horned variety are extremely long. These insects can also be identified by the round holes that they make in a tree. The short-horned, flat-headed borers leave and enter their tunnels through oval-shaped holes.

The whitespotted sawyer is a long-horned, round-headed borer that feeds on pine, spruce, and fir trees. The bronze birch borer is a short-horned, flat-headed borer that is a serious threat to birch trees throughout North America. There are many other kinds of wood-boring insects in North America. Some of the most serious of these forest pests are the mountain and southern pine beetles, Ips engraver beetle, Carolina pine sawyer, and the flat-headed apple tree borer.

Figure 6–28 Wood-boring insects cause greater economic damage to trees than any other class of forest insect. *(Photo courtesy of Boise National Forest)*

Figure 6–29 Wood is seriously damaged by wood-boring insects due to the holes that are made by these insects. Lumber from such trees has little value. *(Photo courtesy of Boise National Forest)*

One of the problems that may impact the control of insect pests is the vast amount of territory and the rugged terrain that some forests occupy. It is sometimes very difficult to get equipment to insect-infested trees. Another problem forest managers face is the tendency of some citizen groups to file legal proceedings to prevent the use of pesticides and other control measures on public lands. Some people believe that nature will seek its own balance, and that the results will be superior to management efforts by humans.

Management of the forest environment to minimize insect damage to trees can be done in several ways. One form of forest insect control that is practiced is removal of dead or dying trees for timber or firewood. This is a form of **mechanical control** of insect pests. It is not very effective at reducing harmful insect populations because most of their damage is done before the tree dies or becomes weakened. It is useful, however, in reducing localized pockets of insect infestations. Perhaps the greatest potential value that can come from removing trees that have sustained serious insect damage is that removal of weakened trees will reduce the risk that a serious disease will gain a foothold in the area. Many serious tree diseases get started in trees that have first become weakened by insect damage. Trees in a weakened condition do not have as much natural resistance to disease as do healthy trees.

Checks and balances are part of nature, and the forest environment is no exception. Birds provide the most valuable form of insect control that is available in forests. Huge numbers of forest insects are eaten every day by birds. Of particular importance as insect predators are the woodpeckers, chickadees, nuthatches, and creepers. These birds live in the forest and eat insects as a major component of their diet.

TREE PROFILE

Woodpeckers

The woodpeckers are important in forest environments as natural predators of wood-boring insects (Figure 6–30). They are well adapted to this role with chisel-like beaks and strong skulls that allow them to hammer out holes in tree trunks as they search for insects or prepare cavities in trees for nesting. Woodpeckers also have long thin tongues with which they extract insects from their holes in the trunks of trees. Most woodpeckers have feet that are adapted to clinging to tree trunks. Two toes face forward and two toes face backward. Their stiff tail feathers are used as props against the tree trunks. This bird has evolved into a highly specialized insect predator.

Figure 6–30 Woodpeckers are important natural predators of wood-boring insects. They use their long tongues to extract insects from their holes.

Some forest insect pests are killed and eaten by other insects. These insect predators may be natural to the environment or they may be introduced to forest environments by forest managers in attempts to control problem insects. The use of this kind of control practice is usually not disruptive or dangerous to other creatures in the environment. Some examples of predatory insects include the ladybugs and ants.

One kind of insect that is helpful in reducing harmful insect populations is the parasitic wasp or fly. These insects lay their eggs within or on the bodies of specific insect hosts. When a natural enemy of a harmful insect can be identified and introduced into a forest environment, some control of the harmful insect species can usually be attained. In most cases this kind of insect control does not harm the populations of beneficial insects. Harmful insects can sometimes be controlled by introducing microorganisms to forest environments that cause diseases. Pathogenic or disease-causing organisms include viruses, bacteria, fungi, protozoa, and others. The use of predators and disease organisms to control insect pests is a **biological control** method.

Chemical control is accomplished by applying chemicals to the environments that are infested by harmful insects. A chemical that is used to kill insects is called an **insecticide.** This type of control is very effective at reducing insect populations. The weakness in this insect control method is that chemical controls often kill the beneficial insects along with the harmful ones.

Some political action groups contend that any use of chemicals is unsafe for birds, humans, and wildlife species that live or work in the treated area. The insecticides that are used in the United States require rigorous testing under a variety of conditions before manufacturers are allowed to market them. The burden of proof that they can be safely used is on the company that owns the patents, and millions of dollars must be invested in scientific testing before these materials are approved for use.

Chemicals that are used according to the directions provided by the manufacturer can be used effectively and safely. This should not be interpreted to mean that insecticides are safe to use and that there are no hazards involved. Almost any substance can be abused when it is used in excess or when it is used under conditions for which it has not been tested. Good judgment and careful attention to detail is required of those who apply insecticides and other chemicals to forest environments. Any real or perceived danger due to the use of chemical insecticides and similar materials generally can be traced to their misuse, not their use.

Despite the known hazards that accompany the use of insecticides, such as the potential destruction of populations of useful insects, the use of insecticides in forests is sometimes the best choice among the insect management options available to forest managers. For example, when populations of pine beetles or gypsy moths reach epidemic levels, the use of insecticides is the only known method of insect control that can immediately reduce the insect population to a level that can be managed in other ways. It is also possible to apply insecticides in remote areas or in forests located on steep or rugged terrain by applying the insecticide with airplanes or helicopters. It is important to make sure that the amount and type of chemical material that is applied to the forest is well within the guidelines for safe use of that particular product.

A relatively new approach to control of insects and other pests is gaining acceptance in the agriculture and natural resource industries. It is called **integrated pest management** or **IPM.** It is a method of controlling harmful insects and providing protection to useful insects. It is proving to be a more practical approach to insect control than mechanical, chemical, or biological control methods when they are used separately.

An integrated pest management program makes use of all of the pest control strategies that are available including the use of limited applications of

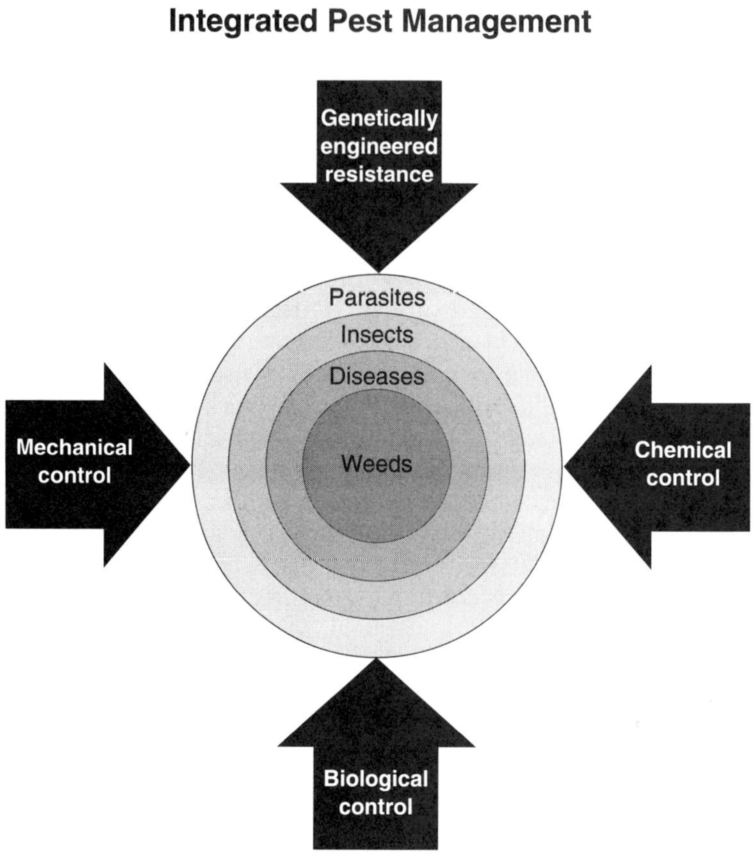

Integrated Pest Management

Figure 6–31 A multipronged attack using a variety of control methods is the most acceptable form of pest control.

chemicals. An IPM program depends mainly on the use of natural insect enemies and other forms of control to reduce harmful insect populations (Figure 6–31). Total destruction of harmful insects is not the objective of IPM. The objective of this kind of control is to keep some of the pests alive as a food source for their natural enemies. An IPM program strives to establish a natural balance between harmful insects and their natural enemies at population levels that are low enough to minimize insect damage.

Future insect control is likely to include the use of genetic engineering techniques to modify the genetics of trees, making them resistant to insect pests. This is accomplished using **recombinant DNA technology.** This procedure makes it possible to transfer a gene that is resistant to a particular kind of pest from a naturally resistant plant to a plant that is vulnerable to the pest. This is done by cutting a desired gene that is resistant to an insect pest from the chromosome of the resistant plant and inserting it into the chromosome of a nonresistant plant. Many years will be required to research the genetics of the different species of

Figure 6–32 Deer are fond of young tree shoots, and they sometimes cause serious damage to tree plantations. *(Photo courtesy of Texas Parks and Wildlife Department)*

Figure 6–33 Tree bark is a main source of food for the porcupine. Trees that lose their bark to this animal will usually survive unless the tree is completely girdled, but the growth rate is significantly reduced. *(Photo courtesy of Boise National Forest)*

trees and to create insect-resistant strains of each different kind of tree. It will also take a long time to produce enough of the genetically altered trees to make large plantings in the vast forests that are found in North America and around the world. The use of science principles and technologies to modify the genetics of a plant, causing it to express genetic resistance to pests and diseases, is part of a growing science known as biotechnology. The process by which plant genetics are changed is called **genetic engineering.**

PESTS

In addition to the insect pests, other creatures sometimes cause serious damage to forests. Among the gnawing mammals that injure trees are such creatures as mice, voles, gophers, rabbits, beavers, and porcupines. These animals are also classed as **rodents** by most scientists. Other animals that sometimes become pests in tree plantations are members of the deer family because they eat the young shoots and twigs found on the growing tips of tree branches and central leaders (Figure 6–32).

Most of the damage that is inflicted on trees by mice, voles, and rabbits occurs as they gnaw on the bark in plantings of young trees. During this stage in its development, the tree can be easily killed if the pests damage the cambium layer beneath the bark. The most common control for small rodents is to poison them, thereby reducing the population. It is important to be as selective as possible in the placement of baits to avoid deaths and injuries to birds and animals other than those for which the poison is intended.

Rodents such as porcupines cause extreme damage to coniferous forests by eating the bark of trees (Figures 6–33, 6–34). A porcupine is a climbing animal that is best known for its protective covering of sharp quills, but it can and does cause serious damage and death to trees. During the summer season, porcupines feed on a variety of plants, but during the winter season, much of their diet consists of the bark of trees.

Figure 6–34 The porcupine is a rodent that causes serious damage to trees by stripping the bark off the trunks of trees and eating it, especially during the winter.

Figure 6–35 Despite the destruction of trees by beaver colonies, it is generally agreed that they make up for most of their damage by raising the water table, thus improving the environment.

The beaver is a rodent that is destructive to the trees in the immediate vicinity of its territory. It is generally conceded, however, that the environmental value of the dams and ponds that it creates more than makes up for the trees it uses for food and structures. The stored water behind a beaver dam raises the water table in the area, contributing to an environment that is favorable to many species of wild plants and animals (Figure 6–35).

Deer and other large game animals can cause problems in nurseries and young tree plantings when they enter sensitive areas in large numbers. In some instances it is possible to discourage these animals from entering or feeding in an area. Small muslin bags of blood meal, tied among the branches of young trees, emit an odor that is offensive to deer. This practice can be effective in plantation plantings when there is an adequate food supply available to the animals in other locations. During times of drought or short supplies of food due to natural disasters such as fires, the only effective control methods for deer are high fences or depredation hunts.

A **depredation hunt** is a legal hunting season established by a state fish and game agency to reduce a herd of browsing animals that is causing serious damage in an area. Fences for deer, elk, and other large animals are usually too expensive to be considered as control measures unless the tree plantings they are intended to protect are high-value species.

CAREER OPTION

Entomologist

An entomologist is a person who specializes in the branch of biology that deals with insects. A career in this field usually requires an advanced degree in biology or a related science, and a specialty in entomology. The work of an entomologist will involve gathering research data through fieldwork, and using the data to learn more about the relationships of insects to the environments in which they live. Entomologists use their knowledge of insect anatomy, feeding habits, and life cycles to discover ways of strengthening populations of useful insects while controlling or reducing populations of harmful insects.

LOOKING BACK

Diseases of trees are of two types known as biotic and abiotic diseases. Biotic diseases are caused by living organisms, and abiotic diseases are caused by non-living factors in the environment. Fungi are the most harmful disease organisms in forest environments because they cause many of the diseases that are found in trees. The worst of these fungal diseases is heart rot. Other important tree diseases include cankers, rusts, rots, wilts, and dwarf mistletoes.

Insects are serious forest pests that cause heavy timber losses. Forest insect infestations are managed using mechanical, biological, and chemical control methods. Integrated pest management (IPM) and the planting of insect-resistant varieties of trees are considered to be the best insect control methods of the future. Other pests that affect forest production include rodents and sometimes deer.

QUESTIONS FOR DISCUSSION AND REVIEW

Essay Questions

1. How are biotic diseases of plants different than abiotic diseases?

2. Describe the symptoms of the heart-rot diseases, and explain their significance to the forest products industry.

3. What roles do fungi play in causing diseases of trees?

4. What are the disease symptoms of canker diseases in trees? How are these diseases controlled?

5. Describe the symptoms of the rust diseases, and identify some control methods that are used in the forest industry.

6. Identify the symptoms of vascular wilt infections in trees, and recommend a method for controlling these diseases.

7. Name and describe some forms of abiotic diseases in trees.

8. What classes of insects are harmful to forests, and what kinds of damage does each class inflict on trees?

9. Describe and give examples of biological and chemical control methods for forest insects.

10. How are genetic engineering and integrated pest management practices used to control insects?

11. What other kinds of pests damage or kill trees, and how can they be controlled?

Multiple-Choice Questions

1. A disease that is caused by a living organism is:
   A. not contagious
   B. abiotic
   C. pollution
   D. biotic

2. Wood rots are diseases that occur in trees due to organisms called:
   A. fungi
   B. cankers
   C. rhizomorphs
   D. rusts

3. Fungi are organisms that lack chlorophyll to make their own food, and some fungi obtain nourishment from other living things. For this reason, these fungi are known as:
   A. endomorphs
   B. rhizomorphs
   C. parasites
   D. cankers

4. When the roots of trees touch each other, they sometimes grow together, creating connections between their roots that are known as:
   A. rusts
   B. root grafts
   C. wilts
   D. rhizomorphs

5. An example of a disease known as a wilt is:
   A. White Pine Blister Rust
   B. Dutch Elm disease
   C. heart rot
   D. Dwarf Mistletoe

6. A fruiting body is:
   A. a structure that produces spores that develop into fungi
   B. a fleshy structure that surrounds the seeds of a plant
   C. a thin strand of fungal tissue that enters tree roots, infecting them with disease organisms
   D. a structure on a tree leaf in which sap becomes fermented to produce honeydew

7. A canker is a plant infection that:
   A. causes red or brown discoloration
   B. affects only conifer trees
   C. kills patches of tissue on the trunk or branches
   D. is a painful infection of the gum of a tree

8. An example of an abiotic disease in a tree is:
   A. canker
   B. Verticillium wilt
   C. Mistletoe
   D. sunscald

9. An entomologist is:
   A. a person who studies relationships between living and nonliving things
   B. a student of the fruiting habits of trees
   C. a hollow beaklike mouthpart with which sucking insects obtain sap for food
   D. a person who studies the branch of science related to insects

10. Which of the following is *not* a destructive type of forest insect?
    A. defoliator
    B. terminal feeder
    C. bark beetle
    D. pollinator

11. To which of the following types of destructive forest insects does the scale insect belong?
    A. sucking insect
    B. defoliator
    C. wood borer
    D. root feeder

12. Which of the following animals is *not* classed as a rodent?
    A. gopher
    B. mink
    C. mouse
    D. porcupine

LEARNING ACTIVITIES

1. Take a walking field trip around the neighborhood, and collect insect specimens that are suspected of causing damage in trees. Identify each of the insects that has been collected, and study how each of them interacts with the trees. Suggest ways that each of these insects might be controlled to prevent damage to trees. Assign students to repeat this exercise by gathering and displaying their own insect specimens.

2. Invite the county extension educator or a forest or park service official to discuss local forest and ornamental tree problems with members of the class. Discuss procedures for gathering plant materials from diseased plants that will prevent the diseases from spreading to new areas.

FOREST MANAGEMENT

Chapter

7

Terms to Know

law
department
agency
regulation
policy
oversight
sustained yield
dominant use

The Role of Government in Forestry

Objectives

After completing this chapter, you should be able to

* relate the early history of the roles that government played in forest management

* describe the impact of the Homestead Act on native American forests

* identify the main purpose of the Timber Culture Act of 1873

* express the two philosophies that influence forest planning in the United States

* explain why a management plan is needed for forest resources

* name the two federal departments that are responsible for managing most of our forest resources

* list the most important forest management agencies and explain their responsibilities

* describe how the Forest Reserve Act of 1891 changed the way forest resources were administered

* explain how the national forests, national monuments, and national parks were established

* discuss some of the differences that are likely to occur in the ways that national, state, and privately owned forests are managed

* distinguish between the multiple-use and dominant-use concepts of forest management

The forest lands of North America constitute a massive renewable natural resource that should be available to every generation of citizens. The federal governments of the United States, Canada, and Mexico along with state and provincial governments can ensure the future of our forests through management that balances present needs with long-term purposes. Forests that are wisely used will always be available to the citizens who depend upon them.

The role of government in conservation and management of forest resources is to manage public forest lands to sustain the yields of forest products in a manner that is consistent with science-validated silviculture practices. Government should also ensure that public forest lands are managed for the benefit of the citizens. Governments are responsible for enacting and enforcing laws and regulations that govern the uses of the land, water, wildlife, timber, and other natural resources (Figure 7–1). These responsibilities of government are sometimes contradictory, and this has contributed to confusion in how forest resources should be managed. Our democratic form of government has resulted in many compromises in forest management laws and policies.

PLANNING

The early colonists in North America probably saw little need for developing a plan for the use of forest resources. Forests were abundant in the new land, and they seemed far too vast to ever be threatened by overuse. Many of the colonists even considered the forests to be barriers to progress. The forests had to be cut, piled, and burned to make way for farms. To these people, a plan for forest management probably would have been considered a waste of time.

Figure 7–1 Most important decisions that affect the forest industry are made in the state capitals or in the United States capital at Washington, D.C.

The attitude of the United States government during the first one hundred years as a nation was to exploit the timber resources. Settlements were moving westward into new territories, and lumber was needed to build cities and towns. The railroads and mines needed timber products, and mature forests were available to supply these materials. Timber harvests were not controlled, and people were allowed to harvest free timber. During this period in American history, large tracts of public lands were converted to private ownership.

The Homestead Act of 1862 granted 160 acres to each settler who would develop the land into a productive farm. This prompted the homesteaders to clear the trees from large tracts of established forest land. The homestead law accomplished the government goal of moving citizens into the western territories, but in the process, it became evident that some of the land claimed under the law was unsuited for agricultural production. The desire on the part of government officials for citizens to take possession of the new western territories took precedence over any need for a comprehensive forest management plan.

The first major attempt toward forest planning occurred with the passage of the Timber Culture Act in 1873. This act was intended to create forests on land that was granted for homesteads in the Great Plains region. This law required 40 acres of trees to be planted on each 160-acre homestead (Figure 7–2). It was hoped that this would create a supply of timber in an area where none existed previously. Other congressional legislation allowed public land to be purchased

Timber Culture Act—1873

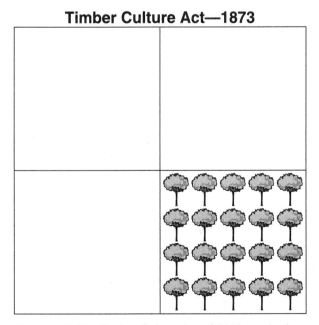

Figure 7–2 The Timber Culture Act of 1873 required the planting of 40 acres of trees on each homestead of 160 acres in the Great Plains region.

in 160-acre blocks for personal use other than agriculture in the four western states of California, Oregon, Washington, and Nevada.

During the mid to late 1800s, large tracts of timber were openly stolen by timber companies from the public forests with little or no penalty. Many of these lands were under the direction of the Department of Interior, but officials did not enforce the laws and regulations dealing with trespass and theft on the public lands. Protection of forest resources became a priority to Carl Schurz, secretary of the Department of the Interior in 1877. The secretary believed that the growing population would soon place greater demands on forest production than the forests could provide. He immediately began to enforce laws to stop the theft of timber, but several more years passed before the official government attitude toward the use of forests changed from one of exploitation to conservation.

The first step in any planning process is to develop a statement of beliefs such as a mission statement or philosophy. This is followed by establishing goals that are consistent with the philosophy. The next step is to develop a plan of action to achieve the goals. Developing a plan for forest management has been difficult. Part of the reason for this is that agreement on a forest management philosophy is split between those who wish to use and conserve forest resources and those who wish to enjoy and preserve forest environments.

The two distinctly different forest philosophies in the United States have evolved into two basic forest plans that affect most forest lands (Figure 7–3). These plans are administered by two federal departments, the Department of Agriculture and the Department of the Interior, and several agencies. Among these agencies are the United States Forest Service and the National Park Service. The Forest Service plan represents the views of the conservationists, and it makes most forest resources available for productive uses. The National Park Service has implemented a plan to preserve the forest lands and resources located within the boundaries of the national parks and monuments. Inside these

Forest Management Philosophies and Uses

Conservation	Preservation
Use the resource:	Save the resource:
—Timber harvests	—Camping
—Livestock grazing	—Hiking
—Public recreation	—Boating
—Mining	—Fishing
—Hunting and fishing	

Figure 7–3 Two different forest management philosophies have emerged in the agencies that manage forest resources, and each of these philosophies influences the management of vast tracts of forest lands.

boundaries, the only forest uses that are allowed are outdoor recreation and the enjoyment of natural forest environments.

The lesson to be learned from the past is that it is important to have a forest plan. A plan that is based mostly on compromises seldom satisfies anyone, but it is better than using the forest resources without a plan. Natural resources that are heavily used before management plans are developed often become abused resources. Evidence of this is the loss of more than 75% of the topsoil from some of the original forest lands located in the eastern region of the United States. Much of the erosion that occurred was the result of removing the trees and other vegetative cover from large tracts of land located on slopes. When the exposed soil surfaces were subjected to the heavy rains that are common to the area, water erosion caused serious damage to the soil.

Congressional action through the National Forest Management Act of 1976 requires the Forest Service to develop comprehensive plans for forest management (Figure 7–4). The plans are expected to consider the effects of timber harvesting on water quality, soil erosion, wildlife populations, and all of the other forest uses. The multiple-use concept of management requires a strong forest planning effort involving the different interest groups that use the forest. Future forest management plans must be flexible enough to function in an environment of changing public opinions. Plans should be based on sound scientific research, and they should take into account the needs of the citizens and the nation.

Figure 7–4 Federal legislation requires the Forest Service to develop management plans that consider the effects of timber harvests on water quality, soil erosion, wildlife populations, and all other forest uses.

ADMINISTRATION AND MANAGEMENT

The management of public forests is assigned to several different agencies at the national level. The United States Department of Agriculture (USDA) administers most of the public forest lands through the Forest Service. It also directs the research activities of the forest experiment stations. The Forest Service is responsible for managing the 155 national forests that have been established in the United States.

THE UNITED STATES FOREST SERVICE

The Forest Service is an agency within the United States Department of Agriculture (Figure 7–5). It is responsible for managing public lands designated as national forests and wilderness. The agency is presided over by the chief forester with headquarters in Washington, D.C. Each of the nine regional forests is under the direction of a regional forester with a regional office located in the region. Each region is organized into national forests with a forest supervisor providing leadership. National forests are divided into ranger districts with a forest ranger serving as the operating officer in each district. The research arm of the Forest Service includes eight experiment stations and a laboratory for forest products research (Figure 7–6). A director and several assistant directors lead each experiment station with each station divided into research units. Leading each research unit is a unit leader and one or more scientists. The Forest Service also includes divisions responsible for state and private forestry located in the regional offices and in one area office.

Figure 7–5 The United States Forest Service is an agency that is part of the United States Department of Agriculture (USDA).

Figure 7–6 Research is an important activity conducted by the Forest Service. In addition to research involving timber, the eight national forest research centers study such interesting topics as relationships between domestic livestock grazing practices and the health of elk herds that graze the same areas.

Figure 7–7 Public forest lands that are considered valuable for their scenic or historical characteristics are often managed by the Department of the Interior. Most of these areas have been designated as national parks or monuments.

The Department of the Interior is responsible for management of extensive public forest areas that have been designated as national parks and monuments (Figure 7–7). These lands are managed by the National Park Service. Additional forest land is mixed in with the vast grazing lands in the western states. Most of these forests are under the direction of the Bureau of Land Management (BLM) (Figure 7–8). Other agencies in the Department of the

Figure 7–8 The Bureau of Land Management (BLM) is a federal agency that manages forest lands that are mixed in with the vast grazing lands in the western desert region of the United States.

Interior that have management responsibilities for forests include the Bureau of Indian Affairs that controls forests located on reservation lands belonging to Native American tribes. The United States Fish and Wildlife Service exercises the right to control forest environments for the recovery efforts of endangered species of animals and plants. The Department of Defense manages the forest lands located on military installations.

FOREST LAWS AND POLICIES

Democratic forms of government operate by passing legislation that bccomes **law.** Once a law has been passed by Congress, it is assigned to an administrative **department** of government that is administered by the executive branch of government. A government **agency** is responsible to implement laws after they are passed by Congress and signed by the president. Agency employees write the rule or **regulation** that will be used to implement a new law. A regulation that is adopted by an agency may be approved as a **policy.** A policy is not a law, but it is a statement of how an agency will fulfill its responsibility to administer a law. The implementation of laws through agency regulations and policies is a major function of government.

One role of government in conservation and management of forest resources is to provide **oversight** or management responsibility for public

---

FOREST PROFILE

North American Free Trade Agreement (NAFTA)

Governments impact the forest industry in many ways in addition to establishing laws, regulations, and policies. An example of a major impact of the governments of North America on the forest and wood products industries is the implementation of the North American Free Trade Agreement by the governments of Canada, United States, and Mexico. Negotiations among these nations resulted in the approval of a "free trade" agreement in 1992 that affected the sale of wood and wood products across the borders of these countries and the elimination of many of the trade barriers that previously existed.

Since the implementation of NAFTA, the United States wood products industry has raised the issue of unfair advantage with regard to the sale of Canadian timber products in the United States. At issue is the ability of the Canadian timber industry to produce and sell timber at relatively lower prices than the prevailing prices in the United States before NAFTA was signed and implemented. In this instance, the reduction of trade barriers had an immediate and noticeable impact on the price of timber products. Since the free trade agreement appears to be here to stay, how should the United States forest products industries react? What options are available to the United States industry to deal with pricing issues such as this?

forest lands. Oversight should be carried out in a manner that is consistent with science-validated silviculture practices. Government should assure its citizens that public forest lands are managed to allow the most and best uses of the resources. The laws that are passed by state and federal legislative bodies define what the best uses of these resources are believed to be. The laws define what must be done, and the policies that are established by government agencies determine how the laws will be implemented. Many different laws and policies have been implemented in the management of our forests.

National Forests

The administration of publicly owned forests in the United States did not become a serious function of government until the mid 1800s. Prior to that time, much of the effort of public land trustees and managers was on producing revenues for the government by selling public lands. The Land Ordinance passed in 1785 called for the sale of public lands to reduce the national debt. Some states had established agencies by the late 1800s to manage forest preserves and to enforce laws and policies concerning timber harvesting, but prior to that time there was little meaningful government administration of forest resources.

The Forest Reserve Act of 1891 is one of the most significant laws regarding forest lands ever to be passed by the Congress of the United States. It gave the president powers to set aside public lands as forest reserves, but it did not provide any directions on how the forest reserves were to be used. The Yellowstone Park Forest Reservation was immediately created by decree of the president, and over the next few years, many other forest reserves were created. Logging and mining became illegal activities within forest reserves, although many illegal harvesting activities continued. The responsibility for management of the forest reserves was assigned to the General Land Office of the Department of the Interior.

The establishment of forest reserves by the executive order of the president of the United States resulted in conflict between the supporters of the two different philosophies of conservation versus preservation of forests. Conservationists believed in wise use of the forests, while preservationists wanted the forests closed to harvesting and mining activities. The establishment of new forest reserves was a victory for preservationists because commercial uses of forest reserves had not been approved by Congress.

President Grover Cleveland set the stage for a new round of conflict between these opposing views when he approved additional forest reserves before leaving office in 1897. Within months, Congress passed legislation including the Organic Administration Act that defined the purposes of forest reserves. These purposes still dominate the ways that public forest lands are managed today. The key points were that forest reserves were for the purpose of protecting the forests while improving the watersheds to provide for conditions that favored water flows in streams and rivers, and forest reserves were expected to provide a supply of timber to meet the needs of the nation.

Gifford Pinchot became the chief of the Division of Forestry in the Department of Agriculture in 1898. With the help of President Theodore Roosevelt, he was successful in getting Congress to pass legislation in 1905 that transferred the forest reserves from the Department of the Interior to the Department of Agriculture. With the transfer accomplished, the Forest Service was organized to replace the Division of Forestry, and the forest reserves became known as national forests (Figure 7–9).

President Roosevelt, through his administrative actions, probably had more impact on forest lands than any other president. Congress passed the American Antiquities Act in 1906 that authorized the president of the United States to set aside public lands for the preservation of sites that were of historic or scientific value. These sites became known as national monuments, and the use of the resources within their boundaries was restricted. President Roosevelt established eighteen national monuments during his term of office, and he transferred vast areas of public forest lands into the national forest system.

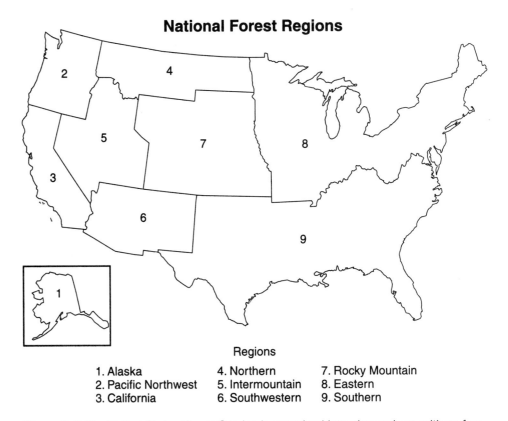

National Forest Regions

Regions

1. Alaska
2. Pacific Northwest
3. California
4. Northern
5. Intermountain
6. Southwestern
7. Rocky Mountain
8. Eastern
9. Southern

Figure 7–9 The United States Forest Service is organized into nine regions, with as few as two national forests (Alaska) and as many as thirty-three national forests in each region.

The issues of the early 1900s were mostly concerned with dividing the public land. Attempts were made to assign management responsibility for national parks to the Forest Service, but the Department of the Interior countered this proposal by assigning the national parks and monuments to a single administrative division. A leader in the preservationist movement, Steven Mather, was appointed to manage these resources. He was successful in obtaining passage of the National Park Act in 1916, and the National Park Service was established under his leadership.

Through much of the twentieth century, the forest conservation and preservation arguments have continued. When the National Park Service pressed for national forest lands to be converted into national parks, the Forest Service opened new roads in the forests for recreation. They even created wilderness areas to preserve the more scenic areas from exploitation. The strategy of the Forest Service in making these changes was intended to demonstrate that recreation and preservation of wild habitat could be accomplished within the national forests, and that no new national parks were needed.

Forest preservation groups have promoted the creation of new national parks from national forest lands, while conservation groups have opposed them. The struggle continues even today with preservation groups lobbying Congress to enlarge the present wilderness areas and to create new ones. Most proposed wilderness areas remain under the jurisdiction of the Forest Service, but converting these lands to wilderness places restrictions on the ways these lands and their resources can be used.

Legislation during the last half of the twentieth century has tended to support the concept that forests under the management of the Forest Service are expected to be productive. The Multiple Use Sustained Yield Act was signed into law in 1960. This law required forest resource management for multiple uses (see Chapter 1), specifically wildlife, recreation, grazing, and timber production. A **sustained yield,** as it was referred to in the law, requires managers to prepare a complete inventory of timber resources that includes the past harvest history and average maturity of each timber stand. Once this information is known, future harvests can be planned to produce a volume of timber products that can be repeated year after year without depleting the timber resource.

The adoption of clear-cutting practices in eastern hardwood forests during the 1960s caused concern for many citizen groups. Court actions followed to stop the practice, and they were successful in several states. The National Forest Management Act, passed in 1976, eliminated provisions in earlier laws that had been successfully used in the courts to stop clear-cutting practices. It also required the development of comprehensive forest plans.

Several bills that have affected forest management practices were passed in the 1970s. They include the Clean Air Act of 1970, the federal Environmental Pesticide Control Act of 1972, and the federal Water Pollution Control Act of 1972. This was followed by passage of the Endangered Species Act of 1973, the Toxic Substances Control Act of 1976, and the Clean Water Act of 1977 (Figure 7–10). All of these acts provide for use of science-validated management practices in solving environmental problems.

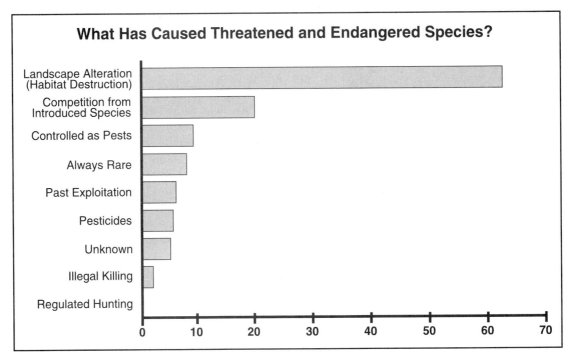

What Has Caused Threatened and Endangered Species?

Figure 7–10 Many different bills have been passed by Congress that affect the ways that forests are managed. One of the most significant legislative actions is the Endangered Species Act of 1973 that prescribes management practices that do not intrude in the environments occupied by endangered or threatened species.

FOREST PROFILE

Endangered Species Act

The United States Congress passed legislation in 1969 that protected animal species that were declining in numbers. The act was expanded in 1973 to require the United States Fish and Wildlife Service to identify species of animals and plants that might become extinct. The act identifies two classes into which those species that are found to be at risk may be placed. Those in immediate danger of extinction are classed as "endangered species." These are the plants and animals that have small numbers in their populations. In many cases, the population is also becoming smaller throughout most or all of the range that is occupied by the species. Species that are in less danger of extinction, but which are at risk of becoming endangered, are classed as "threatened species." These are organisms that can reasonably be expected to survive if immediate steps are taken to protect the environments in which they live.

The act protects both the species and its habitat. Habitat is defined as the environment in which an organism lives and from which it obtains its nutrition. Species that are protected under this legislation cannot be killed, hunted, or disturbed without heavy legal penalties being assessed. The act also protects the organism's habitat from development or other disruptive uses by people. The access to areas where protected species live is restricted under the provisions of the law to prevent human interference. This is considered to be necessary for the purpose of protecting the delicate balances of nature.

Examples of ways that this law has affected the forest industry are easy to find. The protection of two species of birds has heavily impacted the forest industry since the Endangered Species Act was implemented. In the Pacific Northwest, the Spotted Owl has been identified as a threatened species (Figure 7–11). Its favored habitat is mostly composed of "old-growth" forest; however, recent scientific studies have confirmed that some Spotted Owl populations also live and hunt in "new-growth" forests. Harvest of timber in the old-growth forests has nearly been eliminated since the provisions of the Endangered Species Act have been enforced.

In the South-central and Southeastern regions of the United States, the Red-cockaded Woodpecker has been listed on the endangered species list (Figure 7–12). The listing of this bird has impacted the forest industry in those forests in much the same way as the listing of the Spotted Owl has affected the industry in the Northwest. Several court actions have been filed by interested groups to ensure compliance with the act, and the result has been a decline in the timber industry in regions where threatened and endangered species are found.

Figure 7–11 The Spotted Owl has been identified as a threatened species. *(Courtesy of United States Fish and Wildlife Service. Photo by Randy Wilk)*

Figure 7–12 The Red-Cockaded Woodpecker is listed as an endangered species.

Forest Health and Protection

Mike Dombeck, a former acting director of the Bureau of Land Management, was appointed by the Clinton administration to be chief of the United States Forest Service. He immediately issued directives to Forest Service employees in early 1997 declaring forest health and protection to be priorities of his administration. "We cannot meet the needs of the people if we do not first conserve and restore the health of the land. Failing this, nothing else we do really matters," he said. Chief Dombeck also declared his willingness to do battle with anyone who seeks to weaken environmental laws.

In 1996, President Bill Clinton stirred controversy on the preservation issue when he used his presidential powers under the American Antiquities Act of 1906 to create a new national monument in southern Utah. Much of this land was known to be rich in coal reserves and minerals, and the action created an uproar that demonstrated once again that the struggle is still alive between those who wish to practice conservation while using resources and those who wish to preserve resources through nonuse.

State Forests

Some public forest land is owned by the states, and each state has assigned the management of these resources to a state agency or bureau. The state forest agencies are managed under state laws and policies, but the state forests are also subject to national laws, such as the Endangered Species Act and the Clean Water Act.

The state forest lands in some states provide income for the public schools. Where this is the case, many of these state forest agencies are expected to maintain a sustained harvest plan that brings a steady flow of income to the state. This is possible only when forest health is maintained and steps are taken to ensure that the forest is regenerated each time it is harvested. Each state forest agency has access to the USDA cooperative extension forest resource specialists located in the state. In many instances, these specialists are located at a state land grant university, and they are available to consult with public and private forest managers on forest management issues.

Private Forests

Sometimes we think of private forests as small plots of trees located on ground that is unsuitable for farming. While private forests of this type are part of the privately held forests, many privately owned forests belong to huge corporations. In many states, the forest holdings of large corporations make up much of the forest land within the state boundaries. Private forests, especially in the eastern part of the United States, have much more influence on production of forest products than do national forests.

Owners of private forest lands are responsible for managing their own forest resources, but they have help available to them when it is needed. The Forest Service provides resources to owners of private forest land through its private forestry units located in regional forest offices. Professional expertise is available to help solve forest problems and to help develop management plans. Management expertise is also available through the USDA cooperative extension forest research specialists in each state. When forestry information or expert advice is needed, it usually can be obtained by calling the local cooperative extension office. Information on obtaining the services of the state forester is also available from county cooperative extension offices.

Sharing forestry information is important to both public and private forest interests because problems in forests do not recognize ownership boundaries. Problems that develop on private forest lands will soon extend into forests located on public land. This role is often reversed with national forest problems spilling over into private forests. The best solution to such problems is to work together for the benefit of all parties.

Relations between some private forest owners and the state and federal governments have been strained in the past by the efforts of government agencies to regulate private forests. A key issue has been whether personal property rights also guarantee full authority for owners to make forest management decisions on privately held land. Some managers of federal agencies, with encouragement from special interest groups, have wanted to require private forest owners to comply with the same policies that are applied to federal forests. In general, private owners have preserved the right to manage their forest properties.

MULTIPLE-USE MANAGEMENT

Multiple-use forest management is a concept that allows different resources within an area to be used at the same time. It may also involve different uses of the same resource at one time. Several important uses of forest resources are recognized by the Forest Service as appropriate uses of forest resources. Among these are timber production, livestock grazing, recreation, watershed management, and mining (Figure 7–13).

Multiple use of resources is considered to be the best management strategy when the forest uses in a given area are compatible. For example, a person who uses the forest to watch and photograph birds is not likely to be bothered by the person who is fishing at a nearby stream or lake. Considerable conflict could result, however, if motorcycle club members were riding their motorcycles up and down a nearby road. The noise level would likely be so high that few birds would remain in the area. The multiple-use management concept works best when all forest users exercise consideration for others and good judgment in their use of the resources.

Figure 7–13 Among the appropriate uses of public forest lands are timber production, livestock grazing, watershed management, mining, and recreation. *(Photo courtesy of Carolyn Miller)*

It can be difficult to find ways to make some types of mining compatible with maintaining the high water quality that is required in streams and lakes to keep fish alive and healthy. Campers seldom have high-quality camping experiences if livestock are concentrated in or near public campgrounds. Logging activities frequently detract from the enjoyment of an expected environment of quiet and solitude for people who have come to the forest to enjoy nature.

Forest management is sometimes based on a **dominant-use** concept that restricts particular forest uses to isolated areas. The dominant-use approach to forest management assigns greater priority to a particular use of the forest than to alternative uses. This approach to management restricts the use of resources to a single use. It is very different from multiple-use management. The difficult part in implementing the multiple-use management concept on public lands is to identify ways to modify or isolate forest uses that tend to detract from the forest environment for other users.

Examples of proper use of resources are evident in many management units. Sheep that are grazed in units where pine tree seedlings have been planted can reduce the competition from weeds and other vegetation without damage to the trees. Recreational water sports are completely compatible with the need to store irrigation water in reservoirs. Migrating waterfowl and resident wildlife populations also benefit by activities such as the storage of water in reservoirs. Livestock grazing can be compatible with the needs of wildlife. Grazing must be managed, however, to allow late growth of forage plants and to preserve the plant populations in riparian zones (Figure 7–14).

Figure 7–14 Livestock grazing and wildlife needs appear to be compatible as long as grazing is managed to allow late regrowth of forage and to preserve plant populations such as willows and brush in the riparian zones.

The multiple-use concept of forest management may be the most significant development to emerge in the history of forest management in North America. It is a resource management strategy that requires forest resources to be managed in such a way that people with different interests and needs can use the same resources with minimal conflict and without depleting them. To do this, all users must assume full responsibility for understanding and using the resources wisely.

CAREER OPTION

Extension Forester

An Extension Forester works for a state land grant university as an employee of the Cooperative Extension System. He or she is a specialist in forestry who provides consulting services to private and public forest owners within each state. Other extension educators in the counties depend upon the expertise of the extension forester to diagnose problems and to suggest ways to solve them. A graduate degree in forestry or in a closely related field is a prerequisite for this career. Other career preparation should include a strong educational background in the biological and social sciences.

LOOKING BACK

Government plays a leading role in management and conservation of both public and private forest lands. Government officials enact and enforce laws and regulations that control the uses of land, water, wildlife, timber, and other natural resources. One role of government is in forest planning. Numerous forest planning laws have been implemented since 1873 when the Timber Culture Act was first passed by Congress. The United States Forest Service and the United States Bureau of Land Management have become the two federal agencies that provide oversight for most federal forest lands. Congress has given United States presidents the power to set aside lands for forest reserves, national parks and monuments, historical sites, and sites of scientific value. Some forest land is owned and managed by states, and vast areas of forest lands are privately owned and managed. Multiple-use forest management is generally practiced on public lands, making these areas available for timber production, grazing, recreation, watershed management, and mining.

QUESTIONS FOR DISCUSSION AND REVIEW

Essay Questions

1. What roles did the government play in forest management in the colonial period?

2. How did the Homestead Act impact native American forests?

3. What was the main purpose of the Timber Culture Act of 1873?

4. State the two different forest management philosophies that influence forest planning in the United States.

5. Why is it important to develop management plans for forest resources?

6. Name two federal departments and the key agencies that are responsible for the management of most public forest lands.

7. Describe how the Forest Reserve Act of 1891 changed the way that forest resources were managed.

8. What events led to the classification of some forest lands as national forests, national monuments, and national parks?

9. Name one of the ways that a national law may affect forest lands whether they are under federal, state, or private ownership.

10. What are the differences between managing forest resources for multiple-use purposes in contrast with dominant-use management?

Multiple-Choice Questions

1. Forest management in colonial times could best be described as forest:
   A. conservation
   B. exploitation
   C. dominant-use management
   D. preservation

2. Which of the following effects did the Homestead Act of 1862 have on forest resources?
   A. forest acreage decreased
   B. forest acreage increased
   C. erosion of forest soils declined
   D. the quality of surface water improved

3. The Timber Culture Act of 1873 required that:
   A. trees could no longer be harvested
   B. timber products could not be exported
   C. timber lands could be purchased in 160-acre blocks
   D. 40 acres of each homestead in the Great Plains region had to be planted in trees

4. In general, the philosophy for management of forest resources administered by the United States Forest Service could be described as:
   A. exploitation
   B. preservation
   C. conservation
   D. dominant-use management

5. Which of the following government organizations is *not* involved in managing forest resources?
   A. United States Fish and Wildlife Service
   B. Department of Defense
   C. United States Forest Service
   D. Social Security Administration

6. National parks and monuments are administered by:
   A. Department of the Interior
   B. Bureau of Indian Affairs
   C. United States Forest Service
   D. Department of Agriculture

7. Multiple-use forest management, as it appears in federal laws, approves each of the following forest uses with the *exception* of:
   A. grazing
   B. farming for crop production
   C. wildlife habitat
   D. recreation

8. The ability of a forest to produce a volume of timber products that can be repeated year after year is referred to as:
   A. gross annual production
   B. low input/high yield
   C. diminishing returns
   D. sustained yield

9. The USDA Cooperative Extension System employs forest resource specialists who provide information and consulting services to:
   A. managers of privately owned forests
   B. United States Forest Service
   C. state forest managers
   D. all of these

10. A forest management strategy that requires forest resources to be managed for several different purposes is known as:
    A. multiple use
    B. dominant use
    C. resource preservation
    D. competitive advantage

LEARNING ACTIVITIES

1. Obtain a map of a forest region on which private, state, and federal ownership of the land is identified. Invite a forest manager to talk with the class about ways that laws and government regulations and policies affect the management of the forest resources for which he or she is responsible. If you do not have a forest near your community, you might do the same activity using a speaker phone or through an Internet connection.

2. Assign students to search the Internet for evidence of the two competing philosophies of forest management. Classify each document according to its orientation toward conservation or preservation of resources, and discuss the findings of the class members.

Chapter

8

Terms to Know

silvics
stand
natural regeneration
artificial regeneration
direct seeding
germination
seedling
nursery
bare-root stock
allelopathic effect
herbicide
controlled burn
prescribed burn
clear-cutting
even-aged stand
selection cutting
uneven-aged stand
sapling
pole
mature
overmature
senescent
girdling
rodenticide
cleaning operation
stand improvement
liberation
intermediate cutting
salvage cutting
sanitation cutting
pruning

Silviculture

Objectives

After completing this chapter, you should be able to

* define silviculture
* list some important silviculture management practices
* distinguish between natural and artificial methods of regenerating forests
* discuss the advantages of direct seeding or planting seedlings in comparison with natural methods for regenerating forests
* explain the most common methods of producing seedlings for forest regeneration
* describe the steps that should be followed in transplanting a tree seedling
* describe the characteristics of the different growth stages of trees such as seedling, sapling, pole, and mature tree
* explain why it is necessary to control populations of rodents, especially during the seedling and sapling stages of tree development
* describe some intermediate treatments that are applied to forests
* describe some silviculture practices that are used to improve the growth and quality of trees
* explain how the final use of a tree affects the harvesting method that is used

Silviculture is the art and science of tree production. It is a specialized area of study within the larger field of forestry. It is based on an understanding of **silvics,** which is the study of forests and forest relationships. It includes plant, soil, and animal interactions with trees. Silviculture is practiced on the assumption that a forest environment can be manipulated to make it more favorable to the growth of trees than the natural environment.

Tree farming is becoming much more common as the need for forest products increases faster than the capabilities of natural forests to produce them. Increased production of wood products can be realized when ideal conditions are provided for the growth of trees. Silviculture applies the principles of science to the production of trees using modern technologies as tools (Figure 8–1).

Silviculture that is practiced in an intense management system becomes very dependent on the use of specialized machines. In every sense, tree plantations become highly specialized farms where management is as intensive as it is for any other agricultural crop. Competition from weeds is eliminated, soils are tested, fertilizers are added, diseases are treated, and harmful insects are controlled.

Silviculture can also be practiced in less intensive ways. For example, a stand of naturally seeded young trees may be growing so close together that their growth is restricted by competition among the trees. Thinning the trees manipulates the environment to allow for timber production. Any cultural practice that manipulates the forest environment to achieve specific goals in the forest is a form of silviculture.

Figure 8–1 Tree farming is an important business that requires the application of scientific farming practices. This type of forest production is called silviculture.

FOREST REPRODUCTION AND REGENERATION

The forest reproduction processes known as sexual reproduction and vegetative reproduction were discussed in Chapter 4. It was noted that sexual reproduction results in the production of seeds that are capable of producing young trees. Vegetative reproduction, also known as asexual reproduction, occurs when young trees are produced from leaf, root, or stem tissues (Figure 8–2). Both kinds of reproduction are important in the regeneration of forests.

A population of trees that has been established in a forest environment is called a **stand.** When a population of trees is few in number and widely scattered, it is considered to be a poor or weak stand, while a population of healthy trees that are properly spaced in the forest is considered to be a strong or vigorous stand. These terms will be used throughout this text to describe the characteristics of specific populations of forest trees.

Figure 8–2 Some species of trees such as the poplars are regenerated vegetatively by planting "suckers" or young stems and branches in moist soil where they generate roots.

PROFILE IN NATURE

Seed Distribution by Squirrels and Birds

Squirrels and birds play important roles in distributing the seeds of trees and other plants to new locations (Figure 8–3). Squirrels actively harvest seeds by gathering acorns, pine cones, and other kinds of nuts to be eaten during the winter season. Some of the seeds that are hidden in the debris of the forest floor are forgotten, and they may eventually germinate and grow. Most of these seeds are not distributed very far beyond the trees on which they grow, but squirrels play a role in planting them beneath the vegetative cover on the forest floor.

Birds distribute seeds over wide areas. Some of the seeds are carried in flight and dropped in distant locations. Other seeds may be overlooked in the shell or husk materials that surround most seeds from trees. Some seeds may have a hard enough seed coat to survive the digestive process of a bird. These seeds are distributed in the feces of birds.

Figure 8–3 Squirrels and birds play important roles in distributing seeds of trees to suitable new locations. *(Courtesy of United States Fish and Wildlife Service. Photo by Jon Nickles)*

Forests are usually considered to be naturally renewable without any need for human intervention. While this is generally true, forests do not always produce the kind of trees that are wanted or needed. Some kinds of trees compete well with other forest plants, and some trees do not. For example, the eastern white pine forests tended to be replaced by oak forests after they were harvested. The stage in the biological succession on these sites favored the climax species of trees such as oak.

When a particular kind of tree is desired in an area, it may be necessary to change the environment to favor its growth. It is for this reason that silviculture is practiced so widely on private forest lands. Some species of trees reproduce and regenerate naturally when conditions are created that are favorable to the species. For example, regeneration of aspen forests in the region of the Great Lakes can often be accomplished by clear-cut harvest methods. This removes the shade from the soil surface, allowing natural growth of young aspens. Shaded areas favor the growth of other kinds of forest plants and trees.

Two types of forest regeneration occur following timber harvests. **Natural regeneration** occurs on a forest site when young trees begin to grow there without having to be planted. Sometimes seeds have been dispersed in the area by the wind or by wild animals. Some hardwoods grow from the roots or stumps of the harvested trees. In some instances, advance regeneration occurs due to seedlings and saplings that were already there when the harvest took place. Natural regeneration of forest trees depends on several important growth conditions such as the availability and dispersion of fertile seed in the area, the availability of soil moisture, warm temperatures, the condition of the soil, favorable weather, favorable light intensity, and freedom from diseases and harmful insects.

Artificial regeneration is forest renewal that occurs when seeds or seedlings are planted at the harvest site (Figure 8–4). This method often results in a uniform stand of trees that are evenly dispersed throughout the area. The forest manager also has control over the species of trees that make up the new

Figure 8–4 Forest renewal is often accomplished by planting seeds or seedlings soon after a mature forest has been harvested.

forest planting. Seedlings can be selected from superior parent stock that has the potential to increase yields. Trees that are regenerated artificially are all of the same age and growth stage. This makes it easier to manage them. Tree plantations are usually planted in rows. This makes it possible to use mechanical equipment within the plantation for weed control, thinning, pruning, and other purposes.

DIRECT SEEDING

Planting tree seeds to generate new forest growth is called **direct seeding.** Seeds can be planted directly into the soil surface using mechanical equipment, or they can be dispersed in the area by aircraft. Aerial seeding is a good method to use following fires. The soil surface in a burned area is usually free of debris, and large areas can easily be planted using this method. It is important when direct seeding methods are selected to take advantage of good moisture conditions at the time of planting. Drought conditions will prevent germination of the seeds. Seeding should be timed to coincide with adequate soil moisture to ensure a good stand of trees.

The use of high-quality seed is important in forest regeneration. Tree seeds should be collected only from superior trees (Figure 8–5). The seed must be collected at the right stage of maturity, and it must be carefully cleaned to eliminate damaged or shrunken seed. It should be stored in a darkened location that is cool and dry to avoid untimely germination of the seed.

Seed is sometimes damaged by extreme temperatures, drought conditions, and other environmental factors. Such conditions can reduce the fertility of

Figure 8–5 Seed tree farms are often maintained to ensure a supply of high-quality seed for forest regeneration purposes.

the seed. Seed should always be tested before it is used to make sure that the seeds will sprout and grow. The process by which this occurs is called **germination.** Seed laboratories are available in most states to provide seed testing services (Figure 8–6). These laboratories issue certification tags that are attached to each container of seed that was part of the tested seed lot. Using certified seeds reduces the risk of establishing an inadequate population of young trees in the plantation. This is one of the greatest problems associated with direct seeding of trees.

Adequate amounts of seed must be planted to ensure that the population of seedlings that becomes established is properly spaced in the rows. It is important that enough young trees survive to require thinning at a later time. This allows for the removal of weak, damaged, or deformed trees from the stand. Allowances must be made in the seeding rate for seed that is eaten or destroyed by birds, rodents, squirrels, and insects. Some seeds are killed by the molds and fungi that are nearly always present in forest environments. Seeds can be protected from seed-eating insects, birds, and animals by coating the seeds with protective chemicals such as fungicides, insecticides, and repellants. Repellant-coated seeds have been used with some success to discourage rodents and birds from eating the seeds.

Not all of the planted seeds will grow. This makes it necessary to plant more seed than you expect to need. The seeding rate is different for each kind of tree. This is because the seeds of different kinds of trees are of different sizes. The volume or weight of the seed that is needed to plant each acre is much less with

Figure 8–6 Seeds should be tested for germination prior to planting in order to ensure a good stand.

small seeds than it is for large seeds, since there are more seeds per pound for small seeds in comparison with large seeds. The seeding rate is also affected by the way the trees will be used at harvest time. Trees that are used for pulpwood can be spaced more closely between and within the rows than is desirable for timber production.

PRODUCTION OF SEEDLINGS

A **seedling** is a tree that is in the early stages of development. Seedling production for forest plantings usually takes place on a massive scale. Sometimes the seedlings that are raised for forest regeneration are planted in cultivated fields. A large outdoor planting of tree seedlings is called a **nursery** (Figure 8–7). Tree seeds are usually planted in wide rows called beds. When the seedlings are ready for transplanting to forest sites, they are removed from the beds. Sometimes the seedlings are placed in individual plastic bags with soil on the roots. This method can be used when it is expected that planting may be delayed. Another method that is used is to lift the seedlings from the beds leaving the roots free of soil. Seedlings in this condition are called **bare-root stock.** Bare-root stock must be planted right away or placed in a cool, damp storage area with special care to keep the roots moist. Large numbers of tree seedlings can be produced in a relatively small area in an outdoor nursery.

Figure 8–7 A tree plantation that is used to produce seedlings for transplanting is called a nursery. Seeds are planted in rows with very little space between seeds, and the seedlings are removed before they become too crowded in the row.

Figure 8–8 Some seedlings are raised in individual containers in greenhouses. These seedlings are more expensive to produce than those raised in a nursery, but they also have a higher survival rate when they are planted in the forest. *(Photo courtesy of Potlatch Corporation)*

Seedlings that are raised in greenhouses are usually planted in plastic trays or individual containers (Figure 8–8). Containers are filled with potting soil that is kept damp with frequent applications of water. In most cases, seedlings require several months to a year of growth before they are ready to plant in a forest environment. During the time that seedlings are under intensive management in a greenhouse or nursery, they are more susceptible to insects and diseases than at any other time (Figure 8–9). This is because large numbers of

Figure 8–9 A greenhouse environment promotes rapid seedling growth, but it also subjects seedlings to insects and diseases due to large numbers of plants being concentrated in a small area.

plants are concentrated in a small area. Some diseases and insect pests are capable of reproducing rapidly under such favorable conditions. For this reason, great care must be taken to observe the plants frequently and to respond quickly to insect and disease problems.

Two seeds are usually planted in each compartment of a multiple-plant container. This is done to ensure that a healthy plant is produced in every compartment. Thinning is required to reduce the seedling population in the containers to one plant per section. Seedlings that are produced in greenhouses are usually planted with the potting soil on their roots. They are more expensive to produce than seedlings raised in outdoor nurseries, but less root damage is likely to occur as these seedlings are transported and planted in forest sites.

PLANTING SEEDLINGS

The most critical period in the life of a tree is during the seedling stage when it is tender and succulent. For this reason, careful attention must be given to new plantings of seedlings. Healthy seedlings that are planted in damp soil that has been well prepared usually can be expected to have a reasonable rate of survival. Site preparation is very important in establishing new plantings of trees. This is because excess organic matter on the soil surface makes it difficult to be sure that seedlings are planted in the mineral layer of the soil. This is especially true when the seedlings are planted by a machine.

Many methods of site preparation are used. Planting sites are sometimes prepared by gathering the surface debris into piles for burning, or by cutting it up mechanically with a brush cutter or similar machine (Figure 8–10). A distinct

Figure 8–10 Before seeds or seedlings are planted in a harvested area, it is important to prepare the seedbed by cutting up plant debris or burning it. The smoke in the distance pinpoints the location of a recent timber harvest.

advantage of mechanical site preparation is that competition from existing plants can be reduced. This is important, especially when conifers are planted in an area with native hardwood shrubs and trees. Hardwood species of trees and shrubs tend to grow faster than pines in the early growth stages. Because of this, they gain early dominance and overshade the pines, eventually causing many of them to die. For this reason, it is important to reduce the populations of hardwood shrubs and trees in an area where pine seedlings are to be planted. When it is impractical to use mechanical site preparation methods, it is important to at least clear a small site as each seedling is planted.

Live plants on the soil surface present another threat to seedlings besides competition for water, nutrients, and sunlight. Evidence is accumulating indicating that some kinds of plants, such as grasses, maintain their dominance over other plants by releasing chemicals into the area close around them that provide small doses of poison to invading plants. A plant activity of this nature is an **allelopathic effect.** It is believed that some grasses have a greater allelopathic effect on hardwood seedlings than they have on pine seedlings. This allows pines to become established in relatively pure stands in areas where hardwood trees were once dominant.

Chemical control of undesirable shrubs, brush, and trees can be achieved using chemicals that kill plants. A chemical of this type is called an **herbicide.** Herbicides can be applied in several different ways. One method of chemical application is to spray the chemical preparation on the leaves and stems of the plants that are to be controlled. Great care must be exercised to prevent the chemical mixture from drifting to the foliage of trees that are intended to survive. Herbicides are also available in dry form. This material is broadcast on the soil surface and absorbed through plant roots. Some undesirable forest trees and shrubs, such as those that send up new growth from their roots, can be controlled only by completely uprooting the plant or by killing the roots with chemicals.

The use of fire in site preparation is a proven practice, but special care must be taken to control the fire to burn only the desired area. Use of fire in this manner is called a **controlled burn** or a **prescribed burn.** Fire effectively removes debris from the surface and kills or weakens plants that might compete strongly with the seedlings for moisture and nutrients. A common practice in site preparation in areas of low rainfall is to gouge the surface of the soil and plant the seedling at the bottom of the depression that is formed. This concentrates most of the moisture that is available in close proximity to the seedling, increasing its chance of survival.

The best time to plant seedlings is in the spring season before it gets too hot. Excessive heat tends to dry the soil, reducing the chances that the seedling will survive. The spring season is a time when precipitation is abundant in many areas, and seedlings require a supply of moisture to establish their roots in the soil. As each seedling is planted, its roots should have soil firmly packed around them. Mechanical planters use one or more packing wheels to firmly press the soil into place. This ensures that the roots will contact the moist soil particles, making it possible for them to absorb moisture and dissolved nutrients.

Large numbers of seedlings are still planted by hand on steep sloping areas, but mechanical methods of planting seedlings are also widely used. Both planting methods require the roots of the seedlings to be placed in the soil in such a manner as to avoid bending them over. Bending or twisting the roots interferes with root development, and it weakens the tree later on. The seedling should be placed as closely as possible in a vertical position with the soil packed firmly around the roots.

DEVELOPING NEW STANDS

The characteristics of a new forest stand are determined to a large extent by the methods that were used in the previous harvest. For example, **clear-cutting** is a harvest method in which all of the trees in the stand are cut (Figure 8–11). The new stand of trees that follows will be an **even-aged stand** in which most of the trees are approximately the same age (Figure 8–12). This method of harvest leads to plantation tree farming. Even-aged stands also occur naturally in areas where natural disasters such as forest fires, snowslides, or blowdowns have destroyed all of the trees in an area.

Figure 8–11 The clear-cut harvest method leads to the production of timber stands that are even-aged, making it easier to manage the forest. The disadvantages of clear-cut harvesting include the potential for greater erosion and water pollution problems and unsightly landscapes.

Figure 8–12 An even-aged stand of trees consists of trees that are all approximately the same age.

A harvest method called **selection cutting** removes a sustained yield of wood from the forest at regular intervals. Only the most mature trees are selected, although damaged or diseased trees are removed at the same time. The use of this harvest method maintains an **uneven-aged stand** in which trees of all ages are found in the forest (Figure 8–13). Other harvesting methods are discussed later in this chapter.

The forest regeneration method that is used also has an effect on the age of the trees that are found in a stand. Seeding and planting seedlings on ground that is free of live trees results in even-aged stands, while natural reforestation tends to occur over a period of several years resulting in trees of mixed ages.

The first stage of tree growth is the seedling stage (Figure 8–14). It occurs after a seed has germinated, and it continues until the tree has grown to approximately 3' in height. After that, the young tree is referred to as a **sapling.** It will continue to be known as a sapling until the lower branches begin to fall or until it reaches a diameter of 4". In the next stage of development, the young tree is called a **pole.** The diameter of a tree in the pole stage ranges from 4" to 10". Trees in the size range from 10" to 24" in diameter are described as being **mature.** The final development stage for trees occurs when they begin to decay. Trees at this stage of development are described as being **overmature.** A high proportion of trees in this stage of development are becoming **senescent,** meaning that they show evidence of heart rot, decay, and other defects due to age.

Figure 8–13 An uneven-aged population of trees is the result of selecting only the most mature trees in a forest for harvesting.

Stages of Tree Growth

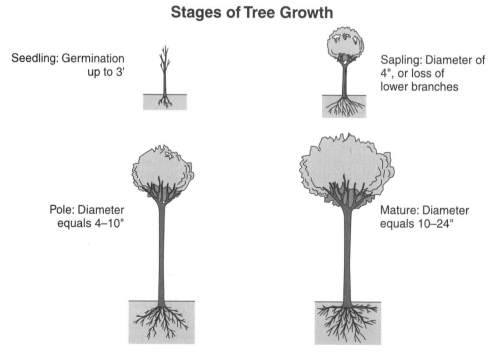

Seedling: Germination up to 3'

Sapling: Diameter of 4", or loss of lower branches

Pole: Diameter equals 4–10"

Mature: Diameter equals 10–24"

Figure 8–14 The growth of trees is described in four phases or stages.

Rodents

Rodents make up the most diverse group of mammals. A rodent can be identified by the four large incisor teeth in the front of its mouth. These teeth never stop growing, and rodents must gnaw on wood or other materials just to keep their teeth worn down. The front edge of a rodent's teeth is composed of harder material than the back edge, causing the back edge to wear faster than the front edge. The result is that the incisor teeth become chisel shaped, and they are sharpened as they wear down. A rodent must chew to continue living. Rabbits have some of the same gnawing habits as rodents, but their tooth structure eliminates them from this group of animals.

During the sapling growth stage, the young trees remain quite vulnerable to damage from insects, diseases, and animal pests. The major diseases and insect pests were discussed in detail in Chapter 6, but the discussion of damage to trees due to the gnawing activities of rodents is expanded here because rodents constitute such a major threat to trees in this stage of growth.

Mice, rabbits, gophers, and other gnawing animals can easily kill a sapling by chewing through the bark all around the base of a tree. This condition is known as **girdling.** It results in the death of the tree because the flow of water and nutrients between the tree roots and the foliage is blocked by loss of vascular tissue. Rodents are known to experience massive population increases when their food supplies are abundant and natural predator populations are small. During these periods of abundant rodent populations, new stands of trees are in danger.

Rodent control is hard to achieve through poisons because new rodent populations move into the area almost immediately from adjacent areas. It is possible, however, to achieve some control over rodent populations with chemicals designed specifically to kill them. Such a material is called a **rodenticide.** The most effective control method for rodents living in or near tree plantings is to control the buildup of dead plant materials such as grass or brush on the forest floor. These materials provide cover in which rodents can hide from their natural enemies. They also provide shelter to rodents from climate- and weather-related hazards.

INTERMEDIATE TREATMENTS

Forests require care and management during the years that pass between planting and maturity. They may benefit from a **cleaning operation** to remove vegetation that competes with young trees. A cleaning operation may be needed to provide control of brush or any other plants that interfere with the development of a stand of young trees. Any silviculture practice that is applied between the seedling and sapling stages of development for this purpose is a cleaning operation. An operation of this kind that is done when the forest stand is older than the sapling stage is called **stand improvement.** A cleaning operation may consist of the removal of undesirable older trees in the stand to make sunlight available to the young trees. This kind of stand improvement is called **liberation.**

Control of brush may be necessary in stands of young trees, especially when moisture for the trees is in short supply. Some control of brushy plants can be done using the mechanical methods described in the first part of this chapter for site preparation. Care must be exercised to avoid damage to the outer bark of trees that could lead to fungi infections or other diseases.

A prescribed burn may be an effective control method in some kinds of trees if the brush and fuel on the forest floor are not too abundant (Figure 8–15). Trees must be sufficiently mature that the bark of the trees will provide protection from fire. They must also be free of low-hanging branches that may catch fire, causing the foliage of the trees to burn. An abundant fuel supply on the forest floor can cause fire to be very dangerous to the stand of trees that the prescribed burn is intended to protect.

Chemicals have only limited uses in established stands of trees, because they are often as dangerous to the crop as they are to the plants they are intended to control. They can be used in spot treatments of undesirable trees or shrubs when great care is exercised in applying them, and they can be injected into the trunks of problem trees. In all cases where herbicides are used in the forest, all of the regulations for their use must be obeyed.

Figure 8–15 Fire is an important tool in preventing forest fires later. Debris that accumulates on the forest floor usually can be removed by prescribed burns without causing serious injury to the larger trees.

An **intermediate cutting** is a silviculture practice that is intended to improve the forest by removing some of the trees. This cutting may occur anytime between the time of planting and harvesting. Early in the life cycle of a forest planting, it is often necessary to reduce the tree population in the planting to provide enough space for each tree to grow. This practice is called thinning. Trees that are planted in rows can be thinned by removing every second or third tree in the row. The trees should be carefully inspected as they are thinned to make sure that the most valuable trees are saved.

It is not always possible to protect forests from damage due to fire, insects, or diseases. When any of these destructive agents causes widespread damage in a forest, it may be necessary to remove some or all of the affected trees. The timber obtained from a cutting of this kind may have commercial value if the trees are harvested in a timely manner. This type of cutting is called a **salvage cutting** (Figure 8–16). An emergency cutting caused by disease or insect problems

Figure 8–16 Salvage logging is a tree harvest that is conducted to remove diseased or damaged trees from the forest. In this kind of logging operation, only the damaged timber is removed.

for the purpose of preventing the spread of the problem to other vulnerable trees is sometimes called a **sanitation cutting.** Salvage or sanitation cuttings may be combined with a complete intermediate harvest of marketable trees to improve the health of the forest.

MANAGING FOR GROWTH AND QUALITY

Several silviculture practices are known to improve the rate of growth and the quality grade of the timber that is produced by a stand of trees. One of these practices is thinning. The successful forest manager expects to have some of the seedlings die, so an adequate number of trees are planted to ensure a good stand. Once the trees fill in the forest canopy, they begin to restrict the growth of adjacent trees. As they grow, some of them must be removed to make room for those that will provide the final harvest. Failure to thin reduces the final production of a stand of trees.

Pruning is the practice of removing the lower branches of a tree. This results in fewer knots in the lumber that is cut from the logs of pruned trees. The cost of pruning is generally considered to be too high to justify pruning most commercial trees, but research has proven that lumber quality can be greatly increased through the use of pruning. In terms of economics, however, it may be wise to restrict pruning to those species of trees that are high in value. One method of achieving some self-pruning in trees is to delay thinning until the trees are in the pole stage of development. The close proximity of trees to one another will result in some natural pruning of the lower branches.

The availability of water is often a limiting factor in the production of trees. Adequate soil moisture can advance the maturity of a tree more quickly than any other factor in its environment. Irrigation is gaining acceptance on forest plantations in many regions where natural precipitation cannot be depended upon to be adequate. This is not a common practice on most plantations, but it can be a valuable production tool, especially in the production of Christmas trees and pulpwood.

Managing a forest to keep it healthy includes the use of all of the practices that have been discussed in this chapter. A healthy forest is capable of rapid growth and the production of high-quality wood products. An unhealthy forest is not. Evidence of this can be seen by examining the annual growth rings in a tree stump. The thickness of each annual ring is evidence of the health and vigor of the tree during the year when a particular growth ring was formed. Drought conditions, insect problems, and diseases can all affect the health and production of a tree.

Among the modern tools that are becoming available for forest management is the use of satellites to observe large areas. This technology can be used to detect differences in the surface temperatures of plants. The foliage of stressed plants tends to radiate more heat than does the foliage of healthy plants. This heat can be measured and recorded on a map using satellite technology to identify locations where trees are stressed. This makes it possible for a forest manager to monitor large areas of forest land. Areas on a satellite map that show evidence of stressed trees can be checked by people on the ground to identify and treat the problem.

HARVEST CUTTING

The harvest cutting is the final event in a cycle of forest production. In even-aged stands, the trees tend to mature at about the same time, whereas uneven-aged stands can be managed for sustained yields through selective-cutting practices. Unlike agricultural crops, mature trees can be harvested over a period of several years without losing their value. This makes it possible for managers of private forests to base the time and method of harvesting on timber prices. They can delay harvests when timber prices are low, or they can choose to accelerate harvesting when timber prices are high.

Most forest plans call for an intermediate timber harvest as part of the intermediate cutting. This improves the forest stand by releasing the crop trees from competition with highly competitive species that are suppressing their growth. The first intermediate harvest is not usually very profitable for lumber, because it consists mostly of inferior crop trees that are relatively small in size, or trees of a species for which demand is generally low. In areas where a market for pulpwood exists, an intermediate harvest of these trees can be profitable, because much of the wood that is harvested can be sold.

Trees that are harvested for pulpwood usually range in size from a minimum size of 5" dbh to large logs. Some pulp mills process trees into chips at the pulp mill, but many large mills now process trees in the forest and haul the chips to the mill in large trucks (Figure 8–17). This makes it economical to process most of the wood in the tree into wood chips.

Figure 8–17 Pulpwood vans are used to haul large loads of wood chips from the harvest location to the pulp mill for processing.

As a stand of timber matures, the products that are obtained from intermediate harvests become more valuable due to the increased size of the trees. As the trees grow in size, the stand must be thinned from a density of approximately seven hundred trees per acre to fewer than seventy trees per acre at the time of the final harvest. All of the other trees, with the exception of those that die during the production cycle, are removed during early thinning operations or harvested as part of an intermediate harvest.

Intermediate harvests that occur in the later stages of the production cycle provide trees that can be used for poles, fenceposts, and lumber. Many forest owners and managers depend upon intermediate harvests to provide business income and to pay the costs of improvements in the timber stand. A very real value that is gained through an intermediate harvest is the growth surge that occurs in crop trees after they are released from competition for sunlight, moisture, and plant nutrients.

Harvest cutting methods depend to some degree on how the harvested trees will be used. Large logs are usually cut by loggers using chainsaws. Next they must be retrieved from the sites where they have been felled to roadside areas where they are loaded on trucks (Figure 8–18). Most logging operations retrieve logs using specialized all-terrain vehicles called skidders. These machines drag the logs along the ground. Some logging operations use winches equipped with

Figure 8–18 Transportation is a major part of the logging process. Most logs are hauled from the harvest site to the mill on large tractor-trailer rigs.

long cables to retrieve logs. Timber harvests that occur in areas that are inaccessible by road or that are highly susceptible to soil erosion frequently retrieve logs by airlifting them with industrial helicopters (Figure 8–19). This method is expensive, but it is also very efficient, and it reduces the need for building roads into sensitive areas.

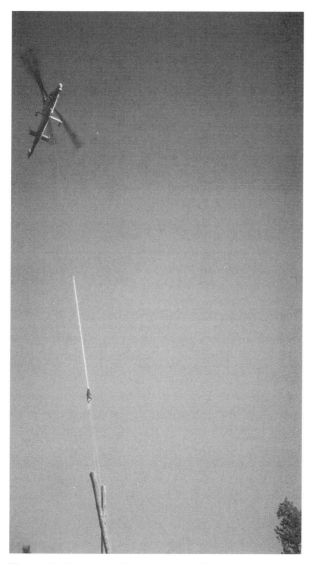

Figure 8–19 Forests that are located in environmentally sensitive areas can be successfully harvested using industrial helicopters to transport the logs to the yard with minimal disturbance to the soil surface.

Whole-tree harvesting is a method that is used to harvest small to moderate-sized trees. This method is especially useful for pulpwood harvests on tree plantations, because these harvests usually consist of moderate-sized trees. Large machines grasp the trunks of trees while they are sawed or sheared off at the base. After each tree is severed from the stump by one of these machines, it is stacked in a pile to be chipped or hauled intact to the pulp mill.

PROFILE ON FOREST SAFETY

Safe Thinning Practices

The importance of thinning a stand of trees was discussed earlier in this chapter. In most cases, thinning is done by cutting down enough of the trees to reduce competition for sunlight and nutrients for those that remain in the stand. This operation is usually accomplished with a chain saw. This is one of the most useful pieces of equipment available to a forester, but it is also one of the most dangerous. This is because it is capable of rapid cutting, and it cuts legs and feet just as well as it cuts trees.

Safe operation of a chain saw begins by keeping the saw sharp. This allows it to cut rapidly through a tree without applying undue pressure to the blade. The operator has a tendency to increase the pressure on the blade as the cutting edges become dull. By increasing the pressure on the blade, the operator loses some of his or her control over the machine and actually increases the possibility that the blade will buck or jerk before the cut is complete. This can cause the blade to rebound from the cut toward the operator, putting the operator's feet and legs in danger. A chain saw should never be operated with the throttle locked in the high position. The only time the locking device should be used is to start the saw. The operator is not able to operate the saw as safely under full power as when the trigger is used to adjust the power to meet the changing needs of the job.

It is always recommended that the operator of a chain saw should wear boots with steel toes. This will help to deflect the blade of the saw away from the operator's toes if the saw gets out of control. Saplings and poles can be difficult to cut safely due to the flexibility of trees that are in this stage of growth. The greatest hazard to a chain saw operator occurs when the small tree leans in the wrong direction, pinching the blade of the saw. As the operator struggles to extract the saw blade from the tree, it is easy to lose caution by stepping closer to the blade. Should the blade suddenly buck or jerk as it comes out of the tree, the operator is in danger.

Another important chain saw safety practice includes the use of a spark arrester on the exhaust of the saw to prevent fires. It is always wise to keep a fire extinguisher and a shovel nearby while working in the woods to put out a fire if one should start. A hard hat is always a must while working with falling trees, and rugged work clothing is needed to protect the skin from falling debris.

CAREER OPTION

Forest Worker

Forest workers perform the tasks related to reforestation and protection of stands of trees. They also do maintenance work on forest roads, trails, buildings, and campsites. Some of these tasks include planting seedlings, removing diseased or damaged trees, pruning trees, preventing and suppressing fires, performing insect and weed control, and constructing improvements in forest facilities. Training in the use of hand and power tools is required, along with a basic understanding of silviculture practices. Work of this kind is sometimes seasonal in nature. Many professional foresters obtain summer employment or serve internships as forest workers during the period when they are obtaining university and college educations.

The information in this chapter concerning forest harvesting practices is intended to provide only basic information about the subject. A detailed account of timber and pulpwood harvesting practices is provided in Chapter 10 of this textbook.

LOOKING BACK

Silviculture is the art and science of tree production. When silviculture is practiced, the natural environment is modified to make it more favorable to the production of trees. The production of forest products increases when growing conditions are ideal. Natural regeneration of a forest occurs when trees begin to grow in a harvested area without being planted as seeds or seedlings. Artificial regeneration of a forest occurs when seeds or seedlings are planted. Seedlings may be raised in greenhouses or in outdoor plantations. Young trees are vulnerable to damage from insects, diseases, and animals. Cultural practices are used to control and limit damage to trees from these sources. The practice of silviculture makes it possible to manage forests for growth and quality. Trees should be harvested when they reach maturity to prevent them from becoming vulnerable to diseases and insect damage. Trees are harvested using methods that are compatible with the ways the timber will be used.

QUESTIONS FOR DISCUSSION AND REVIEW

Essay Questions

1. What is silviculture?

2. What are some important silviculture practices, and how do they improve forest production?

3. How is artificial regeneration of a forest different from natural regeneration?

4. What advantages are gained from direct seeding or planting seedlings in comparison with natural methods of reforestation?

5. Describe two common methods used to produce tree seedlings, and list the advantages and limitations of each method.

6. How should the site be prepared for transplanting tree seedlings, and what transplanting procedures should be followed?

7. Name the stages of growth that trees go through from planting to harvest.

8. What are some intermediate treatments that are applied to forests, and what advantages do they contribute to forest health and production?

9. What are some silviculture practices that are used to improve the growth and quality of trees?

10. How does the final use of a tree affect the method that is used to harvest it?

Multiple-Choice Questions

1. The art and science of tree production is known as:
   A. silvics
   B. forest regeneration
   C. silviculture
   D. forestry

2. Reproduction of trees from the leaves, stems, or root tissues is called:
   A. vegetative reproduction
   B. sexual reproduction
   C. silviculture
   D. suckering

3. The natural growth of a young forest following the harvest of mature trees is called:
   A. spontaneous combustion
   B. arboriculture
   C. artificial regeneration
   D. natural regeneration

4. The process by which seeds begin to sprout and grow is known as:
   A. germination
   B. gymnosperm
   C. regeneration
   D. angiosperm

5. Young trees that have been removed from the soil in preparation for shipping are known as:
   A. seedlings
   B. saplings
   C. bare-root stock
   D. poles

6. Some kinds of plants release chemicals into the area close around them that provide small doses of poison to young plants that invade their territory. This defensive plant response is known as:
   A. allelopathic effect
   B. germicide
   C. herbicide
   D. chemical warfare

7. A harvest method in which the only trees that are harvested are the mature trees and diseased or damaged trees is:
   A. clear-cutting
   B. selection cutting
   C. salvage logging
   D. girdling

8. A young tree with a diameter of 4" to 10" is a:
   A. sapling
   B. seedling
   C. pole
   D. log

9. Removal of undesirable older trees from a stand to make sunlight available to young trees is known as a:
   A. liberation
   B. cleaning operation
   C. stand improvement
   D. sanitation cutting

10. Pruning trees is a cultural practice that is performed for the purpose of:
   A. harvesting firewood
   B. improving lumber quality
   C. attracting wild animals
   D. harvesting damaged timber

11. A mechanical timber harvesting method that is used to harvest small- to moderate-sized trees is known as:
   A. intermediate harvest
   B. aerial harvesting
   C. cleaning operation
   D. whole-tree harvesting

LEARNING ACTIVITIES

1. Obtain seeds and fresh cuttings for a commercial tree species, and generate seedlings. Compare the growth of seedlings obtained from vegetative cuttings with the growth of seedlings generated from seeds. Assign students to keep a log book in which they record their observations and their work throughout the duration of the project. Assemble the combined data of the entire class, and assess the two methods for generating seedlings.

2. Take a field trip to a tree farm, and assign the students to prepare a written report that contains at least the following information: (a) the final product or products of the farm, (b) specific cultural practices, (c) handling procedures for each product, (d) product markets and marketing plans, and (e) future plans.

Chapter

9

Terms to Know

metes and bounds
rectangular survey
initial point
baseline
principal meridian
standard parallel
guide meridian
township
range
section
quarter section
forty
acre
hectare
are
forest type map
stereoscope
planimeter
dot grid
photogammetry
Biltmore stick
hypsometer
chain
increment borer
basal area
timber yield
cruise
systematic sampling
line-plot cruising
forest growth
ingrowth
mortality
cut
survivor trees

Terms to Know continued

Measurement of Forest Resources

Objectives

After completing this chapter, you should be able to

❋ identify the types of information that are needed to develop a long-term forest management plan based on sustained yields

❋ describe the features of the two types of land surveys that are used in the United States

❋ explain the relationship of baselines and principal meridians to the initial point location from which each rectangular survey begins

❋ explain how instruments such as the stereoscope and the planimeter are used in preparing a forest type map

❋ name some tools that are used to estimate the diameter and height of a standing tree, and explain how each tool is used

❋ define the role of a timber cruiser

❋ contrast the differences between a 100% cruise and a partial cruise, and explain when each is appropriate to use

❋ list some assumptions that apply to partial cruises that may influence the accuracy of the results

❋ explain the formula for measuring forest growth, and describe each of the formula components

❋ describe the most commonly used methods for scaling logs

survival growth	Doyle's rule	Smalian's formula
scaling	International log rule	standard cord
Scribner's rule	board foot (BF)	timber cruiser

Management of a renewable resource for a sustained yield requires the ability to measure production. All of the measurements that are included in this chapter are needed to estimate the amount of wood that a forest can produce each year. Accurate records obtained from actual forest measurements are used by forest managers to create timber growth and yield estimates. Tables and graphs, prepared from actual forest measurements, can provide reasonably accurate estimates of future growth and predicted harvest volume. Computer models are available to model alternate management practices and to predict the outcomes.

LAND SURVEYS

Two types of land surveys were used to measure and record the land area in the United States. The **metes and bounds** system used natural physical features of the land as starting points for land measurements. Markers such as streams, roads, rocks, and other features were used. Many of these markers are difficult to locate because streams change their courses during flood seasons, roads are rerouted, rocks and boulders are moved, and features of the land change over time. This system did not prove to be very accurate.

Most areas of the United States, except for portions of the southeast, have been surveyed using the **rectangular survey** system. This system uses an **initial point** as the beginning point for each survey (Figure 9–1). More than thirty

Figure 9–1 The rectangular survey system identifies a prominent physical feature in the landscape from which all land measurements begin. This location is called the initial point.

Figure 9–2 The exact location of each initial point in the United States is marked by a brass marker.

initial points were established as the land survey was conducted. Each initial point is a prominent and permanent physical feature in the landscape, and the exact location of each initial point is marked with a brass marker (Figure 9–2). This type of land survey was developed by Thomas Jefferson for the purpose of establishing property lines for the sale of public lands.

The rectangular survey system established a system of **baselines** that run east-west, and **principal meridians** that run north-south (Figure 9–3). These survey lines are marked with permanent markers, and they are the reference points for defining property lines. A **standard parallel** line is established running east-west for each interval of 24 miles on either side of the baseline. A similar line called a **guide meridian** extends north to the next standard parallel. Because this line converges on due north, the curvature of the earth causes the northern border of each 24-mile tract to be slightly less than 24 miles long.

Each tract of land, measuring approximately 24 miles along each side of a square, is divided into sixteen townships. A **township** is shaped like a square, measuring approximately 6 miles long on each side. Each township is identified using consecutive numbers beginning at the baseline and proceeding north or south. For example, the second township north of the baseline is designated by the code T2N. The third township located south of the baseline would be identified with the code T3S.

A **range** is the east-west location of a township from a principal meridian. For example, the first township west of a principal meridian is identified by the descriptive code T1W; the third township located east of a principal meridian is coded T3E. The location of a township that is in the second township location west of a principal meridian and the third township location south of the baseline would be described as T3S, R2W (Figure 9–4). Each township consists of 36 sections. Each **section** is approximately one mile square with a surface area of 640 acres (1 acre = 43,560 square feet). A section consists of four quarter sections

Rectangular Land Survey

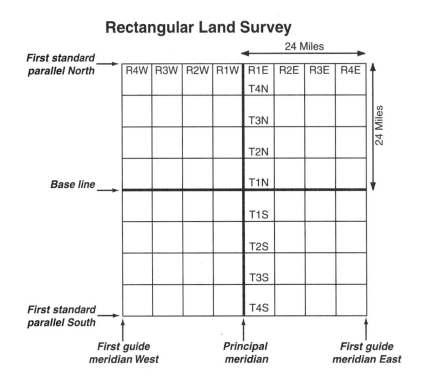

Figure 9–3 The rectangular survey system is based on parallel survey lines, with north-south lines and east-west lines that intersect at regular intervals.

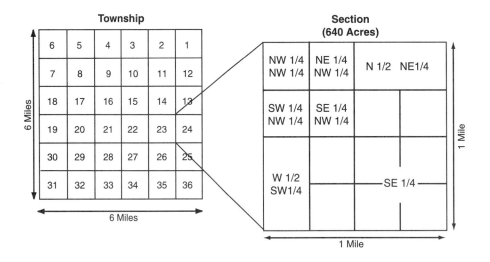

Figure 9–4 Land locations are described by the coordinates that they occupy in a land survey.

with each **quarter section** approximately equal to 160 acres. Quarter sections are further divided into four subdivisions, each of which contains slightly less than 40 acres. Each of these parcels of land is called a **forty.**

LAND AREA

Land area is a measurement of the amount of surface area within established boundaries. The land area measurement used in forestry can be expressed two different ways (Figure 9–5). An **acre** is a land area equal to 43,560 square feet. When metric measurements are used, **hectare** is the unit of measurement for land area. One hectare is equal to 10,000 square meters, or 100 ares. In the metric system, one **are** is equal to 100 square meters. One hectare is also equivalent to 2.471 acres.

Land area is necessary information for those who develop management plans for forests. For example, it would be nearly impossible to determine the amount of saleable timber in an area without first knowing the amount of surface area or measuring every tree. Once the land area has been determined, and a few of the trees have been measured, the amount of timber in a forest can be estimated with a reasonable degree of accuracy.

One way to determine the land area is to measure it while taking compass readings each time the boundary changes directions. A much easier method for determining land surface area is to prepare a **forest type map** from aerial photographs. The technician uses an instrument called a **stereoscope** to view two different aerial photographs at the same time. Two photographs of the same area that are taken from slightly different angles are merged by a stereoscope to convert the two views into a single view with three dimensions. Technicians who are trained to use this instrument are able to prepare maps that accurately record the locations of forests of different types using photographs taken from aircraft or from satellites. A technician who is skilled in the use of this instrument can even estimate the height of trees with accuracy (Figure 9–6).

Maps must be prepared that are proportional to the actual land area that they represent. This is known as drawing a map to scale. For example, a map scale of 1:20,000 means that each inch on the map represents 20,000 inches or 1,667 feet in the forest (Figure 9–7 on page 243).

A forest type map that has been prepared to scale can be accurately measured to determine land area. This is done through the use of an instrument called a **planimeter** by tracing the perimeter boundary of an area on a map. The

---

Land Area Measurements

1 acre = 43,560 square feet
1 hectare = 10,000 square meters = 100 ares
1 hectare = 2.471 acres

Figure 9–5 Forest land area is measured either in hectares and ares or in acres.

Figure 9–6 A stereoscope is used to prepare forest type maps by combining the images of two different aerial photographs of the same area taken from slightly different locations.

instrument accurately measures the area within irregular boundary lines such as those found on a topographical map (showing elevations) or a forest type map.

Another method that is used to determine land area is the **dot grid** method. A transparent plastic sheet on which dots have been uniformly placed is overlaid on a map (Figure 9–8 on page 244). The dots that lie within the boundary lines are then counted. A dot grid with ten rows of ten dots per row (one hundred dots) for each square inch of surface area has a conversion factor of 0.638 when a map with a scale of 1:20,000 is used:

$$20,000 \times 20,000 = 400,000,000 \text{ square inches forest surface area}$$
$$400,000,000 \text{ / } 144 \text{ square inches (1 square foot)} = 2,777,777.78 \text{ square feet}$$
$$2,777,777.78 \text{ / } 43,560 \text{ square feet (1 acre)} = 63.77 \text{ acres per square inch}$$
$$\text{of map surface}$$
$$63.77 \text{ / } 100 \text{ dots per square inch} = 0.638 \text{ acres/dot}$$

If 73 dots are counted within the boundary lines of a timber tract using this grid, the total acreage of the tract is approximately:

$$73 \times 0.638 = 46.57 \text{ acres}$$

The ability to produce aerial photographs of a known scale makes it possible to determine area directly from the photographs. Such photographs are becoming widely used for the purpose of mapping fields and forest lands. This method of measurement is called **photogammetry.** It is used to measure the total area enclosed within boundary lines that are defined on the photograph.

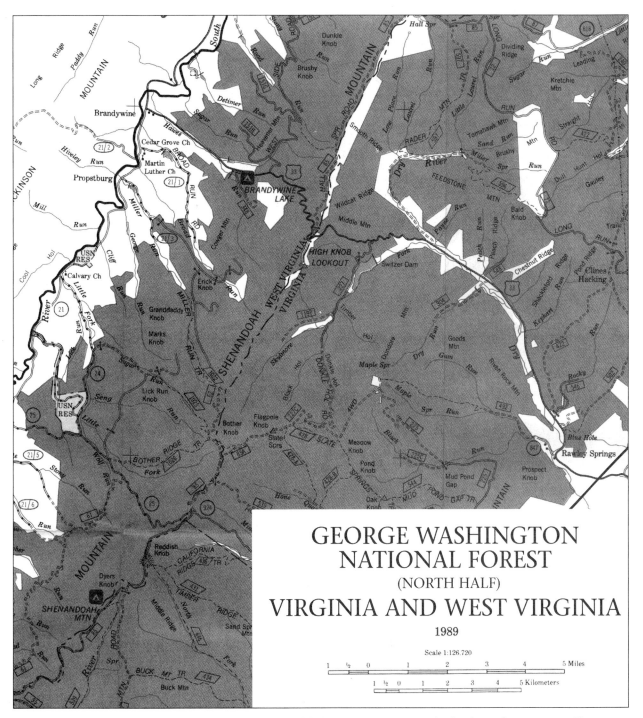

Figure 9–7 A map that is drawn to scale is proportional in its measurements to the land area it represents. (*Courtesy of the USDA Forest Service*)

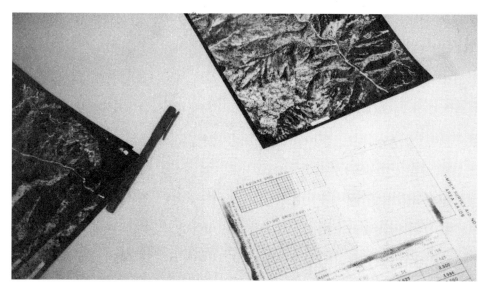

Figure 9–8 One method of measuring the surface area of a plot of land is to use a map overlaid by a clear plastic sheet on which dots have been uniformly placed according to a particular scale. The dots are counted to determine the amount of surface area.

TREE DIAMETER AND HEIGHT

The diameter and height of a tree must be known before it is possible to calculate the amount of wood that it will produce. These measurements are taken while a tree is still standing. For tree measurements that require a high degree of accuracy, instruments should be used that take direct measurements from the tree, and adjustments must be made to compensate for such things as thickness of bark, and slope. The dbh measurement, for example, should be taken on the upper side of a tree that is growing on a slope.

The diameter of a tree is measured outside the bark of the tree at breast height (4.5'), as was discussed in Chapter 2. This measurement is called dbh, and it is used to estimate timber volume. The dbh measurement can be taken in several ways, with the method of choice determined by the degree of accuracy that is required. The most accurate instrument for taking this measurement is the tree caliper. It adjusts to make contact on either side of a tree, and the diameter reading is taken directly from the scale on the instrument.

The dbh measurement can also be made with the use of a steel tape that is calibrated to provide dbh data by measuring the outside distance (circumference) around the tree (Figure 9–9). This method is accurate when a tree is perfectly round, but since most trees are not perfect cylinders, some accuracy is lost when this method is used. A third method for taking dbh measurements is to use a **Biltmore stick.** The instrument is used by holding the stick against the tree in a horizontal position at a distance of 25" (arm's length) from the eye and sighting across it. The left end of the stick is lined up on the outer edge of the tree, and the reading is taken from the instrument at the point where the sighting on the

Figure 9–9 The diameter of a tree can be accurately measured by a steel tape that converts a measurement of the circumference of a tree to the diameter.

other edge of the tree intersects the stick. This method is not as accurate as the caliper or the tape, but it is quick and easy to use, and it is accurate enough to provide reliable information for some forest management purposes. A variety of other devices is available to foresters that can be used to make rough estimates of timber volume. Most of these involve taking sightings through an instrument that is held at a fixed distance from the eye.

The height of a standing tree can be measured in different ways. The most accurate method for measuring height is to use a sectioned or telescoping height pole that is extended upward through the foliage to the top of the tree. This instrument is used in forest research plots, but it is useful only on small and moderate-sized trees. Tree heights of more than 60' must be measured another way.

Indirect measurement of trees can be done using a type of measuring device called a **hypsometer.** This instrument is available in two types. One type uses trigonometry to calculate tree height based on angles and known distances. The other type of instrument is based on geometry, and it is calibrated to use ratios of known fixed distances on similar triangles to calculate the height of the tree (Figure 9–10). Most hypsometers are designed to be held at arm's length (25") at a distance of one **chain** from the tree. A surveyor's chain for measurement is 66' long, and this is the unit of measurement used in the forest.

MEASUREMENT OF AGE

The determination of age in an individual tree can be made either before or after it has been harvested. Age is determined by counting the annual rings in the woody xylem tissue at the base of the tree. The environmental and physiological

Figure 9–10 A hypsometer is an instrument that is used to indirectly measure the height of a standing tree.

changes that accompany the different seasons of the year produce wood of slightly different color and different density. These changes can be seen distinctly in the wood that is deposited beneath the cambium of the stem each year (see Chapter 3).

The age of a standing tree can be measured using an **increment borer** to extract a core of wood from the cross-section of a living tree (Figure 9–11). The data that is obtained using this instrument is used to determine age, growth rate, and soundness in a tree. The instrument is equipped with a cutting tip, and the wood is deposited in the center of the hollow bit as it bores into the center of the tree. The wood cores that are obtained must be carefully removed from the tube to avoid disturbing the alignment of the tissue in the wood core, and the annual rings are counted. A single annual ring of new growth is deposited each year.

Determining the age of harvested trees is easily done by counting the annual rings in a cross-section of the base of a tree (Figure 9–12). Wood that is deposited during a favorable growth season is generally much thicker and lighter in color than the wood deposit from a year when the tree was subjected to drought conditions or other stressful effects.

MEASURING TIMBER STANDS

Timber stands are measured for a variety of reasons. They are measured regularly to determine the timber volume. Timber volume for a single tree can be expressed as **basal area.** This is the area of the cross-section of a tree in square meters or square feet. Basal area can also be used to describe the timber volume in an entire forest when it is expressed as square meters per hectare or as square

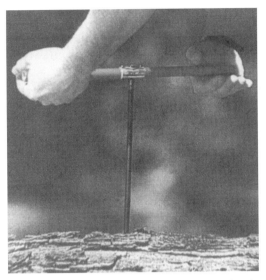

Figure 9–11 An increment borer is a tool that is used to extract a core of wood from a living tree for the purpose of counting the annual rings to determine the age of the tree.

Figure 9–12 The annual rings that are observed in the cross-section cut of a log can be counted to determine the age of a harvested tree.

feet per acre. The timber volume in a forest on a particular date is expressed as **timber yield.** It is a timber product inventory that can be used to measure how quickly the forests are renewed in comparison with the rate at which forest products are used. Forest managers need to know how much timber can be harvested each year to maintain a sustained yield.

A common method that is used to estimate timber yield is to **cruise** or survey the timber in an area. Cruising is the activity of gathering forest data from the sample plots. It is seldom necessary to measure every tree in a forest to determine timber yield because measuring sample plots that are similar to the rest of the forest and applying their yields to the entire area will usually provide data that is very close to actual measurements of all of the trees. A 100% cruise is used only in cases where a high degree of accuracy is required. A forest research plot may require this degree of accuracy, but in most instances, a partial cruise is adequate. Some basic assumptions apply to partial cruises to ensure the accuracy of the results:

1. The sample must be representative of the entire forest.
2. Enough samples should be tallied to reduce the chance that errors will occur.
3. The plot size should be large enough to tally fifteen to twenty trees per plot.

Establishing the sample sites for a partial cruise is important in determining the accuracy of the measurements that are taken. One method that is used to establish the sample sites is random sampling. This sampling procedure selects sample sites in each stand using a process that allows every possible site to have an equal chance of being selected. Random sampling is considered to be a fair way to select data collection sites.

One way to ensure that sample plots are selected that are similar to all of the other sample plots is to establish straight lines that extend across the entire area from which forest measurements are taken. The lines are spaced evenly, and sample units are identified at equal intervals along each line. Lines should be oriented to run up and down slopes instead of around slopes because timber of similar quality tends to be located in similar positions on a slope. Trees representing the different quality and type characteristics found in the entire forest are most likely to be encountered in the trees that are tallied when lines are oriented in this manner. This sampling method is called **systematic sampling** (Figure 9–13).

When systematic sampling methods are used, the process of gathering data is known as **line-plot cruising.** Two cruisers usually work together in teams with one team member taking measurements and the other recording data on the tally sheet or in a small portable computer (Figure 9–14). They refine field maps and make adjustments to type lines located on aerial photographs that identify forest types. Experienced cruisers also learn to make minor adjustments in values as they work. This may be necessary, for example, when one of the sample sites includes a pocket of old-growth timber that is not typical of the stand that is being measured. Cruisers must keep in mind that the reason for doing a

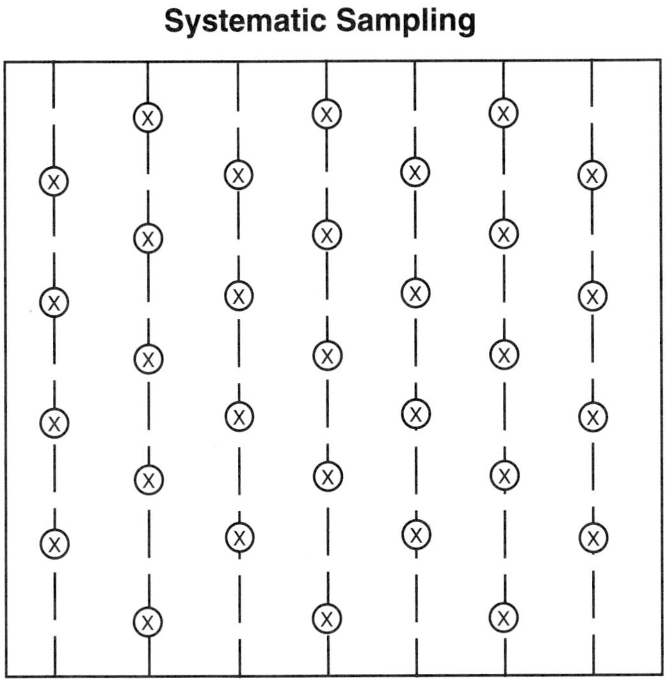

Systematic Sampling

Figure 9–13 Systemic sampling is an attempt to accurately estimate the timber yield of a forest by measuring only a representative sample of the trees.

Figure 9–14 Timber cruising is made easier by using a small portable computer that is designed specifically for recording the cruise data.

partial cruise is to gather data for the purpose of making an accurate estimate of the forest yield for the entire stand.

Several other methods of sampling a forest are used to determine the amount of marketable timber. One of these is to determine the average number of trees in each dbh class per acre of forest. This is done by tallying the number and dbh of all of the trees located in sample sites. Sample sites are distributed in different locations throughout the stand, and data from these sites are used to predict the forest yield for the entire stand. Another method is to use a cruising instrument that gauges the dbh of trees by sight. While standing at the center of the selected site, a person records only those trees that are larger than a certain dbh while turning slowly through a full circle. Values derived from these and similar sampling techniques can be used to estimate the amount of timber in the stand.

DETERMINING FOREST GROWTH

Forest growth is defined as the increase that occurs in the volume of wood in a forest over a specific period of time. Data that are used to determine forest growth are obtained by doing forest surveys. Second surveys must be conducted at each site following an interval of several years to determine forest growth. The difference in volume between the original survey and the second survey represents growth.

Conditions at a particular survey site in a forest tend to change during the time interval between forest surveys. New trees that were not present when an earlier survey was conducted, plus trees that were not big enough to be tallied in the original survey, add forest volume to the total that inflates forest growth beyond the amount of growth that occurs in individual trees during the same time period. This growth is known as **ingrowth.** Ingrowth is partly offset by the loss of trees that were measured during the first survey that were dead when the second survey was conducted. The data that are collected from these trees are recorded in a category of losses called **mortality,** and attempts are made to determine what caused the deaths of these trees.

Trees that have been harvested prior to the second forest survey are tallied using data from their stump measurements. This category is called **cut.** Trees that were measured in both surveys for the time interval during which growth is evaluated are **survivor trees.** The total difference in volume of these trees between the two surveys is called **survival growth.** The net forest growth is calculated according to the following formula:

Ingrowth + survivor growth – mortality – cut = forest growth

The growth rate of a tree or group of similar trees can be measured by using an increment borer to remove a core of wood from the tree and measuring the width of the most recent growth rings. The average width of the annual rings multiplied by 2 is equal to the annual increase in the diameter of the tree.

MEASURING FOREST PRODUCTS

The final forest measurements and the most accurate ones are the measurements that are taken after the timber has been harvested. Raw timber products include sawlogs, bolts, and chips. Each of these products is measured in a different way. For example, sawlogs are sold by the board foot. Bolts that are sold for firewood are usually measured in stacks and sold by the cord, while bolts that are sold for pulpwood are measured by the cord or by weight. Most pulpwood and paper products are sold by net weight. Measurement of forest products to determine the quantity or amount of product is called **scaling.**

Sawlogs include logs that are at least 8' long with a small-end diameter not less than 6". These logs are scaled by first measuring the diameter of the small end of the log inside the bark. This value in combination with the log length is used to determine the number of board feet in a log. This value is usually obtained from a table constructed for this purpose. The tables in most common use are the **Scribner rule** and the **Doyle rule.**

The unit of measurement known as the **board foot (BF)** is equivalent to the volume of wood in a board that measures 1' long × 12" wide × 1" thick (Figure 9–15). A more useful way to calculate a board foot is to express the formula as

$$BF = L \times W \times T / 12$$

LOG SCALING SYSTEMS

Scribner, Doyle, and International Rules

Three common methods for measuring logs are used in North American forests. The Doyle method tends to favor the buyer of logs that are less than 28" in diameter because more lumber can usually be cut from trees of this size than the formula indicates. The Scribner rule is quite accurate in its lumber estimates, and a modified version called the Scribner decimal C is used to measure timber sales by the United States Forest Service.

Many different rules for scaling logs have been used in the past, but the most accurate of these is the **International log rule.** Unlike the Scribner and Doyle rules for log measurement that are still widely used in the timber industry, the International log rule takes into account the amount of wood in the taper of the log from the small end to the large end. This system was developed by Judson Clark, a forester who mistrusted the log rules that were in use at the time. Using the International log rule, the volume of a log is determined by calculating the value of each 4' section of the log and adding those values together.

Doyle	$V = (D-4)^2 L / 16$
Scribner	$V = (.79D^2 - 2D - 4) L / 16$
Scribner decimal C	Round Scribner value to nearest 10 board feet; drop the last 0.
International	$V = .22D^2 - .71D$; Calculate each 4' section separately, then add them all together for final total (adjusts for ½" taper per 4' section of log).

where L (length) is expressed in feet, and W (width) and T (thickness) are expressed in inches. For example, a piece of lumber with dimensions of 8' × 8" × 2" has a volume of 10.67 board feet:

$$8 \times 8 \times 2 / 12 = 10.67 \text{ BF}$$

Board feet may be more easily calculated in odd-sized lumber by changing the formula to express all measurements in inches. For example, the formula for a board measure of 10'6" × 10" × 2" would be changed to:

$$126 \times 10 \times 2 / 144 = 17.5 \text{ BF}$$

Another log scaling method that has proven to be dependable is **Smalian's formula.** This measurement estimates the volume of solid wood in a log. Measurements are taken from both ends of the log to determine the area of the cross-section surface inside the bark. These measurements are combined with the length of the log to determine the volume of solid wood. Despite its high degree of accuracy, the wood industry continues to use the board foot as the most common unit of measurement instead of converting to a volume system of measurement. This is probably because more time is required to make the measurements that are needed to use the volume system. In comparison, the Doyle and Scribner methods are faster and easier to use. The formula for calculating total volume of solid wood in a log using Smalian's formula is:

Examples of One Board Foot

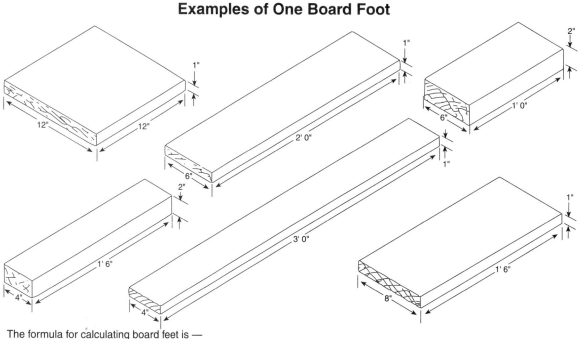

The formula for calculating board feet is —
bd. ft. = number of pieces × thickness in inches × width in inches × length in feet ÷ 12.
Calculate the board feet in the following:

Problem 1. 5 pieces 2" × 4" × 8'.

Solution: $\dfrac{5 \times 2" \times 4" \times 8'}{12} = 26^2/_3$ bd. ft.

Problem 2. 6 pieces 1" × 8" × 10' *or*, the same pieces dressed would be 6 pieces 3/4" × 7 1/2" × 10'. (Fractions from 1/2" to 1" are considered as 1".)

Solution: $\dfrac{6 \times 1" \times 8" \times 10'}{12} = 40$ bd. ft.

Problem 3. 8 pieces 2" × 6" × 38".
(If the length is in inches, divide the product by 144 instead of 12. Why?)

Solution: $\dfrac{8 \times 2" \times 6" \times 38"}{144} = 25^1/_3"$ bd. ft.

Figure 9–15 The basic measurement of lumber is the board foot. One board foot is equal to the volume of wood in a board that measures 1' long × 1' wide × 1" thick. (*Visual courtesy of Elmer Cooper*)

$$V = (A1 + A2) \, L \, / \, 2$$

where V = volume, $A1$ and $A2$ = areas of log ends, and L = length.

Pulpwood and firewood scaling is usually done using a unit of measurement known as a **standard cord.** One cord of stacked wood measures 4' × 4' × 8', or 128 cubic feet in volume (Figure 9–16). The following formula can be used to calculate the number of cords in a stack of wood when each measurement is expressed in feet:

$$(\text{Length} \times \text{Height} \times \text{Width}) \, / \, 128 = \text{Number of Cords}$$

For example, how many cords of firewood are there in a stack of wood 12' long, 5' high, and 4' wide?

$$12' \times 5' \times 4' \, / \, 128 = 1.88 \text{ cords}$$

Figure 9–16 One cord of stacked pulpwood or firewood is equal to the volume of wood measuring 4' wide × 4' high × 8' long, or 128 cubic feet.

ESTIMATING BIOMASS

Wood has long been used as a form of energy. Even in the modern world, more wood is used for fuel than for all other uses combined. Biomass is defined as the weight of all of the material in a tree that can be used as a source of energy. In more realistic terms, it is the portion of the tree that is above the ground that can be used as fuel. Intensive culture of fast-growing trees has the capability of producing high tonnages of biomass when trees are planted in dense stands and harvested after twenty or thirty years of carefully managed growth.

Biomass production is measured as net weight of dry matter. One of the contributing factors to its potential as an energy crop is that whole-tree harvest methods convert the entire tree into biomass. This includes the branches, bark, and even the leaves. All of this plant material is used, and none is wasted. The trees are usually harvested mechanically and chipped in the field. In most instances, they are hauled to electrical power plants where they are dried and burned for fuel. Wood chips can even be converted to gas for more efficient generation of electricity.

Most of the biomass that is used for power production today comes from waste products. Biomass is now the second leading renewable source of energy for power production behind hydroelectric power. The production of biomass is a rapidly growing industry. As the production of biomass as an energy crop increases, it is estimated by the United States Department of Energy that up to 50,000 megawatts of electricity will be produced in biomass-powered generating plants by the year 2010. This is enough electricity to light approximately 7,000 football stadiums both day and night for a year. The biomass industry is also expected to create 120,000 new jobs in rural areas by 2010.

The weight of biomass that is available in a stand of trees can be very difficult to estimate because it involves several variables. Examples of these are differences in tree varieties (hardwoods tend to be higher in density than soft-woods), and maturity of the trees (immature wood tends to be higher in water content than mature wood). Biomass yield can be estimated by first measuring the volume of wood in a similar manner to the way cruising is done in a stand of timber. These weights must be adjusted upward, however, because the entire portion of the tree that is above the ground contributes to the yield of a biomass crop. Adjustments also must be made to convert the weight of the biomass that is available to a dry matter basis.

PROFILE ON FOREST SAFETY

Safe Use of Forest Vehicles

The forest work environment involves the use of trucks and off-road vehicles that are operated on rough and uneven terrain and narrow logging roads. The conditions under which trucks, tractor-trailers, and other vehicles are operated are often far from safe, and it is important for the workers who operate these vehicles to learn safe operating procedures. Recognition of the hazards involved in operating forest vehicles is the first step in developing safe work habits.

The combination of speed and top-heavy loads can be extremely dangerous. Most of the driving that occurs in the forest involves loads of logs that sit high above the road, so the driver must learn to slow down on curves to avoid tipping the vehicle over. Quite often, it is difficult or impossible to see far enough ahead to anticipate problems with other vehicles or blockages on the road. The key to safe handling of forest vehicles is to drive according to the road conditions and avoid speeding. Forest vehicles must be well maintained to ensure that they are mechanically sound and to avoid such problems as faulty brakes or loss of power on steep grades. It is important to secure the load by chaining it down and to check the chains enroute. Great caution is also needed during the loading and unloading processes to make sure that people and equipment are not crushed by heavy logs.

CAREER OPTION

Timber Cruiser

A **timber cruiser** is a person who estimates marketable timber volume by taking sample measurements from sites throughout a stand of trees. These measurements are used to predict the total volume of timber in the designated area. Estimated losses are subtracted from the total volume to determine the total volume of marketable timber. A cruiser also gathers data about forest conditions that can be used to make forest management decisions related to logging, timber sales, sustained yields, land use, and many other purposes.

A timber cruiser spends a great deal of time in the forest. He or she must be skilled in sampling techniques, use of various measuring instruments, and mathematical calculations. Skills in careful observation are required in order to notice such things as symptoms of decay or insect infestations.

LOOKING BACK

Forest management for sustained yields requires the ability to measure living forests and forest products. Land surveys are used to define property lines and to describe locations of timber stands on maps. Maps are also used to record forest types. Timber cruisers refine maps derived from aerial photographs, and measure timber stands to determine the management practices that are needed. They do this by measuring representative samples of the trees in the stand and calculating volume estimates based on data obtained from the sample sites. Timber yield charts are used to calculate timber yields. The Scribner decimal C rule is the method that has been adopted by the United States Forest Service to measure timber sales. Timber stands are measured and marked using measuring devices of several types to estimate forest yield. Timber products are measured after the trees have been harvested by using direct measurements on the trees. Timber products are sold mostly on the basis of volume and weight.

QUESTIONS FOR DISCUSSION AND REVIEW

Essay Questions

1. Explain the management concept of sustained yield, and identify the kinds of information that are needed to develop a sustained yield forest management plan.

2. What are the two types of land surveys that are used in the United States? What are the key differences between them?

3. How is the rectangular survey system organized to identify the locations of property lines and boundaries?

4. Describe how instruments such as the stereoscope and planimeter are used to create a forest type map.

5. How does a timber cruiser determine the height and diameters of live trees?

6. How does a forest worker determine the age of a live tree?

7. What is the difference between a 100% timber cruise and a partial timber cruise? When is each method appropriately used?

8. Suggest some sampling and measuring practices that can influence the accuracy of a partial timber cruise.

9. Explain the formula for determining forest growth, and describe each of the formula components.

10. List the steps and explain the procedures that are used to scale logs.

Multiple-Choice Questions

1. The type of land survey system that is used in most areas of the United States except in some areas of the southeast is:
   A. metes and bounds
   B. rectangular survey
   C. Scribner's rule
   D. Gannet survey

2. Which of the following is a survey line that runs east-west?
   A. principal meridian
   B. section
   C. guide meridian
   D. baseline

3. Each section of land has a surface area of approximately how many acres?
   A. 640
   B. 320
   C. 160
   D. 40

4. A tract of land approximately six miles square is called a:
   A. section
   B. township
   C. quarter section
   D. range

5. A procedure called the dot grid method uses a clear plastic sheet with evenly spaced dots to:
   A. measure land area
   B. measure tree heights
   C. make forest type maps
   D. calculate forest basal area

6. A method that uses aerial photographs to make field maps to scale is:
   A. photogammetry
   B. altimetric scaling
   C. planimetery
   D. fotometry

7. An instrument that is used to measure the circumference of a tree is called a:
   A. Biltmore stick
   B. steel tape
   C. hypsometer
   D. timber cruiser

8. The age of a living tree is determined using an instrument called a:
   A. increment borer
   B. planimeter
   C. Biltmore stick
   D. stereoscope

9. The activity of gathering data for the purpose of estimating the volume of standing timber is called:
   A. slumming
   B. scaling
   C. cruising
   D. Scribner's rule

10. The timber volume table that is used by the United States Forest Service to measure timber sales is:
    A. Doyle's rule
    B. Scribner's decimal C
    C. Scribner's rule
    D. International log rule

11. The activity of measuring piled logs to determine their yield in board feet is called:
    A. slumming
    B. scaling
    C. cruising
    D. scribbing

12. The standard of measurement that is used to express biomass production is:
    A. cubic feet
    B. board feet
    C. cubic meters
    D. net weight

LEARNING ACTIVITIES

1. Cut some cross-sections that are 2–3" thick from several cut trees of different sizes. Assign students to work together in small groups with each group receiving one of the cross-sections of a tree. Assume that the tree was harvested last year. Determine the age of the tree at the time it was harvested. Determine some years when drought or other stressful factors affected the tree, and determine some years when growth conditions were favorable. Use pins to mark the annual ring that was deposited when significant events occurred, such as the end of World War II, or the year men walked on the moon.

2. Obtain some aerial photographs or maps of agricultural fields or timber sales areas from a government agency such as the Farm Service Agency or Forest Service. Also obtain some dot grid sheets from the same agency or purchase some from a forestry supply catalog. Invite a professional forester or soil conservation officer to instruct the class on the methods and purposes for measuring land area. Make sure that each student successfully calculates the land area on his or her field photograph or map.

Chapter

10

Terms to Know

logging
seed tree method
shelterwood method
coppice method
environmental impact statement
Environmental Protection Agency
(EPA)
chain saw
sawhead
limbing
bucking
yard
landing
bunching
skidder
prehauler
forwarder
siltation
silt load
deadwood
choker
litigation

Harvest and Reforestation Practices

Objectives

After completing this chapter, you should be able to

* identify factors that influence decisions affecting timber harvests

* describe some important components of a timber harvest plan

* distinguish between harvest methods leading to even-aged and uneven-aged forests

* explain how the planned method of forest regeneration affects the selection of a harvest method

* relate the volume of timber harvests to forest growth as they affect the forest management concept of sustained yield

* speculate on the reasons the National Environmental Policy Act included a requirement for an environmental impact statement to be filed as part of each timber harvest plan

* list and explain each of the steps that are involved in harvesting timber

* explain the historical relationship between road construction in forests and surface water quality

* evaluate the practice of salvage logging

* describe some methods that are used to minimize litigation related to timber harvests

Timber harvest practices in North America have changed in remarkable ways since the first logs were harvested along American shores. Early European explorers found a dependable supply of tall, straight trees to replace broken ship masts and timbers in the coastline forests. There was no need to obtain timber permits, and no environmental impact statements were required. Early settlers in North America enjoyed these same privileges. They could harvest trees as they wished, and the seemingly endless timber supplies, though abundant, were discovered to be a finite resource. Timber harvests today are highly regulated and controlled, and the average citizen in the United States is unlikely to ever harvest a tree in his or her lifetime.

Many steps are involved in the process that leads to a timber harvest. Forest management decisions are influenced by many factors besides science-based research. Among these influences are the legislative actions that have been taken by Congress and the state legislatures. Legislation has led to regulation of many aspects of the forest industry (see Chapter 7). Environmental impact statements, occupational safety regulations, and forest management rules have all changed the way timber harvests are planned and conducted in both private and public forests (Figure 10–1).

PLANNING THE HARVEST

A timber harvest, also known as **logging,** is a natural outgrowth of planning. Once a long-range management plan for a forest has been implemented, it is expected that proven silviculture practices will lead to a harvest of quality timber on a timetable that is consistent with maintaining sustainable yields. Most forest managers can predict, with reasonable accuracy, how long it will take for an even-aged timber stand to mature. They also know that once the trees have matured, a harvest must be planned and initiated to avoid timber losses (Figure 10–2). Heart rot and decay become serious problems in overage forests. The increase in timber volume due to growth is more than offset by losses due to rotting.

Planning for a timber harvest begins long before any trees are cut. One of the first decisions that must be made is the type of harvest that will be conducted. The different types of timber harvest methods were discussed briefly in Chapter 8 with reference to the ways that harvest methods affect forest regeneration. This chapter will consider ways that the different timber harvesting methods are implemented.

Clear-cutting

One of the most important considerations in planning a timber harvest is to determine what method will be used to regenerate the forest. If an even-aged stand is desired, it can be achieved by planning the way the current stand of trees is harvested. For example, an even-aged stand of Eastern White Pine can best be established by the clear-cutting method of harvesting. This is followed by site preparation and planting either seeds or seedlings to regenerate the desired stand.

Figure 10–1 Environmental impact statements evaluate proposed forest uses in terms of how the use will affect the resources such as water, wildlife, soils, and so on. Timber harvest planning is intended to reduce environmental impacts on other resources and uses.

Figure 10–2 Mature trees are valuable natural resources that supply raw products to the wood industry. Once a tree has matured, however, its health usually declines until it dies. Failure to harvest in a timely manner results in dead or decaying trees that have no commercial value.

Clear-cutting is an unpopular harvest method among some political action groups due to the loss of scenic value and dangers to the environment, both real and perceived. If clear-cutting is the method of choice, harvests in adjacent areas must be timed to ensure that forests of different ages are established (Figure 10–3). In the political world of the twenty-first century, it will be important to restrict the size of clear-cut harvests to reduce the potential for erosion. Harvest planning should also provide for using selection-cutting methods near roads and leaving unharvested buffer zones of trees near streams to protect water quality and to provide wildlife habitat (Figure 10–4). These kinds of compromises will be necessary if clear-cutting harvest methods are to be used in the future.

Seed Tree Method

The **seed tree method** of timber harvesting is part of a management plan in which mature trees of the desired species are protected from cutting in scattered locations throughout the forest (Figure 10–5). The purpose of such protection is to provide seed for regeneration of the forest. In most instances, site preparation

Planning for Clear-cuts on a Seventy-Year Rotation

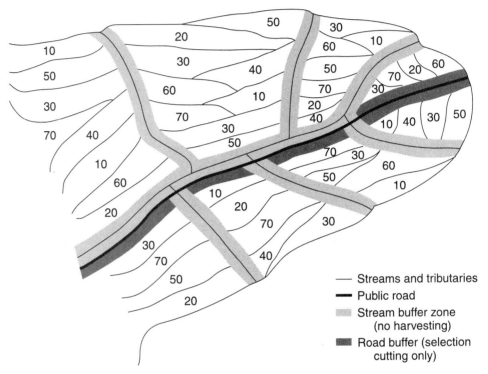

	Streams and tributaries
	Public road
	Stream buffer zone (no harvesting)
	Road buffer (selection cutting only)

Figure 10–3 Clear-cuts should be set up parallel to contour lines of the terrain and isolated in relatively small pockets. This harvest plan allows trees to be harvested for sustained yields every ten years on a seventy-year rotation.

Figure 10–4 Unharvested buffer zones near roads and streams provide wildlife habitat and preserve the beauty of the forest environment.

Seed Tree Regeneration Method

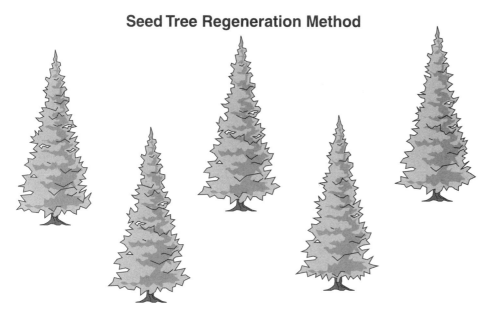

Figure 10–5 The seed tree method of timber harvesting protects mature trees that are scattered throughout the harvest area to provide seed for the next crop. The seed trees may be harvested once the young trees have become established.

is necessary to reduce competition from shrubs and established seedlings of less desired species. This method is effective in establishing even-aged timber stands with Southern pine and Western larch. The seed tree method has limitations in situations where site preparation is not practiced.

Shelterwood Method

The **shelterwood method** is a modified seed tree harvest method in which mature trees are left in the harvested area in sufficient numbers to provide shade and protection for seedlings (Figure 10–6). Once seedlings have become established, some of the mature trees are harvested, leaving an overstory that provides partial shade on the forest floor. The number of these mature trees that are harvested in the first cutting depends on the shelter needs of the seedlings. These needs vary from one species to another, and the species that eventually become dominant in the stand may be strongly influenced by the availability of shelter during critical periods of development.

When the new stand has become well established, the remaining mature trees are harvested. This harvest method has proven to be effective where harsh environmental conditions exist, making it difficult for young trees to survive. This method for establishing even-aged timber stands also tends to be viewed more favorably than clear-cutting because the landscape is never completely stripped of trees.

Shelterwood Regeneration Method

Figure 10–6 When the shelterwood method of harvesting is used, enough mature trees are left in the harvested area to provide seeds along with shade and protection for the young seedlings.

Coppice Method

The **coppice method** of forest regeneration is a silviculture system in which trees are clear-cut and the forest is regenerated from stump sprouts (Figure 10–7). It is included here because of its connection with clear-cutting as a harvest method. As with any clear-cut harvest, care must be taken to protect against soil erosion by planning the sequence of clear-cuts. Most of the trees that are generated by the coppice method are fast-growing trees that are managed on short rotation periods between harvests. Oaks and aspens are among the trees that are managed with this system. In most cases, these trees are used for fuel or pulpwood. This method also has great promise in the production of biomass.

Selection Cutting

Selection cutting as it is described in this textbook is not the same as the selection cutting that occurred as native forests were harvested. In that instance, selective cutting was a harvest system in which only the high-value trees were harvested. Eventually, the only trees that remained were the least desirable varieties. Selection cutting is a timber harvest method that is used to identify and harvest trees near the end of their productive lives (Figure 10–8). It is important to harvest these trees while they are still vigorous, and before they become victims of decay or disease.

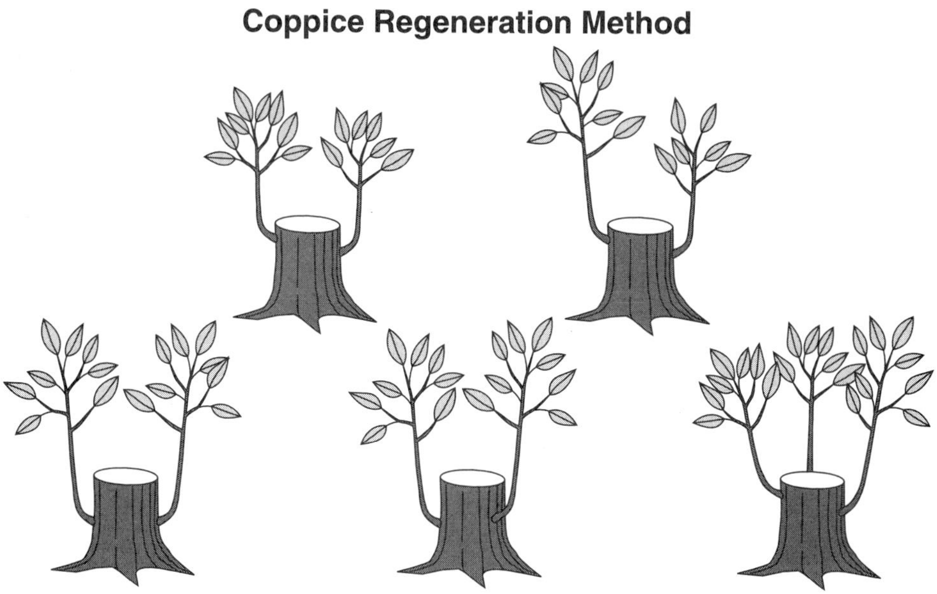

Coppice Regeneration Method

Figure 10–7 A new generation of trees is produced from sprouts arising from the stumps of harvested trees when the coppice harvest method is used.

Figure 10–8 When the selection-cutting harvest method is used, only the mature trees that are near the end of their productive lives are marked for harvesting.

Harvest planning is a continual process when selection-cutting practices are followed. This is because trees are removed from the forest at regular intervals as they mature. Regular inspections and forest inventories are required with this kind of harvest method to ensure that the correct amount of timber is harvested. When too much timber is harvested, the ability of the forest to produce at a sustained level gradually decreases. Eventually the size of the trees diminishes, and larger numbers of trees are required to maintain the same harvest yields.

Sustained yields can be established and maintained by determining the forest growth, and harvesting only as much timber as the forest can replace. When selective cutting is practiced, a plan with a regular harvest rotation or a minimal tree size is needed. Prior to each harvest, individual trees are evaluated, and a harvest decision is made. The harvest decision is based on the vigor and the potential future production of each individual tree. Growth rate is a good indicator of vigor, and individual trees may be retained in the forest instead of being harvested because they are still healthy and their potential for growth is high. Individual trees must be evaluated and marked for each harvest when selection cutting is practiced.

ENVIRONMENTAL IMPACT STATEMENTS

The passage of the National Environmental Policy Act (NEPA) in 1969 required the filing of an **environmental impact statement.** This is a science-based study of the harvest area that specifically details the expected effects of human activity on the environment and wildlife in the area. Such studies must be completed before timber harvests may be conducted on federal lands. This piece of legislation has had major impacts on forest management practices in the western area of the United States where large tracts of federal forest lands are concentrated.

An environmental study on the impacts of logging includes the effects of road construction in the area. This activity has the potential of causing pollution of streams with runoff water laden with silt. Among the issues that such a study will address are soil types and the expected effects of logging activity on soil

---

ENVIRONMENTAL
PROFILE

The Environmental Protection Agency

The **Environmental Protection Agency (EPA)** was established in 1970. This agency is charged to protect and maintain the environment for future generations. The EPA is responsible for the enforcement of environmental laws designed to reduce pollution of air and water due to noise, radiation, pesticides, and harmful chemicals and materials. This agency has established the water quality standards, and it monitors the disposal of toxic waste. The EPA monitors chemical residues in humans, wildlife, and food.

stability. Effects of logging on native plants and wild birds and animals are included in the environmental impact statement. The presence of an endangered or threatened species in the area may result in major modifications in a timber harvest plan or even stop all logging activities. The environmental impact statement also addresses the ways that environmental damage and pollution problems will be prevented (Figure 10–9).

HARVESTING PRACTICES

Timber harvests involve a number of separate operations that take place in the forest. The first of these activities is felling the tree. This can be done mechanically when the size of the trees and the slope of the terrain will allow it. Trees up to 24" in diameter at the butt can be felled mechanically, but larger trees are usually felled with a gasoline-powered cutting tool called a **chain saw.** Trees that are harvested for pulpwood are often sheared off at the base with hydraulic shears. Concerns about compression damage to sawlogs using this method have resulted in the development of a **sawhead** that is mounted on a mechanical feller, replacing the more common shears.

The process that follows the felling of a tree is **limbing** or removal of the tree limbs (Figure 10–10). Limbs have little value except for pulpwood or biomass. They are removed from the tree to facilitate handling. Despite the development of large mechanical machines called limbers that break limbs off the tree trunks,

Figure 10–9 Senior level forest managers spend a great deal of their time researching and writing forest plans that limit negative environmental impacts on forest resources.

Figure 10–10 Power saws are still very much in use in the forest for felling, limbing, and cutting trees into appropriate lengths.

Figure 10–11 Logs are transported from the harvest locations to yards or landings where they are piled for storage until they can be transported to a mill for processing.

much of this work is still done with a chain saw. It is followed by the **bucking** process during which the tree is cut into segments of the proper length for hauling or processing. Much of this work is done with chain saws, although mechanical methods are available for performing the limbing and bucking processes.

Logs for which the intended use is lumber production must be moved to a site where they can be loaded for transport to a sawmill. This site is called a **yard** or **landing** (Figure 10–11). Logs in close proximity to one another are assembled in small piles. This is called **bunching,** and its purpose is to allow several logs to be skidded to the landing at the same time. This is done with a machine such as a **skidder** that drags logs to the landing. Some logging operations use mechanical winches or cable yarding systems. These yarding methods involve pulling or lifting the log with a cable from the place where it was felled to the yard. On sensitive or inaccessible terrain, it is a common practice to use helicopters to carry logs to the yard or landing (Figure 10–12). A **prehauler** or **forwarder** is a machine equipped with a grappler or knuckleboom loader that is used at the landing (Figures 10–13, 10–14). Its best use is to pick up logs for the purpose of sorting, stacking, or loading them for transport.

Most logs are scaled at the landing or at the mill before they are processed. When logs are felled by different workers, it is a common practice to stack the logs in individual piles. This makes it possible for a logging contractor to pay individual workers according to the amount of timber that they cut. Most logging contractors pay their workers according to the amount of timber products that they deliver. This creates incentive for workers to work quickly, safely, and efficiently. Safety is a key factor in efficiency in the forest because no money is earned by workers who are injured so badly that they cannot work.

In instances where trees are to be used for pulpwood, they are usually debarked and chipped in the forest. Where whole tree-chipping operations are practiced, the entire tree may be chipped for biomass or pulpwood, and the

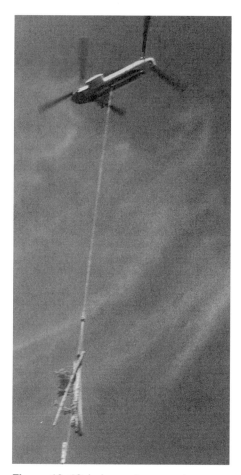

Figure 10–12 In harvest areas where terrain is steep or where damage is likely to occur to soil surfaces, commercial helicopters are often used to transport trees to landings.

Figure 10–13 Large industrial machines equipped with grappling devices are used at the landing and at the sawmill to handle heavy logs.

Figure 10–14 A knuckleboom loader is a specialized machine that is used to sort, move, stack, and load logs.

chips are usually transported to pulpmills or generating plants in large transport vans (Figure 10–15). The basic operations outlined here may be performed in a different order or modified in any logging operation.

ROAD CONSTRUCTION

Logs are usually transported by truck, although railroad cars may be used once the logs are out of the woods (Figure 10–16). It is sometimes cost-effective to use the railroad for long transports while reserving trucks for short hauls. The advantage of using trucks to haul timber products is that they are very adaptable to different transportation needs, such as hauling logs or hauling chips. They also can be readily maneuvered to logging sites.

Figure 10–15 When whole tree-chipping operations are conducted at the harvest site, large transport vans are used to haul the wood chips to the mill or electrical generating plants where they are used.

One of the biggest environmental problems associated with timber harvests is the construction of logging roads to gain access to the harvest area. Road construction is a major part of many logging operations. It must be done properly, especially in mountain areas, to avoid serious **siltation** problems in streams caused by silt particles settling out of runoff water. Silt contamination destroys

Figure 10–16 Large log trucks are usually the most effective and efficient means of transporting logs from the forest to the mill.

fish habitat and fills lakes and reservoirs with soil (Figure 10–17). Many problems of this nature can be avoided when logging roads are constructed to channel excess water and the **silt load** it carries into areas where the rate of flow is slowed down enough to allow the silt particles to settle out. When this is done, damage from siltation is greatly reduced.

Other logging activities that contribute to silt contamination of surface water include skidding and yarding of logs. All of these activities expose soil directly to rainfall, and they cause the soil to become highly susceptible to erosion.

SALVAGE LOGGING

Salvage logging is an emergency harvest practice. It is practiced when large numbers of trees are killed by insects, diseases, fire, or other natural disasters (Figure 10–18). A standing tree that has died before it is harvested is called **dead-wood.** It still contains good lumber, but wood quality begins to decline rapidly if the tree is not harvested within two or three years. The actual length of time that can elapse before deadwood is harvested depends on the tree species. Deadwood begins to dry out after a few months, and the wood in the center of a log does not become dry as quickly as the outer layers of wood. This often causes the surface of the log to crack as the outer shell shrinks.

Trees that have been blown down by wind bursts must be harvested soon afterward to avoid damage to the wood caused by a combination of soil moisture, molds, and mildews [Figure 10–19(a) and (b)]. Insect damage sometimes

Figure 10–17 Siltation problems have always existed in nature, but they are increased when careless timber harvesting practices are used. Silt that gets into the surface water ends up at the bottoms of lakes, reservoirs, and oceans.

Figure 10–18 Salvage logging is an important forest management practice that is used to remove dead, dying, or damaged trees from the forest.

extends deep into the heartwood of a tree. Some insects continue to damage the tree even after it has died. It is important to perform salvage logging operations within a few months to preserve the quality of the wood.

Timber that is killed by fire often remains standing for an extended time. In many instances, the soil is very fragile following a fire, and special measures

Figures 10–19(a) and **(b)** Natural disasters such as wind bursts or fires sometimes make it necessary to plan special harvests to remove the damaged trees. *(Photos courtesy of Boise National Forest)*

must be taken during the harvest operation to avoid causing serious damage to the soil. In some instances, salvage logging operations are conducted during the winter while the soil is frozen to minimize damage to the soil. In other instances, salvage logging operations are conducted by cutting damaged trees with chain saws, and using helicopters or even balloons to carry logs to landings. This is done by fastening a steel cable called a **choker** around the log and lifting it to a landing where it is loaded on a truck or train to be transported to the mill (Figure 10–20).

Figure 10–20 When helicopters are used to transport logs, high-strength steel cables called "chokers" are used to connect the logs to the large cable that extends downward from the underside of the helicopter.

PROFILE ON FOREST SAFETY

Safe Use of Equipment

The key to working safely in a forest environment is to be observant and anticipate unsafe situations and practices. For example, logs are round and heavy, resulting in a tendency to roll down slopes and hills. Forest workers who think about what might happen before cutting off the last branches will be less likely to be crushed by a rolling log than workers who hurry through their work with little thought of the consequences of their actions. This line of thinking, relating actions to consequences, could be called a "safety mind-set." The most important safety practice in the forest is to develop a safety mind-set.

Many different kinds of equipment are used in the forest industry. One of the most common types of injury to forest workers occurs when the worker is pinched between two moving parts. Examples include movable parts that are powered by hydraulic rams or pneumatic (air pressure) devices. In each instance, tremendous pressure is brought to bear on anything that gets in the way, including human body parts. It is important to identify "pinch points" before attempting to operate equipment. Once they have been identified, the worker must consciously avoid exposure to the moving parts.

Forest workers often work with cutting equipment such as saws and blades. Any cutting edge should be considered hazardous to workers, even when it is protected by a safety guard. Many of the workers who receive cuts on the job are aware of the cutting surfaces, but they choose to ignore safety practices with which they are familiar. Cuts that are received from logging equipment are capable of easily slicing through body parts, leaving the worker dead or maimed for life. There is no forgiveness by a machine when a worker places his or her body at risk of injury. The machine goes on performing its function without regard to hazards to workers.

MINIMIZING LITIGATION

Litigation is the act of filing a lawsuit with a court. Legal actions that are filed with the courts to prevent logging operations have become favorite tools of some special interest groups. Many of the timber sales that are proposed on public lands must be defended in court before the sale can proceed. Care must be exercised by agencies to anticipate legal issues during the harvest planning period, and the issues must be dealt with as part of the environmental impact statement. Anticipating legal arguments allows the United States Forest Service and other public agencies to deal with potential legal issues by gathering data before the timber sale is advertised.

Forests are important to both living and nonliving components of an ecosystem. Trees contribute to healthy watersheds, and they improve the habitats of fish and other life forms that are found in water and soil environments.

Most people in North America believe that forests should be managed in ways that will preserve them for future generations while allowing them to be used. The multiple use concept of forest management provides for a sustained yield of forest products while allowing for uses such as recreation, wildlife habitat, mining, and livestock grazing.

Young animals such as this newborn elk calf depend on forest environments for safety. Wildlife is most vulnerable to predators in the first few days following birth. This young calf depends on forest cover and camouflage coloring to help it hide from its natural enemies. *(Photo courtesy of Robert Pratt)*

Debris on the forest floor provides shelter for wildlife, and it contributes to the buildup of humus on the surface of the soil. However, too much debris on the forest floor also has negative effects because it contributes to destructive forest fires.

Forests also provide materials with which beaver colonies construct dams and lodges. It is generally believed that a beaver colony contributes more to the forest environment than it takes. Beaver contribute to forest environments by raising the water table in the area, thereby improving conditions for forest growth.

Poorly constructed forest roads are easily eroded, and they contribute to serious pollution of streams, rivers, and lakes with silt during seasons of heavy rainfall. Silt is the most widespread form of water pollution in forest environments.

The larvae of bark beetles and some other insects cause considerable damage to the phloem and xylem tissues of trees. Sometimes they completely girdle and kill the trees they feed on, but their greatest damage to the forest is the loss of production. *(Photo courtesy of Boise National Forest)*

Some harmful forest insects feed on the foliage of trees. Such insects are known as defoliators. They are capable of removing all of the leaves from a tree causing it to be weakened. Such trees are highly susceptible to diseases. *(Photo courtesy of Boise National Forest)*

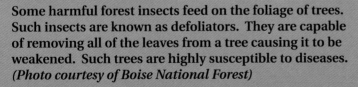

Some insects feed on terminal tissue, damaging the growing tips of branches and central leaders. Some insects eat the meristem tissue while others destroy the phloem and xylem tissues of immature trees. *(Photo courtesy of Boise National Forest)*

Tree leaves may be damaged by sucking insects that feed on plant juices. These trees often become weak, and they may be unable to resist disease organisms in their weakened state. *(Photo courtesy of Boise National Forest)*

Some insects burrow through valuable wood causing serious damage to the trees. The lumber of such trees has little commercial value due to the insect holes that are present in the wood. *(Photo courtesy of Boise National Forest)*

Birds such as the woodpeckers are natural enemies to insects. The holes in the bark of this tree were drilled by woodpeckers in search of grubs located beneath the bark. They are able to extract insect larvae from these holes with their long tongues.

The forces of nature sometimes cause devastation of large tracts of trees. Wind shear occurs when high speed wind gusts break the trunks of the trees or uproot them. In such cases, the wood is often recovered from the forest by conducting a "salvage logging" harvest operation. *(Photo courtesy of Boise National Forest)*

Timber harvests are accomplished with the use of large industrial machines that are used at the landing or yard to move or load heavy logs. These machines are equipped with hydraulic grappling devices that are capable of grasping and manipulating heavy logs.

Trees that are harvested in the forest are transported to yards or landings where they are piled for storage until they can be transported by truck to a mill for processing.

Fire can be a very destructive force in a forest, but it can also be a useful tool for forest management. A controlled burn is a fire that is ignited on purpose to burn trash and debris on the forest floor. When such fires are used regularly, the threat of wild fires is diminished. *(Photo courtesy of Boise National Forest)*

The health of a forest stand is improved by thinning the tree population to reduce competition between individual trees for sunlight and nutrients. Thinning may be accomplished by cutting some of the trees or by making regular use of controlled burns.

A surface fire usually lacks sufficient fuel to generate the intense heat that is required to kill trees of moderate size or larger. This type of fire consumes brush and grasses, but seldom burns through the bark of a live tree. *(Photo courtesy of Boise National Forest)*

Destructive forest fires can roar out of control when sufficient fuel is present to destroy mature trees. Once a decision has been to suppress a fire, many resources can be brought together to fight it. Aircraft are often used to drop fire retardant materials on the flames. *(Photo courtesy of Boise Interagency Fire Center)*

When a surface fire increases in intensity until it ignites aerial fuels such as the upper foliage of trees, it becomes a dangerous crown fire. Such fires are of such intensity that the trees explode, throwing burning materials ahead of the fire.

Sometimes the most effective way to fight a fire is to build a backfire in the path of a wildfire. The backfire is used to burn the fuel from the path of the large, uncontrolled fire. A backfire is ignited by the fire crew along the inner edge of a firebreak where it burns toward the wildfire consuming the fuel in its path. *(Photo courtesy of Boise National Forest)*

Wood is an ideal construction material for buildings because it is capable of flexing when a heavy load is placed upon it. It is especially useful for rafters and trusses that may have to sustain heavy snow loads or the stress of high winds.

A growing market for wood products is the biomass industry. Entire trees are chipped and dried for use as fuel. Biomass is obtained from tree plantations and waste materials that are recovered from lumber mills and manufacturing plants. Large amounts of biomass are used as fuel to generate electricity.

Large amounts of wood are required for the manufacture of paper products such as tissues, paper towels, writing paper, cardboard, and packaging materials. Wood fibers are separated and then reconstituted in a thin layer on heated metal drums. Large rolls of paper are produced for later use. *(Photo courtesy of Marilyn Parker and Potlatch Corporation)*

CAREER OPTION

Litigation Specialist

The United States Forest Service and other state and federal agencies depend upon people who are well educated in the legal issues affecting forests and natural resources to research and defend the agency against lawsuits and to bring court action against people who violate laws. A license to practice law is required of those who defend or prosecute cases in court. Some litigation specialists work as researchers and are not lawyers; however, a law degree and a license to practice law are generally considered to be prerequisites for this career.

LOOKING BACK

Harvest management of forest lands is influenced by public pressures in the form of legislation, occupational safety rules, environmental impact studies, and forest regulations. Each of these influences has changed harvesting practices on both private and public lands. All of these influences have caused the planning component to be one of the most important steps in a timber harvest. A decision to harvest a stand of timber is based on the maturity and health of the trees, and the method of harvest depends upon the findings of an environmental impact study and whether an even-aged or uneven-aged forest is planned. Clear-cutting, seed tree, shelterwood, and coppice methods of timber harvesting lead to the growth of even-aged stands. Selection cutting is used where uneven-aged stands are desired.

Sustained yields constitute an important concept in harvest management, meaning that the amount of timber that is harvested cannot exceed the amount of forest growth that occurs without depleting timber resources. The EPA is a federal agency that is charged with making sure environmental impact studies are conducted appropriately and that they are filed before timber harvests are conducted. Trees are harvested using both mechanical methods and human labor. Trees are felled, limbed, gathered, loaded, and hauled as part of the harvesting process. Road construction is done in such a manner that erosion of the soil will be minimized to protect surface water from silt contamination. Legal issues must be anticipated and steps must be taken during the planning process to resolve potential legal problems.

QUESTIONS FOR DISCUSSION AND REVIEW

Essay Questions

1. Name some factors that influence decisions affecting timber harvests.

2. What are the key components in a timber harvest plan?

3. List some timber harvest methods that lead to the regeneration of even-aged forests and uneven-aged forests.

4. How is the seed tree harvest method different from the shelterwood harvest method?

5. How is the selective harvest method applied in a forest?

6. Explain how forest managers determine the amount of timber that should be harvested to establish and maintain a sustained yield.

7. Why do you think Congress passed the National Environmental Policy Act requiring environmental impact statements to be filed as part of timber harvest plans?

8. List the main steps that are involved in harvesting timber, and explain how each step is performed.

9. What effect does the construction of roads for timber harvests sometimes have on the quality of surface water?

10. What are the positive and negative effects of the forest management practice known as salvage logging?

11. What kinds of data do forest managers include in timber harvest plans to minimize the potential for litigation in the court system?

Multiple-Choice Questions

1. The greatest losses of wood in a forest that is past maturity are due to:
   A. decay
   B. flooding
   C. fire
   D. insects

2. Which of the following timber harvest methods results in an uneven-aged forest?
   A. coppice method
   B. seed tree method
   C. selection-cutting method
   D. shelterwood method

3. The timber harvest method that removes all of the trees from the harvest area is the:
   A. seed tree method
   B. shelterwood method
   C. selection-cutting method
   D. clear-cutting method

4. A timber harvest method in which enough mature trees are saved in the harvest area to provide shade on the forest floor is called the:
   A. shelterwood method
   B. seed tree method
   C. coppice method
   D. selection-cutting method

5. A type of forest regeneration in which the new generation of trees arises from the stumps of the harvested trees is the:
   A. seed tree method
   B. coppice method
   C. selection-cutting method
   D. shelterwood method

6. The responsibility to enforce timber harvest regulations requiring environmental impact studies is a duty of a government agency known as:
   A. NEPA
   B. BLM
   C. USFS
   D. EPA

7. A problem associated with logging activities such as log skidding and road construction resulting in contamination of surface water is known as:
   A. bucking
   B. siltation
   C. scaling
   D. skidding

8. A timber harvest practice in which logs are cut into marketable lengths is called:
   A. bucking
   B. siltation
   C. scaling
   D. skidding

9. A timber harvest operation in which entire trees are processed in the forest to produce biomass is called:
   A. debarking
   B. bucking
   C. chipping
   D. chaining

10. A type of timber harvest that is initiated due to timber damage caused by insects, fire, or other natural disasters is called:
    A. selection cutting
    B. salvage logging
    C. coppice harvesting
    D. damage control

LEARNING ACTIVITIES

1. Invite a forest owner or a representative from the Forest Service to make an illustrated presentation and lead a class discussion on forest harvesting practices. He or she might be asked to bring some examples of old saws and harvesting equipment along with some modern harvesting tools and equipment. Discuss the following points and any others that might be important in your region of North America:

   ❋ changes in timber harvest practices
   ❋ changes in forest species in a particular forest
   ❋ harvest issues that often lead to litigation
   ❋ laws and regulations and their effects on timber harvesting

2. Make two identical models of a watershed using real soil in a shallow tray. Design it in such a way that it can be tilted to different slopes. Design a water distribution system to be placed at the top of the model and a water collection system to be placed at the lower end of the model. Place a mulch of leaves and plant material on the soil surface, and apply an adequate amount of water through the distribution system to cause water to flow into the collection system. The amount of water that is used and the rate of flow must always be the same with both experimental models. Collect all of the runoff water and strain it through a filter to collect any silt that it may contain. Dry the filter and compare its weight with that of a dry, clean filter. Repeat the experiment using the second model with the only difference being that no plant cover is used on the second model. Compare the results. Repeat the experiment changing only the slope of the model watershed. Compare results for plant cover versus no plant cover and for differences in slope. Exhibit the model and report your results in a science fair.

Chapter

11

Terms to Know

surface fuel
ground fuel
duff
peat
surface fire
ground fire
aerial fuel
crown fire
blowup
draft
firestorm
spotting
prescribed burn
incendiary fire
incendiarism
wildfire
fire suppression
firebreak
fire line
backfire
indirect attack
direct attack
short-duration fire cycle
smoke jumper

Fire and the Forest

Objectives

After completing this chapter, you should be able to

* describe ways that fire is both beneficial and destructive to forests

* identify three key elements that must be present for a fire to occur

* analyze the differences that exist among surface, ground, and crown fires

* list the major causes of destructive forest fires

* explain the effects of wildfires on forests and forest environments

* discuss ways that prescribed burns may be used to improve the health of forests

* analyze the fire suppression policy of the United States Forest Service as it has been implemented in the past and as it exists today

* calculate the rate of spread for a fire at different wind speeds

* describe the indirect attack method of fire suppression

* explain how direct attack methods of fire suppression act on the key ingredients of fire

* evaluate past and present efforts of government and industry to prevent destructive fires in the forests

* assess the effects of a short-duration fire cycle on the health of a forest

Fire is both the friend and the enemy of the forest (Figure 11–1). It is a friend because fire is the natural force that cleans debris consisting of dead trees and plant materials from the forest floor. It also provides some brush control and thins the stand when it passes through a forest from time to time (Figure 11–2). Some trees sustain very little damage from surface fires, because their bark is capable of protecting them. Fire becomes a deadly enemy to the forest when there is a buildup of fuel on the forest floor. This causes a ground fire to burn with such heat that the foliage of the trees catches fire, and the trunks of trees explode as sap is converted to steam. When a fire of this magnitude erupts, many of the live trees in the forest are killed or burned. Sometimes thousands of acres of forest are destroyed by a single fire. Fire is exceeded only by insect damage as the leading cause of timber losses.

Three key ingredients interact to control the rate and intensity of a fire (Figure 11–3). They include the availability and concentration of flammable fuels, heat energy sufficient to raise the fuel to its combustion temperature, and an adequate supply of oxygen. When all of these elements are combined in the same location, the result is fire. A change in any of the ingredients of a fire will change the nature of the fire. For example, damp fuel requires more heat than dry fuel to burn. Damp fuel also burns slowly, and the intensity of such a fire is reduced in comparison with that of a fire that burns dry fuel. The concentration of fuel material in an area increases the potential for an intense fire that generates a lot of heat. The total weight of the fuel in a burning area is a good indicator of how difficult it will be to control the fire.

Figure 11–1 Controlled fire can be a useful tool for burning trash that accumulates on the forest floor. *(Photo courtesy of Boise National Forest)*

Figure 11–2 The health of a forest stand is improved by a controlled ground fire in which trash and competing shrubs and young trees are burned. *(Photo courtesy of Boise National Forest)*

Three Ingredients of Fire

1. Availability and concentration of fuel

2. Heat energy sufficient to raise the fuel temperature to its combustion point

3. Adequate oxygen supply

Figure 11–3 When all three ingredients of fire occur in the same location at the same time, the result is *fire!*

FIRE TYPES

Fire types are determined largely by the kinds of fuels that are present. Undecayed fuels such as dry leaves and plant materials, including live plants that are located on the surface of the soil, are called **surface fuels** (Figure 11–4). Combustible materials located beneath the surface that have begun to decay are **ground fuels. Duff** is a ground fuel that is made up of decaying plant material located on or just beneath the surface of the forest floor (Figure 11–5). Another ground fuel called **peat** consists of organic matter that has accumulated in the water of a swamp or bog where decay has been limited.

Figure 11–4 Surface fuels consist of dead plant materials that accumulate on the forest floor.

Figure 11–5 Duff or ground fuel is made up of decaying plant material located on or beneath the surface of the forest floor.

A **surface fire** burns surface fuels consisting of the dry layer of twigs, dead branches, grass, and leaves that lie on the soil surface. It moves quickly across the surface of the ground, but it seldom becomes intense enough to generate large amounts of heat (Figure 11–6). Frequent fires of this nature prevent large

Figure 11–6 A surface fire usually lacks sufficient fuel to generate the intense heat that is required to kill trees of moderate size or larger. *(Photo courtesy of Boise National Forest)*

amounts of fuel from building up on the forest floor with high potential for serious damage, and most trees larger than saplings are able to withstand them (Figure 11–7).

A **ground fire** burns ground fuels that are found on or beneath the surface of the ground. During dry periods, these fuels will burn, and a fire in a deposit of peat or duff can smolder for many weeks. Such fires are limited in their rate or intensity by the amount of oxygen that is available to the flames, but they can be very difficult to extinguish.

The conifers and hardwoods differ considerably in the makeup of their foliage. Due to these differences, the foliage of conifers is more likely to catch fire

Figure 11–7 Forest professionals often use a small ground fire as a tool to remove fuel from the forest floor in the prevention of wildfires, and to kill young seedlings that tend to overpopulate the forest. *(Photo courtesy of USDA, Forest Service)*

than the foliage of hardwood trees. Fuels that are located much more than six feet above the ground in the mid to upper canopy of the forest are **aerial fuels.** A surface fire that builds in its intensity until it ignites the aerial fuels including the upper foliage of trees becomes known as a **crown fire** (Figure 11–8). This kind of fire generates a large amount of heat and can move quickly through the canopy of a forest, inflicting heavy damage on the trees and upon the wildlife that depends on the forest for food and shelter.

In addition to these three types of fires, a crown fire sometimes erupts rather suddenly into an extremely intense fire. This sudden increase in the intensity of a fire is called a **blowup.** Following a blowup, a fire sometimes increases even more in its intensity, creating strong winds and whirlwinds, and sending updrafts composed of hot gases and smoke high into the atmosphere. A fire of this magnitude sometimes creates its own **draft.** This is a supply of fresh air that the convection column draws into the base of the fire, causing the fire to intensify. Such erratic behavior by a crown fire is called a **firestorm.** A

Figure 11–8 When a surface fire increases in intensity until it ignites aerial fuels such as the upper foliage of trees, it becomes a dangerous crown fire.

firestorm often hurls burning debris far beyond the leading edge of the fire, causing it to jump across natural barriers such as rivers. A fire of this intensity often jumps across fire lines that have been constructed to halt its forward movement. The process by which a firestorm hurls burning materials across fire

f lightning strikes (Figure 11–9). It is
to generate hundreds of lightning
rest fires. A heavy rainfall prevents
drought conditions exist in the for-
ntly accompanies lightning, and it
hrough the forest and the intensity
ply of oxygen that wind provides to

olcanic activity. This is a rare occur-
t Helen's volcanic explosion in 1980
rce. The trees within a radius of sev-
wn or incinerated by the wind and

l by lightning, are caused by human
hese fires is failure to put out camp-
or discarding them in the forest. The
illegal use of fireworks inside forest boundaries accounts for some forest fires each year. A few fires are also caused by sparks from the friction of train wheels on the steel rails or by car fires alongside public roads. A few fires are caused by sparks from internal combustion engines, or from hot tailpipes making prolonged contact with forest vegetation.

Forest fires are known to start on some occasions when forest workers lose control of a prescribed burn. A **prescribed burn** is a fire that is started in an established forest for the purpose of removing vegetation and fuel from the forest floor, eliminating trash from a logging operation, or eliminating trees that are

---

Causes of Forest Fires

* Lightning
* Human activities: cigarettes, campfires, fireworks, and so on
* Friction (train or vehicle wheels)
* Internal combustion engines (sparks or hot tailpipes)
* Prescribed burns (loss of control)
* Incendiarism (fires that are purposely set)
* Volcanic activity

Figure 11–9 Forest fires are ignited in a variety of ways, and many fires can be prevented by using caution when you are in wooded areas.

Figure 11–10 Forest fires sometimes require fire control measures, and firefighters must enter the fire zone to stop them from spreading. *(Photo courtesy of Boise National Forest)*

afflicted by diseases or insects. This management practice is implemented from time to time to prevent the buildup of fuel materials in the forest. It is also used to control brush and other vegetation that compete with trees for soil moisture and plant nutrients. Fires have also been known to burn out of control in forests when trees and vegetation catch fire as piles of waste materials are burned.

In some cases, forest fires are deliberately started by people. These are called **incendiary fires,** and they account for over one-fourth of all forest fires in the United States each year. **Incendiarism** is the willful setting of fires, and this problem is most common in the southern states. This practice accounts for nearly 40% of all forest fires in that region of the United States. The practice of setting fire to vegetation for insect and pest control and to improve grazing for livestock has been done each year for many years in some southern states.

EFFECTS OF WILDFIRES

A **wildfire** is any fire that burns out of control, or that has not been prescribed for a specific purpose. Such fires often cause massive damage in forested areas due to their size and a tendency to be unpredictable (Figure 11–11). The most serious wildfires usually occur in times of drought following a period of plentiful precipitation. Readily available moisture causes grasses and other flammable plant materials to grow in abundance. This provides a ready supply of highly combustible fuel when these plants become dry during periods of drought.

Wildfires are more abundant in dry desert regions than in humid regions. This is because plant materials absorb moisture from the air in humid regions that helps to reduce the threat of fire. Fires do not burn as freely in humid areas because some of the heat that is generated by the fire is used to drive moisture out of the fuel. Regions with low humidity also tend to support vegetation that dries out by late summer. This causes the threat of wildfires to be high due to the presence of large amounts of highly combustible fuel.

The heat that an intense forest fire generates is capable of causing damage to nearly everything in the forest. Large trees are usually able to survive surface fires, but fire that is intense enough to get into the crowns of trees is usually fatal (Figure 11–12). This is because so much heat is generated by a fire of this type that the cambium and phloem tissues in the stem and branches are damaged, restricting the flow of nutrients and water in the tree.

Figure 11–11 A forest fire leaves the land without its protective plant cover, and the potential for serious soil erosion problems is high.

Figure 11–12 Intense fires generate tremendous amounts of heat, and they often prove fatal to trees because they damage the cambium and phloem tissues.

Seedlings are susceptible to any type of fire because the stems are so slender and the bark is so thin that heat can readily penetrate to the interior of stems and branches. Destruction of seedlings by a surface fire in an even-aged forest that is advancing toward maturity may be beneficial to the forest. This is because a fire of this type reduces the fuel load on the forest floor, reducing the potential for a more damaging fire at a later time. A similar fire in an uneven-aged forest may be considered destructive if it destroys too many of the seedlings that are needed to establish the next generation of productive trees.

Wildfires are known to have both good and bad effects on wildlife. Major fires destroy habitat upon which wild animals depend for their shelter and food. Many wild animals are able to leave fire zones to escape fires, but young birds and animals along with slow-moving creatures are often killed by forest fires. In the big picture, such losses usually have little impact on a population of animals. Such a fire can be devastating, however, when an endangered or threatened species is placed at risk by a fire.

In many instances, wildlife populations may benefit from fires. This is because plant species such as brush, grass, and other forage plants often fill in areas that were once covered by trees (Figure 11–13). This increases the food supplies for many kinds of birds and animals, allowing their populations to increase. This is particularly true for large browsing animals such as deer and elk. They tend to move back into burned areas almost as soon as new growth begins to appear. The succulent young plants that grow in burned areas tend to be high in nutrients in comparison with the plants that formerly grew there.

Wildfires often inflict serious damage to soils. This is because organic matter is often destroyed, causing the soil structure to break down. Combined with the loss of beneficial soil microbes, the effect of intense fires on soils is often devastating. The loss of soil structure and protective plant cover leaves soils vulnerable to severe erosion. Heavy rainfall during this time period sometimes erodes soils down to bedrock (Figure 11–14). One of the first remedial actions that must be taken following a major fire is to protect soils to prevent silt from being carried into nearby streams, rivers, and lakes.

Water quality is often seriously affected by forest fires. Large amounts of ash and soot are generated by a large fire, and some of this almost always gets into the surface water. The most damaging pollutant, however, is silt that has eroded from the unprotected soils in the fire zone. The threat of soil erosion is greatest in areas where the terrain is steep. Some protection can be provided to damaged soils by seeding the area with grasses. Large areas where ground cover has been destroyed by fire are sometimes reseeded using aircraft to distribute the seeds.

FOREST FIRE MANAGEMENT

Forest fire management has always been based on the ability to detect fires in the vast regions where forests grow. As people became concerned with the loss

Figure 11–13 One outcome of a forest fire is that open areas are created in the forest where large grazing animals like elk and deer can find an abundance of food plants.

Figure 11–14 Heavy rainfall following an intense fire often removes topsoil all the way down to bedrock. *(Photo courtesy of Boise National Forest)*

of timber resources due to wildfires, lookout towers were constructed in remote forest regions. A constant watch was maintained during the summer fire season to detect fires as they occurred. Some of these towers are still used, although fires are usually detected more effectively using modern technologies such as remote sensing equipment and satellite surveillance (Figure 11–15).

Forest fire management is based on the concept that it is neither necessary nor cost-effective to attempt to suppress or put out every forest fire that occurs. There is plenty of evidence to suggest that most fires that are allowed to burn will be extinguished naturally before they have burned more than a few acres of trees. This is fortunate, because it is common for a single thunderstorm to generate several thousand lightning strikes. Most of these strikes do not result in large forest fires, but more fires are started by lightning strikes than it would be possible to suppress.

The key to fire management is to quickly make a judgment of the probable effects that a particular fire is likely to have on the forest if it is allowed to burn. An important factor in making this judgment is an assessment of the likely impact that the fire will have on the management plan for the forest (Figure

Satellite Fire Surveillance

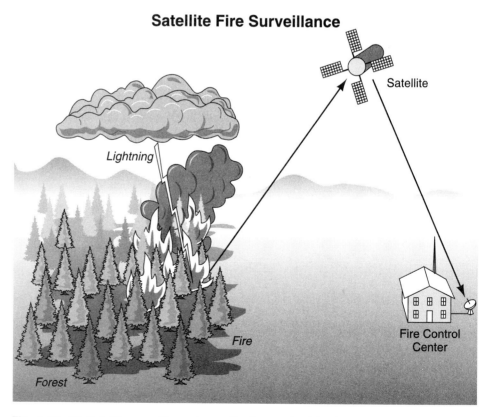

Figure 11–15 Satellite surveillance is an effective method for fire detection in the forest.

11–16). In the case of a surface fire, the goals of the forest management plan may be best served by allowing the fire to cleanse the forest floor.

One of the considerations in deciding how to best deal with a forest fire is to assess the cost of suppressing the fire in comparison with potential losses if the fire is allowed to burn. When a valuable timber resource is located in the area where a fire is burning, it may be cost-effective to suppress the fire. On the other hand, when a fire is burning in an area that has very little high-value timber, the cost of suppressing the fire may exceed the value of the timber that is saved. In some cases it may be wise to watch the progress of the fire for a period of time before making a decision on fighting it or allowing it to burn. Sometimes a decision (to suppress a fire or to let it burn) is based entirely on the availability of people who are trained to fight forest fires. During the 1990s, several different fire seasons have challenged firefighters due to large numbers of fires burning simultaneously in the United States.

Figure 11–16 It is neither practical nor possible to extinguish all forest fires. Forest managers must assess how a fire is likely to impact the forest management plan if it is allowed to burn. *(Photo courtesy of Boise National Forest)*

FORESTRY PROFILE

National Interagency Fire Center

The National Interagency Fire Center is the control center that coordinates fire suppression efforts of several federal agencies that have responsibilities related to management of federal range and forest lands (Figure 11–17). The center is located in Boise, Idaho, where firefighters are equipped, trained, and transported to wildfires in any location where they are needed. Large warehouses filled with fire-fighting equipment, protective clothing, camping supplies, vehicles, aircraft, parachutes, radios, and anything else that might be needed to fight a fire are

Figure 11–17 A wildfire in a forest is a serious threat, and public agencies coordinate their efforts to manage or contain dangerous fires. The National Interagency Fire Center in Boise, Idaho, is a superagency that coordinates fire control efforts throughout the United States.

located at this center. During each fire season the center is filled with activities around the clock. The weather service tracks weather patterns and records lightning strikes. Food vendors provide thousands of meals every day. Workers mix batches of fire retardants and load them in the cargo bays of airplanes. Warehouse workers clean the used gear and equipment and restock it on the shelves. Hundreds of communications between fire controllers and fire crews are handled daily. The National Interagency Fire Center personnel make decisions every day that determine how each fire will be handled.

FOREST FIRE SUPPRESSION

Fire has always been part of the forest ecosystem, and it was not until the Weeks Law of 1911 and the Clarke-McNary Act in 1924 that the United States Congress first funded efforts to control fires (Figures 11–18, 11–19). **Fire suppression** includes all of the work that is done to discover and extinguish fires. An intensive fire suppression policy was implemented in 1935 that was referred to as the 10:00 A.M. fire control policy. The objective of this policy was to completely suppress (by 10:00 A.M. the next day) every fire that occurred on public land during hot, dry weather. This policy proved to be impossible to fully implement in the western United States, because public agencies lacked the equipment and the people to complete the task.

The fire suppression policy evolved into a fire priority approach during the mid 1900s. During this period, the first fires that were suppressed were those that burned high-value timber areas. A strong fire suppression effort was still practiced, but firefighters and equipment were directed to high-priority fires first. Today, the policy is to practice fire management. This policy allows some fires to burn. For example, a surface fire will probably be allowed to burn when it occurs

Figure 11–18 Fire suppression for forest fires was implemented by law in 1924. *(Photo courtesy of Boise National Forest)*

Figure 11–19 Fire suppression includes all of the work and activities that occur to extinguish a fire. *(Photo courtesy of Boise National Forest)*

Figure 11–20 Once a decision has been made to suppress a fire, many resources can be brought together to fight it. Aircraft are often used to drop fire retardant materials on the flames. *(Photo courtesy of Boise Interagency Fire Center)*

in an area where prescribed burns were planned for the purpose of reducing surface litter. Such a fire may be consistent with the long-term objectives in the forest plan.

The first priority in any fire suppression effort is the safety of the fire crew (Figure 11–20). When a decision has been made to suppress a fire, several factors that contribute to the behavior of the fire must be assessed. Among these factors are weather conditions, fuel supply, fire type, and topography of the land. Each of these conditions can affect the way the fire burns through an area.

The effects of weather include the ways that fire is influenced by precipitation, humidity, and wind. Precipitation or high humidity reduces the intensity of the fire and controls the rate at which it spreads. Wind has a major impact on fire because it provides oxygen to the fire, and because it influences the fire type. Wind that blows at a constant speed in a single direction tends to influence fire in predictable ways. For example, the rule of thumb for calculating the rate at which a fire will spread can be expressed as:

$$R = W^2$$

where R = rate of spread, and W = wind speed.

Wind can seldom be depended on to blow in the same direction or in the same pattern over extended periods of time. As wind direction changes, a pattern of turbulence is often evident as well. Radical changes in an up-down

EFFECT OF WIND SPEED ON RATE OF SPREAD

Wind speed of 10 miles per hour (mph) increases to 20 mph. How much does the increase in wind speed affect the rate at which the fire can be expected to spread?

Calculate R for a wind speed of 10 mph:

1. $R = 10^2$
2. $R = 100$

The rate of spread at 10 mph has an index of 100.

Calculate R for a wind speed of 20 mph:

1. $R = 20^2$
2. $R = 400$

The rate of spread at 20 mph has an index of 400 or four times the rate of spread when compared to a wind speed of 10 mph. The rate of spread quadrupled when the wind speed doubled.

---

direction of the wind can cause a simple surface fire to become a crown fire. Wind changes pose serious threats to the safety of fire crews, and people who work in fire zones must be constantly alert to shifts in the direction of the wind.

The effects of fuel supply were discussed earlier in this chapter, but in general, as the fuel supply on the forest floor increases, the threat of a highly destructive fire increases (Figure 11–21). It is for this reason that some fires are allowed to burn in a forest. In some instances, prescribed burns are intentionally conducted in a forest to reduce the supply of fuel on the forest floor. The use of fire as a tool is considered to be a reasonable method of preventing major fires later.

The type of fire that is burning has a major impact on the way it is controlled or suppressed. A surface fire usually can be controlled by creating a **firebreak** or **fire line** in the pathway of the fire. This is done by removing the fuel from a strip of ground to stop the spread of the fire. Mechanized equipment such as bulldozers, blades, and plows is used to create firebreaks in areas where the terrain is accessible to machines (Figure 11–22). On steep slopes and mountainous terrain, fire lines are constructed with hand tools and human labor.

A ground fire is sometimes very difficult to suppress. Such fires do not tend to expand rapidly, but they smolder in underground fuel deposits for several weeks or months. Massive amounts of water are often required to suppress ground fires. A fire in a peat deposit may even require the diversion of a stream to drown it.

A crown fire that races before turbulent winds may require the setting of a **backfire** to gain control (Figure 11–23). This is a fire set along the inside edge of a firebreak to burn the fuel supply back toward the wildfire. This practice

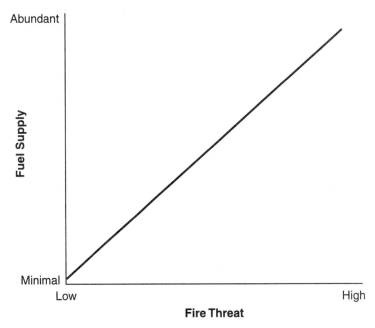

Figure 11–21 As the supply of fuel increases in a forest, the threat of fire also increases.

Figure 11–22 Firebreaks are constructed to stop the advance of a fire using large tractors equipped with blades and plows.

Figure 11–23 Sometimes the most effective way to fight a fire is to build a backfire in the path of a wildfire. The backfire is used to burn the fuel from the path of the large fire. *(Photo courtesy of Boise National Forest)*

removes the fuel on which the wildfire is feeding, and it helps to establish a wide barrier between a major fire and other vulnerable timber. Backfires are also used to force an uncontrolled fire to change the direction of its advance.

It was noted at the beginning of this chapter that three key elements interact to control the rate and intensity of a fire. They include: (1) the availability and concentration of flammable fuels, (2) enough heat energy to raise the fuel to its combustion temperature, and (3) an adequate supply of oxygen. A fire can be extinguished when any one of these three fire ingredients is brought under control. The success of a fire suppression effort depends on the ability to control these three fire elements. The mission of a firefighter is to isolate the fire from fuel, cut off its oxygen supply, and lower the temperature below the kindling point of the fuel (Figure 11–24).

Fire can be isolated from the fuel supply using an **indirect attack** such as establishing firebreaks and setting backfires. The oxygen supply can be interrupted and the temperature can be reduced by applying water, soil, and fire

---

Firefighter Mission

1. Isolate the fire from its fuel supply

2. Cut off the oxygen supply

3. Lower the temperature below the kindling temperature of the fuel

---

Figure 11–24 The mission of a firefighter is to remove or isolate the key elements of a fire.

retardants. Water and fire retardant chemicals are applied to fires from aircraft, trucks, and backpacks. This approach to fire suppression is known as a **direct attack** (Figure 11–25). It is a common practice to make airdrops of supplies for fire suppression to crews on the ground.

Fire crews are highly trained in fire suppression techniques, and they perform their jobs with militarylike precision. A key element in the success and safety of these workers is good communication. Communication systems allow the workers on the ground to talk with each other and with airborne personnel to coordinate their efforts (Figure 11–26). Communications are also very important

Figure 11–25 Fire suppression involving the use of firefighters, equipment, or materials on the fire itself is known as a direct attack. *(Photo courtesy of Boise National Forest)*

Figure 11–26 Firefighters use two-way radios to communicate as they coordinate an attack on a fire. *(Photo courtesy of Boise National Forest)*

# Smokey Bear	**FORESTRY** **PROFILE**

Smokey Bear became the mascot for fire prevention after he survived a forest fire in the 1940s. At the time he was rescued from the fire zone, he was a young cub. After he recovered from his fire injuries, he lived at the National Zoo in Washington, D.C., until his death. An educational program for prevention of forest fires was developed with Smokey as the star. Nearly everyone in America knew Smokey Bear. His one line message was always the same: "Remember, only you can prevent forest fires."

when an injured person must be evacuated, or when a fire becomes erratic, forcing ground crews to retreat.

FIRE PREVENTION

The best possible approach to dealing with wildfires is to prevent them. Fire control is very expensive, and any effort to educate people in methods of fire prevention is worth the cost. The United States Forest Service has adopted many different approaches to educating the public on ways to prevent forest fires. One of these is the ongoing promotion to "Keep America Green." Some states use variations of this theme, but the logo can still be seen in national parks, monuments, and forests throughout North America. Another very successful fire prevention campaign involved the adoption of Smokey Bear as a fire prevention symbol and mascot.

Other methods of educating the public about forest fire prevention include posting signs, promoting fire prevention through the media, printing and distributing brochures, and providing educational materials to schools. All of these educational activities have contributed to fire prevention.

The most important approach to fire prevention is elimination of fire hazards from the forests. The use of prescribed burns may prove to be the best fire prevention tool of all (Figure 11–27). It is interesting that the Native American tribes followed a tradition of burning in forested areas to protect trees that they considered useful. They understood that some kinds of trees are more competitive than others, so they used fire as a tool to give favored trees a competitive advantage. In doing so, they also helped to control the buildup of fuel on the forest floor that might contribute to more destructive forest fires later.

FIRE AS A MANAGEMENT TOOL

It has been noted throughout this chapter that not all fire is bad. It should also be noted that some effects of fire are very harmful to trees. Fire can be used as a tool to help control some kinds of timber losses due to insects and diseases. Forests that are infected by diseases or weakened by insects sometimes benefit

Figure 11-27 A prescribed burn is a surface fire that is used to burn trash and debris before it becomes abundant enough in the forest to support a destructive fire.

PROFILE ON FOREST SAFETY

Fire Containment Safety

Fighting fires is one of the most dangerous jobs in the forest. This is because forest fires do not always "behave" in predictable ways. Perhaps the most important aspect of containing a fire is to be prepared for anything. This means that the firefighter should wear protective clothing in the fire zone. The crew leader should regularly seek the latest information on the velocity and direction of the wind, and he or she must also pay close attention to other fire conditions. Remember, the first priority in fighting a fire is to protect the crew from injuries and death.

Crew members should always be supplied with the best fire equipment that is available. They should always have fire blankets available to them in case a fire turns on them. They should also have fire retardants, shovels, saws, axes, and other modern gear such as Global Positioning System (GPS) equipment to pinpoint their exact locations.

Training is a very important element of fire containment safety. The training component consists of becoming proficient at putting out fires, but it also includes teaching firefighters to deliver an immediate, disciplined response to orders from fire crew leaders. This is necessary for the safety of the entire team, because delayed action in a fire environment could result in injuries and death to crew members. Much of the work that fire crews perform is hand labor using hand tools, and in many cases, the crews work in remote areas. They must depend upon one another and the fire support teams for safety.

CAREER OPTION

Smoke Jumper

A **smoke jumper** is a firefighter who jumps from an airplane and parachutes to a location near a fire zone (Figure 11–28). Smoke jumpers work together as "first response teams" to suppress forest fires that are located in isolated locations or rough terrain. It is a difficult and dangerous job that requires a high degree of physical conditioning. These forest workers are also highly trained in the techniques of fire suppression. They work long hours while they are on the fire lines, and they earn good wages, particularly during the fire season. Smoke jumpers can expect to work in several different locations during a fire season. They are very mobile firefighters, and they are transported to any location where a forest fire threatens to burn out of control.

Figure 11–28 A firefighter who jumps from an aircraft and parachutes into the fire zone is called a smoke jumper. Smoke jumpers are highly trained professionals who play important roles in controlling dangerous forest wildfires. *(Photo courtesy of Boise Interagency Fire Center)*

from a cleansing fire. Unfortunately, trees that have endured a fire may also be injured and weakened to such a degree that they cannot resist invasions of diseases and insects.

The value of prescribed fires in preventing larger fires has been established beyond doubt, and the Forest Service has adopted a policy of using prescribed burns on a much broader scale in recent years. Despite this change in policy, however, it will take many years to return the entire forest to a natural **short-duration fire cycle** where prescribed fires are used regularly to eliminate fuels from the forests. This delay is due to an inability to use fire in this manner in forests where fuel has been allowed to accumulate on the forest floors. In some cases, the fuel buildup has continued for so many years that a serious fire hazard exists. It will probably be best in many of these areas to wait until the next harvest cycle is completed before using prescribed fire to reduce fuel supplies. Such use of fire now is likely to cause major damage to our forests.

LOOKING BACK

Fire is a force that performs both useful and destructive functions in a forest. Three key elements must be present for fire to occur. These consist of a fuel supply, heat energy sufficient for combustion, and oxygen to support burning. A fire can be suppressed by controlling one or more of these key fire ingredients. Surface fires burn fuel that has accumulated on the forest floor. When a fire begins to burn decomposed materials beneath the surface, such as peat, it is called a ground fire. The most destructive fire in a forest is one that gets into the upper canopy of the trees. It is called a crown fire. Prescribed burns use surface fire to reduce fuel on the forest floor in efforts to prevent more destructive fires later. Forests that are subjected regularly to prescribed burns are said to be on a short-duration fire cycle. The intensity and rate of spread of a fire are influenced by the fuel supply, weather factors, fire type, and topography of the land. Fire suppression includes all of the activities that are involved in extinguishing a destructive fire. Effective fire suppression often includes both direct attack and indirect attack methods of fire control. Human activity is a leading cause of fire in forests. Fire prevention consists of reducing fuels in forests and educating people to practice fire prevention.

QUESTIONS FOR DISCUSSION AND REVIEW

Essay Questions

1. In what ways is fire considered to be destructive to forests? How is it beneficial?

2. What are the three elements that must be present for a fire to occur?

3. How are surface fires, ground fires, and crown fires different from one another?

4. List the major causes of destructive forest fires.

5. Explain how wildfires affect forests and forest environments.

6. What are some ways that prescribed fires are used to improve the health of forests?

7. How effective has the fire suppression policy of the United States Forest Service proven to be?

8. Calculate the rate of spread for a fire when the wind is blowing 30 mph in comparison with 10 mph.

9. Give some examples of the direct attack and indirect attack methods of fire suppression.

10. What steps have government agencies and the forest industry taken to prevent destructive fires in the forests?

11. What effects can be expected on the health of a forest from a short-duration fire cycle?

Multiple-Choice Questions

1. Which of the following elements is *not* required in order for a forest fire to occur?
   A. heat energy
   B. a supply of oxygen
   C. fuel
   D. high humidity

2. A partially decayed fuel that is found on the forest floor is called:
   A. duff
   B. charcoal
   C. peat
   D. coke

3. A fire that burns fuel consisting of decayed plant material deposited beneath the ground surface is which of the following fire types?
   A. surface fire
   B. crown fire
   C. ground fire
   D. firestorm

4. A sudden dramatic increase in the intensity of a forest fire is called a:
   A. crown fire
   B. draft
   C. blowup
   D. infernal combustion

5. The greatest single natural cause of forest fires is:
   A. lightning
   B. matches
   C. fireworks
   D. campfires

6. A firestorm is caused by:
   A. two bolts of lightning striking one another
   B. erratic winds that move the fire into the forest canopy where there is plenty of fuel
   C. a thunderstorm
   D. ash that has drifted down from the atmosphere above the fire

7. A fire that is burning out of control is called a:
   A. prescribed burn
   B. surface fire
   C. cleansing fire
   D. wildfire

8. A fire that is intentionally set in a forest for the purpose of burning the fuel on the forest floor is called a:
   A. wildfire
   B. prescribed burn
   C. ground fire
   D. blowup

9. Weather-related factors that influence the intensity of a fire and the rate at which it spreads include all of the following *except*:
   A. topography
   B. precipitation
   C. wind
   D. humidity

10. Which of the following terms best describes the mode of action of a firefighter who shovels soil on the flames at the leading edge of a fire?
    A. fire line
    B. incendiarism
    C. direct attack
    D. indirect attack

LEARNING ACTIVITIES

1. Obtain a training video used by a public agency to provide training for fire crews, and show it to the class. Invite an experienced firefighter to describe some of his or her experiences with fires to the class. Concentrate on the training that is required before a person is qualified to work as a member of a fire crew.

2. Demonstrate the importance of oxygen to a fire by removing flammable materials and preparing an appropriate demonstration site. Light two candles. After both candles are burning, place an inverted glass jar over one of the candles. Blow gently on the base of the flame on the second candle. Discuss what happened in each case, and ask key questions such as: (1) Why was the fire extinguished in the closed container? (2) Why did the fire burn more vigorously when you blew on it?

FOREST TECHNOLOGIES

Chapter

12

Terms to Know

noise intensity
decibel (dB)
noise duration
spiking
sawyer

Safety in the Forest

Objectives

After completing this chapter, you should be able to

* describe the type of clothing and personal gear that should be used by forest workers to provide for their safety

* distinguish between noise intensity and noise duration

* define the term *decibels* and explain how this measurement is related to the safety of workers in the forest and wood products industries

* discuss ways that high-quality tools and equipment contribute to the safety of forest industry workers

* explain how training contributes to the safety of workers in the forest industry

* list some examples of safe work habits that are known to protect forest workers

* discuss the importance of workers' demonstrating respect for the power of the machines and equipment that they operate as they do their work

* describe some safety practices that contribute to the safe operation of a chain saw

* list some work habits that contribute to worker safety as timber is harvested and transported

* identify some safety practices that have been proven to protect workers in timber processing mills

orest safety is a planned activity that requires effort to make it happen. Planning for safety is important in the forest industry, because many of the occupations related to forestry are considered to be hazardous. Safe working conditions are the result of training, good equipment, and a safety mind-set on the part of the worker. This chapter is intended to raise the level of awareness toward safety practices that are recommended for workers in the forest and wood products industries.

CLOTHING AND PROTECTIVE GEAR

Each of the safety principles that is included here should become personal habits of the people who plan careers in forestry. Safety in the forest begins with proper safety clothing and equipment for the workers. Clothing and equipment should be designed to protect the workers from the hazards of falling debris, misdirected cutting tools, and crushing by heavy objects. Each of these hazards causes serious injuries and deaths to forest workers every year.

Proper clothing includes steel reinforced boots, safety glasses, hard hats, gloves, long-sleeved work shirts, and durable work pants (Figure 12–1). It is important for clothing to fit properly because loose clothing can catch on moving machine parts such as shafts and chains. No amount of proper clothing can protect a worker who is careless in the forest; however, good clothing can provide protection from some of the more common injuries.

Steel reinforced boots help to protect a worker's feet and toes from accidental cuts while using an axe or chain saw. One of the hazards of using cutting tools is that the blade can be easily misdirected when it comes in contact with bushes,

---

**Clothing Needed for a
Properly Dressed Forest Worker**

Hard hat

Safety glasses

Hearing protection

Long-sleeved work shirt

Leather gloves

Durable work pants

Steel-toed shoes

Figure 12–1 Proper fitting clothing contributes to the safety of people who work in forest environments.

vines, or other small shrubs that surround the trunks of trees. When a saw blade is moving, its own momentum can sometimes jerk the blade into the feet of the operator. Steel-toed boots sometimes make the difference between serious injuries and protection of the feet.

A good pair of work boots provides good footing for the person wearing the boots, helping to prevent turned ankles and other similar injuries. Boots that are reinforced with steel give some protection from crushing by logs or other heavy objects; however, prevention is the best way to protect the human body from being crushed.

Falling trees are accompanied by falling limbs and debris. A good hard hat is a necessity in this work environment. Even a small branch can cause a serious head injury to a worker whose head is not protected. Even though a hard hat may not protect a worker from large objects, it can certainly reduce the effects of injuries.

Safety glasses should be standard gear for all forest workers. Eye protection is necessary in any environment where flying objects are likely to occur in the work area. During the logging process, there are many sources of flying particles and other debris. For example, the concussion of a large tree with the forest floor and other trees causes rocks, dust, branches, bark, and other debris to become airborne hazards. In addition, the branches of small trees and bushes become dangerous as they are displaced by falling trees. Eye protection and protective clothing contribute to worker safety under these conditions.

Hearing protection is an aspect of forest safety that was not addressed until recent years. Many of the machines that are used in the forest are extremely noisy, and it is likely that unprotected workers will experience hearing loss after prolonged exposure to high levels of noise. Ear protection such as that provided by protective ear muffs or ear plugs should be worn when loud noise is sustained over long periods of time (Figure 12–2).

Two characteristics of noise contribute to hearing loss. They are intensity and duration of sound. **Noise intensity** is the amount of energy that is in the sound waves of a noise source. The unit of measurement for noise intensity is **decibel (dB).** The length of time that a person is exposed to a sound is called **noise duration.** The human ear can tolerate high decibels for short periods of time, but the combination of high decibel levels and sustained noise duration can lead to hearing damage. The Occupational Safety and Health Act administered by the Occupational Safety and Health Administration (OSHA) in the United States Department of Labor established a 90 dB noise level as a maximum safe limit over an eight-hour work period (Figure 12–3). Accurate measurements of noise intensity are possible using a noise or decibel meter.

HIGH-QUALITY EQUIPMENT

The quality of tools and equipment is closely related to safe work practices. Poor-quality tools and equipment usually require extra effort on the part of the user to complete a task. For example, as the blade of an axe becomes dull, the worker tends to swing the axe harder to compensate for the loss of cutting potential. As

Figure 12–2 Workers who are exposed to high levels of noise for extended periods of time should wear equipment that will protect their ears against hearing loss. *(Photo courtesy of Potlatch Corporation)*

Duration of Time Permitted at Various Sound Levels

Duration/day in Hours	Sound Level in dB	Duration/day in Hours	Sound Level in dB
8	90	1	105
6	95	½	110
3	97	¼ or less	115
2	100	none	over 115
1½	102		

Decibel (dB) Levels of Common Sounds at Typical Distance from Source

0	Acute threshold of hearing	90	OSHA limit—hearing damage on excess exposure to noise above 90 dB
15	Average threshold of hearing		
20	Whisper	100	Noisy tractor, power mower, all-terrain vehicle, snowmobile, motorcycle, in subway car, chain saw
30	Leaves rustling, very soft music		
40	Average residence		
60	Normal speech, background music	120	Thunderclap, jackhammer, basketball crowd, amplified rock music
70	Noisy office, inside auto 60 mph		
80	Heavy traffic, window air conditioner	140	Threshold of pain—shotgun, near jet taking off, 50 hp siren (100')
85	Inside acoustically insulated protective tractor cab in field		

Figure 12–3 Noise intensity and noise duration are the two factors that contribute to hearing loss. This chart was developed as a safety guide for workers who are exposed to potentially dangerous noise levels.

the extra effort becomes uncomfortable to the worker, the ability to control the axe is reduced. Loss of control over tools and equipment contributes to accidents and injuries to the workers who use them.

A power tool, such as a chain saw, should have enough power to do the job for which it was designed. Insufficient machine power often results in dangerous work conditions because the operator tends to compensate for the lack of power by increasing the amount of pressure that is exerted on the cutting tools. A similar response often occurs when the saw becomes dull. The operator of a saw with a dull blade often applies excessive force to the blade in an effort to maintain the accustomed cutting rate. This is dangerous because it leads to loss of control and frequently results in serious injuries. A cutting blade should be sharpened as often as necessary to maintain a good cutting edge (Figure 12–4).

Machinery and equipment used in the forest industry must be heavy-duty and high-quality. Only high-quality machines and equipment can hold up under the abuse of heavy logs. Safety for workers and equipment is quickly reduced when machines do not work the way they should. For example, the band saw blade that is used in a head saw is used to cut logs into lumber. Such blades must be removed from the saw for sharpening, and the replacement process closes down the processing line. It is important that the blade should consist of high-quality materials so that it retains a sharp cutting edge for as long as possible. The cost of a high-quality blade is quickly offset by the production time that is saved by less frequent changes of the saw blade. In addition, worker safety is increased through the use high-quality blades.

Figure 12–4 Cutting tools should be kept sharp to ensure that they will cut properly.

TRAINING FOR SAFETY

Training is probably the most important activity contributing to forest safety. This is because safe work habits can be learned. Most people develop safety habits by learning from mistakes of the past. Fortunately, we can and should learn from the mistakes of others. For example, it is not enough to know that a chain saw is dangerous. What the worker really needs to know is that a chain saw or any other machinery or equipment can usually be operated safely once the basic rules of safety are applied. A few simple lessons can eliminate a lot of pain.

Safety training is an important part of preparation for work and careers in every aspect of forestry. It is needed to protect against injuries during the operation of machinery and equipment. Training is needed for workers who operate the mills, producing wood and paper products. Firefighters and smoke jumpers receive extensive training as they wait for their services to be needed, and the lives of the team members often depend on how well each member performs his or her assigned duties (Figure 12–5). Forest workers require training in emergency procedures for injured workers and others who spend time in the forests. They also need training in protecting themselves from the effects of harsh weather conditions and chemical applications for control of pests, water safety, law enforcement safety, and safe operation of vehicles. Safety training takes many different forms in an industry that is as diverse as forestry.

Figure 12–5 Intensive training is required for firefighters because a fire crew must work together as a team to protect their own lives and the lives of others. *(Photo courtesy of Lisa S. Nolan)*

SAFE WORK HABITS

Safe work habits are cultivated by thinking through each work situation in an effort to identify potential hazards. This should be followed by selecting ways to approach dangerous work situations that will reduce the risk of injury. For example, a buildup of trash near a moving chain or blade often alerts workers to the need to clean the area. By thinking through the plan of action, it should be obvious that the power should be turned off before an attempt is made to clean the trash from the area (Figure 12–6). It is never appropriate to work near unprotected chains, blades, and other dangerous areas until the power to the machine has been turned off.

Safety conscious workers learn to recognize where they should position themselves to safely work in hazardous work settings. Falling trees do not always remain where they fall. Sometimes they are very unpredictable, bouncing toward the worker off forest undergrowth or heavy limbs. The best protection against crushing is to carefully avoid high-risk situations. For reasons of safety, a seasoned worker does not stand or place any part of his or her body in a direct line with a moving blade, chain, grinder, or other moving part. This is because these moving parts are likely to throw projectiles such as rocks, metal chips, wood knots, and other materials with such force that they can injure, maim, or kill people.

Figure 12–6 A worker in the forest and wood products industries must develop the habit of turning off the main power source before working near moving belts, chains, shafts, and blades.

An example of such a hazard is the potential for injury that exists in a lumber mill when antilogging protesters are known to be active near a disputed timber harvest. It is well documented that some of the radical extremists in these groups have engaged in the practice of pounding metal spikes into trees with the intention of damaging saws and other milling equipment. This illegal practice is called **spiking.** It is an act of domestic terrorism. A worker who is standing in line with the high-speed saw blade when it strikes a spike is very likely to be injured or even killed by wood or metal debris from the metal spike or the damaged saw blade.

Workers who have developed safe work habits do not place their hands or any other part of their bodies near moving gears, chains, or pinch points of any kind. The high-powered nature of these devices (in processing mills and other uses in the industry) has the potential to cause the loss of arms, hands, fingers, feet, and legs (Figure 12–7). All machines should have protective guards in place to prevent workers from being injured by moving parts. It is important for industry workers to keep in mind that any machine or device that can manipulate a heavy log is capable of easily crushing the human body. Guard rails should be installed to prevent workers from getting into unsafe locations.

Safe work practices contribute to successful careers, but indifference to safety often results in serious injuries to individual workers and those who work with them. No amount of safety gear, high-quality equipment, government and

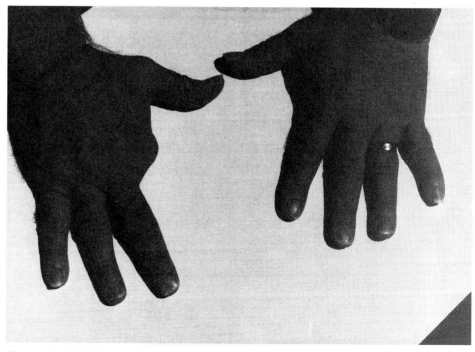

Figure 12–7 Serious injuries occur when powerful machines come in contact with soft human body parts such as fingers, toes, legs, and arms.

company policies, or safety training can protect a worker who does not adopt safe work habits. These are the people who refuse to wear hard hats, safety glasses, hearing protection, and steel-reinforced shoes because they consider them to be unnecessary, uncomfortable, or inconvenient. A glass eye meets all of these criteria. It does not contribute at all to a person's ability to see, and it is uncomfortable, unsanitary, and inconvenient to wear. Take your choice: the final decision to work safely is really up to individual workers as they choose to adopt or ignore safety procedures.

SAFE OPERATION OF EQUIPMENT AND MACHINERY

Workers must have respect for the power of the machines and equipment used in the forest industry if they are to operate them safely. The operation of aircraft in forested areas requires strict attention to safety practices (Figure 12–8). A single mistake is likely to cause serious injuries and deaths to those who pilot or navigate the aircraft. Helicopters are often used in salvage logging operations following forest fires. This is done to protect damaged soil from erosion. Airplanes and helicopters are used to disperse fire retardant chemicals in fire zones to slow the advance of wildfires across forest regions. It is very important to understand safe operating practices for all of the machinery and equipment used in the forest to ensure the safety of the workers, their equipment, and machinery.

There are many different kinds of heavy equipment that are used to transport logs to the landing or to different locations at the mill. Each of these

Figure 12–8 Safety procedures must be followed carefully in order for aircraft to be operated safely in forest environments.

machines must be used with caution to ensure that workers are not injured and equipment is not damaged. For example, logs should be lifted only high enough to keep them from dragging on the ground. If they are carried with the booms extended to their full height, the machine becomes top-heavy and can easily tip over. Extending the booms also places an extreme amount of pressure on the frame of the machine. If a wheel should suddenly drop into a hole while the machine is carrying a full load, the pressure may break the machine in half. Such an accident is likely to injure the operator as well as damage the machine. The key to safe operation is to carry loads with the booms only partly extended.

Log-handling equipment should be equipped with safety cabs and roll bars to protect the operator in case a tree is knocked down or the machine tips or rolls over. Machines such as prehaulers are equipped with grapple loaders attached to overhead booms. Care should be taken to make sure that logs are handled safely to avoid injuries to nearby workers. People who are working near heavy equipment must take responsibility for their own safety because the operator of the machine is unlikely to hear or see them. Workers also must avoid working in areas where they might accidentally be bumped or dragged into a dangerous machine such as a whole-tree chipper.

Chain Saw Safety

The chain saw is one of the most important tools of the forest industry. It is used for many different cutting applications, and workers must learn to safely use this important tool. The most dangerous part of a chain saw is the moving chain that is equipped with evenly spaced cutting teeth. It is important to keep the blade pointed away from feet and legs while the saw is in use. The throttle should never be locked in the on position except during starting. A locked throttle keeps the saw running at high speed, even when it is not cutting. It is best to let the chain slow down or stop between cuts, because this is when the blade is most likely to be in close proximity to the operator's body. Every precaution should be taken to avoid contact with the cutting teeth of a chain saw.

Cuts are not the only injuries that are caused by the careless use of chain saws. The exhaust muffler gets very hot while the saw engine is running, and serious burns can be inflicted on anyone who touches it. For this reason, it is important to avoid touching the muffler and to wear clothing that will provide some protection if a hot exhaust muffler should accidentally be touched. When dry conditions exist in the forest, fire becomes a real danger, and a chain saw must be equipped with a spark arrester for fire prevention. It is always a good idea to keep a shovel and a fire extinguisher nearby while a chain saw is in use. A spark from the exhaust of the chain saw or a spark that is caused by the chain striking a rock can quickly turn into a serious fire.

Allowing a moving chain saw blade to come in contact with soil or rocks not only dulls the blade, but also can throw debris toward the operator, injuring the eyes, skin, and other vulnerable body parts. Proper clothing and safety gear protects against most injuries of this type (Figure 12–9). It is important to carefully instruct inexperienced workers on the safety procedures that should be

Figure 12–9 Proper gear and clothing improve the safety of the working environment for those who operate chain saws.

observed as they learn to operate a chain saw. It is important to remember that the owner's manual contains instructions for the safe use of this important tool.

Safe Harvest Practices

The job of harvesting trees requires workers to pay close attention to safe work habits. A cautious logger will take time to make sure that each step in the harvesting process is performed carefully. Most timber harvesting accidents can be prevented by following some basic safety rules. For example, the first saw cut that is made in felling a tree is a triangular notch at the base of the tree (Figure 12–10). It should be located on the side of the tree where the logger desires the tree to fall. The notch should extend approximately one-third of the way through the tree.

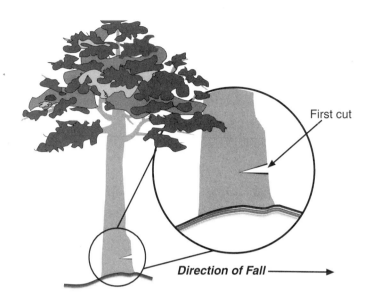

First cut

Direction of Fall ⟶

Figure 12–10 The first step in felling a tree is to cut a triangular notch at the base of the trunk to help control the direction of the tree's fall.

The second cut should be located approximately four inches above the first notch on the opposite side of the tree (Figure 12–11). As the second cut approaches the first cut from the opposite side, the tree will begin to lean. The logger should cautiously continue cutting until the tree begins to fall. The saw should then be extracted from the cut, and the logger should step back and to one side. This location avoids a position that is in a direct line with the trunk of the tree. A logger who is positioned in a direct line with the tree trunk is in a dangerous location because the tree sometimes bounces directly backward toward its stump. The energy that is generated from the fall of the tree can propel it several feet beyond the stump. The logger should look around and have retreat routes in mind before the tree is cut (Figure 12–12 on page 320).

It is important to consider the wind direction and to observe any tendency of the tree to lean when determining which way to fell a tree. An expert logger often can make the cut in a way that pulls the tree, causing it to fall in a predicted direction. Wind, however, can exert enough pressure on a tree to control the direction of its fall. It is usually best to use the wind to help control the direction that the tree falls. This is much safer than attempting to work against the forces of nature.

Transportation Safety

Trucks are the main method of transportation for logs and wood chips. Most large commercial logging operations use tractor-trailer rigs that are capable of hauling large loads of logs or chips from the isolated forests and plantations to processing mills and generating plants. These loads are heavy due to the high

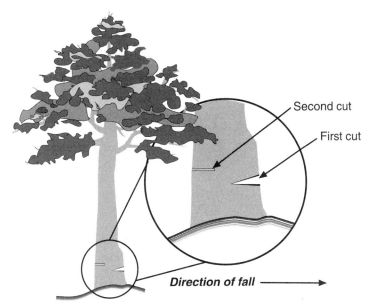

Second cut

First cut

Direction of fall ⟶

Figure 12–11 The final cut in felling a tree is made by cutting toward the notch from the opposite side of the tree trunk.

water content of the trees and fresh wood chips, and drivers must use extreme caution on steep grades. Mountain roads are often narrow and steep, and this combination is dangerous with heavy loads. Drivers must drive cautiously to avoid tipping their loads, and they also must be aware of the safety of other people who use the roads.

Brake failures account for many of the serious trucking accidents that occur as logs and other wood products are transported on steep, winding roads. Drivers should check their brakes regularly and follow a routine maintenance schedule to make certain that their truck brakes are always in proper working order (Figure 12–13). In the event that a brake failure occurs, the driver should keep the truck in gear, and attempt to slow down by keeping the clutch engaged and reducing the engine speed.

Driving speed is a safety factor in the transportation industry. High-speed driving reduces the reaction time that is available to the driver to make driving decisions. When a hazard appears on the road, the driver must react quickly and appropriately to avoid an accident. A second safety hazard related to high speed is that the momentum of the truck and its heavy load is greatly increased. The distance that is required to come to a stop is much greater at high speeds than it is at low speeds. Many accidents can be prevented by slowing the speed of travel.

One of the safety hazards associated with hauling logs and lumber products is that loads can shift during transit. It is always wise to secure loads with chains to make sure they stay in position. If a load should shift while the truck is going around a corner, the movement of the load can exert enough momentum to

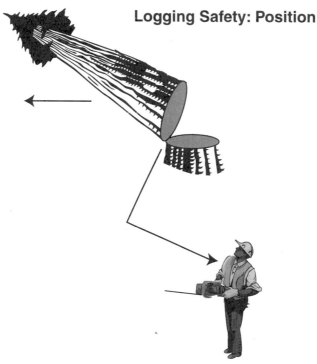

Logging Safety: Position

Figure 12–12 As the tree falls, the logger should take a step back from the stump and move to one side, out of direct alignment with the falling tree.

Figure 12–13 The brakes of logging trucks require frequent safety checks to ensure that heavy loads can be safely transported over logging roads and highways.

carry the truck off the road. The combination of too much speed and a shifting load is a serious safety hazard.

Safety in the Mill

Wood product mills are equipped with heavy-duty machines designed to operate in highly mechanized environments. The "high-tech" nature of such wood product mills undoubtedly prevents some kinds of accidents, but there are so many mechanical devices in a modern mill that workers must use extreme care to avoid being crushed or entangled in the machinery. Rigid safety rules are often developed for hazardous work environments, and signs are frequently posted throughout modern mills to remind workers of the need to follow safe work practices (Figure 12–14).

The view in a modern lumber mill reveals long production lines filled with wood products in various stages of processing. The workers who are present are usually engaged in operating one or more machines and carefully watching each machine to make sure it is working correctly. Power switches at various locations in the mill are capable of shutting off the power to large sections of the mill when emergency repairs are needed or when an injury to a worker has occurred.

Figure 12–14 A modern lumber mill is equipped with power machinery that may expose the workers to dangerous moving parts. Mill workers must develop an awareness of safety hazards in order to preserve their personal safety. *(Photo courtesy of Potlatch Corporation)*

PROFILE ON FOREST SAFETY

The Widow-Maker

Anyone who has worked as a logger for very long is likely to be familiar with the term *widow-maker*. A widow-maker is a situation in which a tree has fallen into another tree where it leans, suspended in the branches of the second tree (Figure 12–15). When the second tree is cut down, the logger is in danger because it is hard to predict where either of the two trees will fall. In too many instances, the worker is injured or even killed by one of the trees as it falls. Frequently the logger leaves a spouse and children behind to cope with his or her death. For this reason, a leaning tree is called a "widow-maker." The wise approach to dealing with this problem is to pull the first tree down with a skidder or other large machine before cutting the second tree.

Widow-maker

Figure 12–15 A widow-maker is a potentially deadly safety hazard in which a cut tree has fallen into the upper branches of a nearby standing tree where it is supported in an upright position. The logger who attempts to cut the support tree is at risk of serious injury or even death.

CAREER OPTION

Sawyer

A **sawyer** is a person whose career is sawing wood. This is a highly skilled specialty that makes use of computers and electronic scanners to determine the most profitable cuts to make in logs. A sawyer is trained to operate the large saws that cut logs into lumber. This is the "key job" in the entire mill because the productivity of the sawyer determines the level of production for the entire mill. The sawyer is expected to keep the main head saw operating at full capacity. Training for this career usually occurs on the job under the direction of an expert sawyer.

Many of the injuries that have been documented in wood product mills involve cuts or crushing injuries to fingers, hands, arms, legs, and feet. Workers must develop safety mind-sets, causing them to consider the possible consequences of every action in terms of safety. As workers begin to think in this manner, safety in the workplace becomes reality.

LOOKING BACK

Careers in the forest and wood products industries are considered to be hazardous. Planning for safety awareness plays an important role in the training programs that prepare people for careers in forestry. Safety training should include instruction on proper clothing and protective gear, such as steel reinforced boots, and protective equipment such as safety glasses, hard hats, and hearing protection. Properly maintained high-quality machinery and equipment is known to reduce work-related injuries. Safety training should include instruction in developing safe work habits and raising the level of safety awareness so that workers will recognize and avoid unsafe working conditions. Safe operation of machines and equipment should include specific emphasis on chain saw safety for all workers who expect to work in logging or yarding operations where chain saws are regularly used. Additional safety training in specific timber-related jobs such as harvest practices, transportation, and processing should be provided where it is appropriate.

QUESTIONS FOR DISCUSSION AND REVIEW

Essay Questions

1. What kinds of clothing and personal protective gear should be worn by workers who work in outdoor forestry careers?

2. What is the relationship between noise intensity and noise duration as they affect hearing loss?

3. What is the unit of measurement for noise intensity? How is it measured?

4. How does the quality of tools and equipment contribute to the safety of forest industry workers?

5. Discuss the importance of safety training as it relates to workers in forestry careers.

6. What is meant by "safe work habits"? List some examples.

7. Why is it important for workers to have respect for the power of the machines and equipment that they use?

8. What does a worker need to know in order to safely operate a chain saw?

9. List some work habits that contribute to the safety of workers as they harvest and transport trees.

10. What are some safety practices that should be used to protect workers in timber processing mills?

Multiple-Choice Questions

1. Which of the following clothing items would *not* be considered as protective clothing for a worker in an outdoor forest environment?
   A. wool hat
   B. long-sleeved work shirt
   C. steel reinforced boots
   D. heavy-duty work pants

2. Which of the following items of personal protective equipment should be worn to protect against a safety hazard that is measured in decibels?
   A. hard hat
   B. eye protection
   C. ear protection
   D. steel reinforced boots

3. Which of the following factors makes a positive contribution to worker safety?
   A. indifference to danger
   B. too little time to do a job
   C. worn or damaged tools
   D. high-quality tools

4. Which of the following approaches to safety training is most likely to help workers learn from the mistakes of others?
   A. reading the owner's manual
   B. watching a training video
   C. attending a safety attitude class
   D. participation in a mentor training program

5. Which of the following is an example of a safe work habit?
   A. turning off the power before making repairs to a machine
   B. trying to get along with a machine that is not working properly to avoid closing the production line
   C. consulting the design engineer before buying new machinery
   D. placing the highest priority on finding ways to do jobs faster

6. Which of the following situations would *not* be considered to be a safe log handling practice?
   A. equipping log handling machines with safety cabs and roll bars
   B. extending the lift to its full height while carrying logs to new locations
   C. extending the lift just enough to lift the log off the ground as it is moved
   D. using helicopters to transport logs from areas that are sensitive to soil damage

7. Chain saw safety is described in each of the following statements with the *exception* of:
   A. Make sure the saw is equipped with a spark arrester.
   B. Keep the blade of the saw pointed away from the legs and feet of the operator.
   C. Lock the throttle in the "on" position except during starting.
   D. Do not allow the blade of the saw to come in contact with rocks or soil.

8. A logger who is harvesting a tree should move to a position of maximum safety as the tree begins to fall. Where is the safest working position in relation to the stump?
   A. back from the stump and to one side
   B. to the immediate left of the stump
   C. directly behind the stump
   D. at least 20' away from the stump

9. Which of the following conditions is a major contributor to accidents related to transporting forest products?
   A. driving too fast for road conditions
   B. top-heavy loads
   C. faulty brakes
   D. all of the above

10. Which of the following classes of injuries is most likely to occur in a wood processing facility?
    A. cuts and bruises
    B. chemical poisoning
    C. head injuries
    D. burns

LEARNING ACTIVITIES

1. Obtain a safety training video for forestry from the National Safety Council, a state industrial commission, state library, or another source in the public domain. Show the video to the students and have them identify each unsafe situation or condition that might result in an accident. Discuss ways that the accidents might have been prevented.

2. Invite a guest speaker to come to the school to speak on the topic of "worker safety in forestry careers." The speaker should be someone who has had experience in the industry, such as a logging contractor, aircraft pilot, firefighter, or mill worker. Suggestions for speakers also could include an extension forester, a representative of a wood product industry, a supervisor from a wood processing mill, or someone from another forestry career. Invite the speaker to spend most of his or her time on positive suggestions for creating safe working conditions with just enough good examples of consequences (for unsafe practices) to hold the interest of the students.

FOREST PRODUCTS

Chapter

13

Terms to Know

pores
resin
resin duct
tensile strength
head saw
head rig
gang saw
debarking
slab
cant
lumber
board
plank
timber
beam
square
loading jack
pneumatic power
planer
green chain
warp
flitch
veneer
block
plywood
cross-banding
particleboard
fiberboard
hardboard

Wood Construction Materials

Objectives

After completing this chapter, you should be able to

❋ distinguish between processed woods of the hardwood and softwood varieties

❋ list some distinguishing characteristics that are useful in identifying woods of different species

❋ identify some characteristics of wood that contribute to its value for construction purposes

❋ identify some characteristics of wood that detract from its value for construction purposes

❋ describe the general process by which logs are processed into lumber

❋ classify the different cuts of processed wood according to their dimensions

❋ distinguish between lumber and timbers

❋ identify some characteristics of hardwoods that contribute to their usefulness

❋ explain the source and methods of processing wood veneers

❋ distinguish the differences among the different types of reconstituted boards

Wood has long been considered to be an excellent material for construction purposes. It has qualities of durability, and it is easily shaped using modern tools and equipment. Wood that is used as construction material should be carefully chosen. The different kinds of wood have different qualities and construction capabilities. They should be used for the purposes for which they are best adapted. Wood is a durable and beautiful construction material when it is properly chosen and used.

IDENTIFICATION OF WOODS

The different kinds of wood are divided into two broad classifications: hardwoods and softwoods. Hardwoods are obtained from broad-leaved trees, and softwoods are obtained from conifers which bear needles or scale leaves. The different kinds of trees were discussed at length in Chapter 2 of this textbook, and information was provided that is useful in identifying live trees. Woods that have been processed also have characteristics that are unique. For example, one distinguishing feature between hardwoods and softwoods is the presence of pores in hardwoods and the absence of pores in softwoods (Figure 13–1). **Pores** are the specialized vessels through which dissolved nutrients moved when the tree was alive.

Porosity in Wood

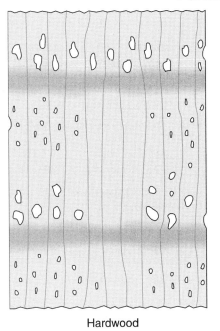
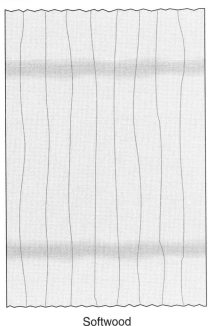

Hardwood

Softwood

Figure 13–1 A distinguishing characteristic between hardwoods and softwoods is the presence of pores in hardwoods and the absence of pores in softwoods.

Resin is a viscous, sticky substance that is clear or translucent in appearance. It is not soluble in water, and it is exuded by a tree as a defense mechanism to protect the tree in places where injuries have occurred (Figure 13–2). **Resin ducts** are openings in the wood through which resin moves within a live tree. They are present in softwoods, but not in hardwoods.

Some woods can be identified by their color or odor. For example, the odor of wood obtained from cedar trees is distinctive. Another aromatic wood comes from the sassafras tree. Among the woods that are distinctive in color are mahogany, rosewood, and cherry; however, woods are found in nearly every color. Color differences between sapwood and heartwood are also evident in many kinds of trees. Sapwood tends to be light in color, while heartwood is dark. Differences exist between species in the proportions of sapwood and heartwood.

The density of wood is a quality indicator, but it is also a method of wood identification. The weight or density of wood depends on the ratio of cell wall material to the volume of the wood. Wood density is usually expressed as pounds per cubic foot or per thousand board feet. Density is also a measure of hardness (Figure 13–3). The pignut hickory is the most dense of all North American hardwoods.

The texture and grain of woods are also useful in identification. Some woods are fine-grained, while others are coarse-grained (Figure 13–4). Many woods have distinctive grain patterns that aid in their identification. Some woods even appear to have no visible growth rings, and the grain is nearly invisible. The texture of these woods appears to be fine, while the texture of conifers where growth rings

Figure 13–2 Resin is a viscous, sticky substance that is exuded by a tree to protect places where injuries have occurred.

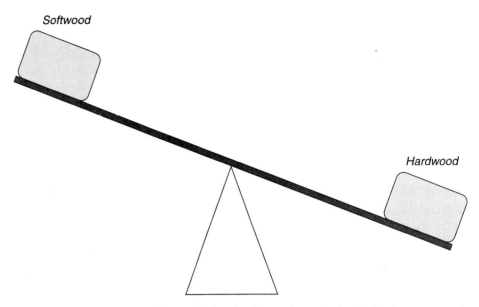

Softwood

Hardwood

Figure 13–3 Hardwoods are higher in density than softwoods due to the larger amount of cell wall material in a given volume of wood. This results in hardwoods being heavier than softwoods.

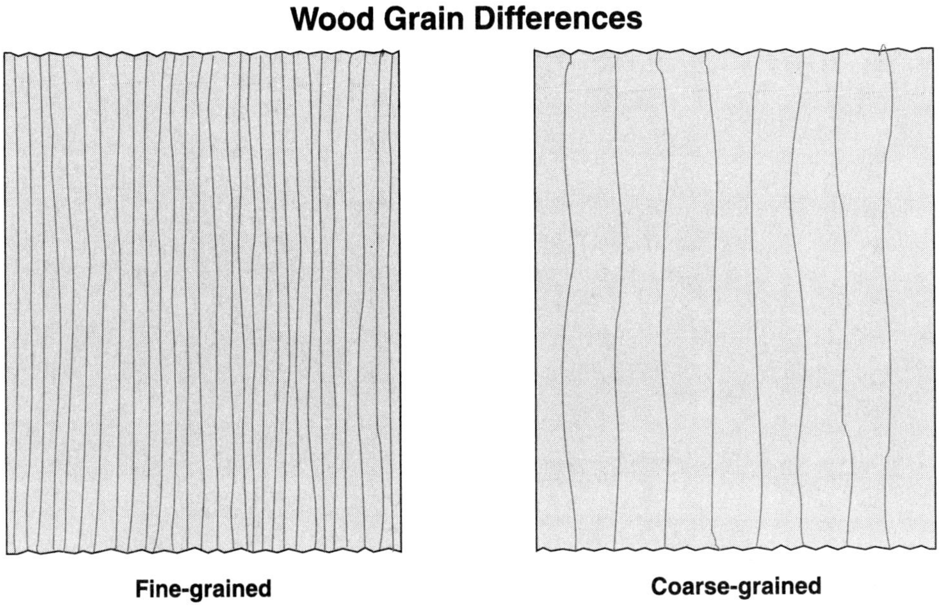

Wood Grain Differences

Fine-grained

Coarse-grained

Figure 13–4 The texture of wood can be used to identify it. Conifers tend to be coarse-grained, while most deciduous woods are fine- to medium-grained.

are prominent is described as coarse. When distinct differences appear in the growth rings between the springwood and the summerwood, the texture is considered to be uneven.

PROPERTIES OF WOOD

Wood has many qualities that make it desirable for constructing useful things. Hardwoods have great durability and toughness, making them desirable in uses where the structural material is expected to endure high stress loads or abrasive wear. Softwoods are used extensively in single-family home and apartment construction for structural uses.

Strength along the long axis of the tree is much greater than it is across the thickness of a tree (Figure 13–5). This measurement of strength is called **tensile strength.** It is attributed to the type of wood cells that are found in the trunk of a tree. Most of them are tracheid cells (discussed in Chapter 3) that are much longer than they are wide. They are aligned lengthwise in the trunk of the tree, making the wood up to ten times stronger along the length of the tree than across it. This is the source of strength of the wood that is used for timbers and structural beams.

Wood has aesthetic properties that make it desirable in the construction of paneling, doors, cabinets, and furniture. Highly polished wood that has been

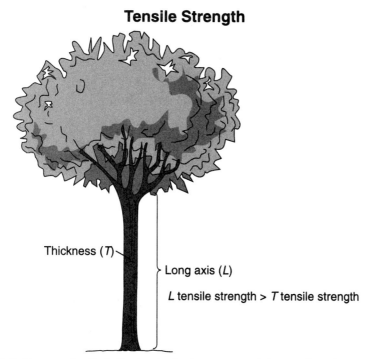

Figure 13–5 The wood cells in the trunk of a tree are much longer than they are wide. The wood is strongest along the length of the tree.

crafted into useful things is highly regarded for its beauty (Figure 13–6). Some of the woods are also highly durable, and furniture constructed of these materials is often valued by the owners for several generations.

One reason that wood is desirable for construction of homes and buildings is because it has insulating qualities against the transfer of heat. In comparison with metal or concrete, wood conducts very little heat. This is because heat moves very slowly between wood cells. Wooden wall studs and wall coverings actually resist the movement of heat. This becomes evident when fire consumes a structure. Quite often, the wooden beams remain in place long after the metal structural beams have been warped and bent by heat.

Another quality that makes wood useful as construction material is its flexibility. Buildings that are constructed using wood structural materials are capable of some movement at times of stress. Flexibility is also useful in the manufacturing industry. Wooden devices, such as sporting goods, require flexibility. The ability of wood to flex without breaking is an important characteristic that favors its use as a construction material (Figure 13–7).

Some of the properties of wood restrict its uses for some purposes. One of these properties is the tendency of wood to shrink when dry conditions prevail and to swell when conditions are damp. Among the problems caused by this property are loose joints and warped members. This can be a serious problem with hardwood floors and wooden furniture.

Another property of wood that interferes with its usefulness is that it cannot be fused or melted together. Wood can be glued together, but its fibers cannot be welded. Wood has elastic qualities that allow it to bend, but wood also springs

Figure 13–6 Some kinds of wood have aesthetic properties that contribute to the use of such woods in constructing fine furniture.

Figure 13–7 Most wood is capable of flexing when a heavy load is placed on it. This makes wood an ideal construction material for rafters and trusses that may have to sustain heavy snow loads during winter months.

back to its original shape once pressure is released. It can be shaped by cutting or grinding, but it is not easily molded to new shapes.

Other qualities that detract from wood as a material for construction include its lack of uniformity. Variability exists even in lumber from the same tree. Wood also breaks down over time, especially when damp conditions prevail.

LUMBER MANUFACTURING

Modern lumber mills are equipped with high-tech equipment and machinery that is capable of high-speed operation. Logs are sawed into marketable dimensions using a **head saw** or **head rig** (Figure 13–8). This is the main cutting platform in the mill where logs are sawed into boards. Many modern lumber mills cut timber with a band saw that has teeth on both edges. This allows the saw to cut a board as the moving carriage carries the log in either direction. The double-cutting band saw blade has greatly contributed to the efficiency of modern sawmills. Nearly twice as much timber can be cut using this blade in comparison to a band saw blade that cuts in a single direction. Band saws are also capable of cutting very large logs. The only size restriction for a band saw is the distance between the upper guide for the blade and the floor of the carriage.

Some sawmills have continued to use circular blades to saw logs, but most mills use these blades to trim boards. Circular blades also may be used in a multiple-blade combination called a **gang saw.** A gang saw is capable of making several cuts at the same time in processing small logs or in cutting boards to standard lumber sizes. Circular blades are effective, but they can cut in only one direction. They also have fewer cutting teeth than band saw blades, which makes it necessary to sharpen them at more frequent intervals, and they are limited in the size of log they can cut by the size of the blade. The diameter of the largest

Figure 13–8 Logs must be processed into dimensions that are useful for manufacturing wood products. This is done by cutting the logs to the proper sizes using a head saw or head rig. *(Photo courtesy of Potlatch Corporation)*

log that can be cut on a circular saw cannot exceed the cutting radius of the blade.

Each log that enters a modern lumber mill is subjected to **debarking.** The bark is removed by large machines that apply pressure and friction to the log surfaces to remove the bark (Figure 13–9). Large lumber mills often operate more than one processing line. When this occurs, logs are often sorted by size after they enter the mill. The largest logs are moved to the main head saw where they are cut into boards and timbers.

Many modern lumber mills now use X-ray machines to scan logs. The X-ray profile of the log is evaluated by the computer to determine how the log should be positioned and which cuts should be made to obtain the highest value from the log. A computer link to the carriage and the head saw allows the log to be cut automatically to maximize its value. The person who operates the head rig is called a sawyer. The sawyer is responsible for carefully watching the automated operations and overriding the system by operating the machinery manually when problems arise.

The first cut on each of the four sides of a log removes an exterior piece of wood called a **slab.** This wood is usually chipped for use as paper pulp or biomass (Figure 13–10). A log from which all four slabs has been removed is called a **cant.** Logs are often exported in this form. Each board or timber falls from the

Figure 13–9 Many modern sawmills remove the bark from logs before they are processed. This is done by large machines that apply pressure and friction to the surfaces of the log. *(Photo courtesy of Potlatch Corporation)*

Figure 13–10 Wood that is recovered from slabs and other waste materials is frequently chipped at the mill for use as pulpwood or biomass material. *(Photo courtesy of Potlatch Corporation)*

saw blade to a set of rollers that carries it to the next processing area. A wood piece that is less than 5" × 5" in its dimensions is generally called **lumber.** A cut of wood that is less than 2" thick and greater than 4" wide is called a **board.** A piece of wood that measures 1⅞–4" in thickness and greater than 11" in width is a **plank.** A wood piece that is greater than 5" × 5" in its dimensions is called a **timber.** A **beam** is a timber that is greater than 8" × 8" in its dimensions. Any timber that is square-cut with equal dimensions on all four sides is called a **square.** Standard wood sizes are recognized by the timber industry (Figure 13–11).

Boards and timbers are sorted by thickness, using mechanical sorting gates located throughout the mill. Lumber of 1" thickness passes under the mechanical gates, while lumber of thicker dimensions strikes the gate, causing a mechanical arm known as a **loading jack** to be raised from between the rollers, lifting the timber to another set of rollers located on a higher plane. The same process is repeated as 2" lumber is separated from 4" lumber and larger timbers. Most mills use compressed air to provide **pneumatic power** for the operation of loading jacks and sorting gates.

No two lumber mills are designed exactly the same way, but in general, boards move along rollers until they come to a gang saw. Here they are positioned, trimmed, and cut to standard widths. All of the lumber products smaller than timbers and beams are run through a machine called a **planer.** This machine reduces the rough lumber to standard lumber dimensions that are less than their nominal sizes, and it makes the surfaces smooth. It does this by shaving or planing off the outer surfaces.

Standard Lumber Sizes

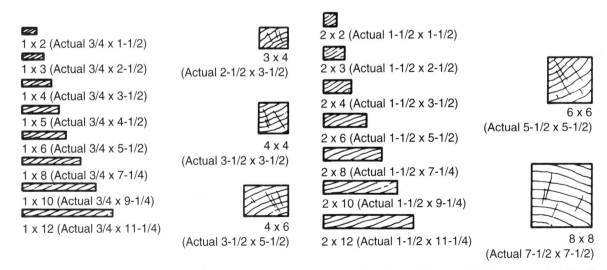

1 x 2 (Actual 3/4 x 1-1/2)

1 x 3 (Actual 3/4 x 2-1/2)

1 x 4 (Actual 3/4 x 3-1/2)

1 x 5 (Actual 3/4 x 4-1/2)

1 x 6 (Actual 3/4 x 5-1/2)

1 x 8 (Actual 3/4 x 7-1/4)

1 x 10 (Actual 3/4 x 9-1/4)

1 x 12 (Actual 3/4 x 11-1/4)

3 x 4
(Actual 2-1/2 x 3-1/2)

4 x 4
(Actual 3-1/2 x 3-1/2)

4 x 6
(Actual 3-1/2 x 5-1/2)

2 x 2 (Actual 1-1/2 x 1-1/2)

2 x 3 (Actual 1-1/2 x 2-1/2)

2 x 4 (Actual 1-1/2 x 3-1/2)

2 x 6 (Actual 1-1/2 x 5-1/2)

2 x 8 (Actual 1-1/2 x 7-1/4)

2 x 10 (Actual 1-1/2 x 9-1/4)

2 x 12 (Actual 1-1/2 x 11-1/4)

6 x 6
(Actual 5-1/2 x 5-1/2)

8 x 8
(Actual 7-1/2 x 7-1/2)

Figure 13–11 The wood and timber industry recognizes standard lumber sizes. The actual thickness and width of planed lumber is less than its nominal size.

Lumber now moves into the sorting sheds on the **green chain,** which consists of parallel chains that carry the boards [Figure 13–12(a)]. In some mills, the ends of the lumber pieces are trimmed to standard lengths at this stage of processing. Once the lumber products arrive in the sorting sheds, they are inspected, sorted by grade, and stacked in layers [Figure 13–12(b)]. Modern equipment is available that inspects, sorts, and stacks the lumber according to its size and grade. Some machines even test lumber and certify it as a stress graded product.

Green lumber still contains much of the moisture that was in the live tree. Lumber that is allowed to dry before it is stacked is likely to **warp** or become distorted in shape. This is because loss of moisture causes the wood to shrink. When moisture is lost at different rates from different sides of a board or timber, the lumber becomes crooked or warped. Green lumber is stacked in layers separated by narrow strips of wood called stickers to allow air to circulate over the board surfaces. This allows the damp lumber to dry in the stacks, or it can be dried in kilns to speed up the process.

TIMBERS AND STRUCTURAL PRODUCTS

Timbers and beams consist of large-dimension lumber that is used to construct the frames of buildings, bridges, and other structures. This lumber is also used extensively by the mining industry to shore up mines, preventing cave-ins.

Figure 13–12(a) The green chain is a conveyor system that is used to carry boards to new locations in the mill for additional processing. *(Photo courtesy of Potlatch Corporation)*

Figure 13–12(b) Boards are carried by "J" hooks to the proper sorting bins. They are sorted by length and width by the trimmer scanner. *(Photo courtesy of Potlatch Corporation)*

Timbers and beams are used in large numbers as the crossbeams that are used as railroad ties to which rails are fastened in the construction of railroad tracks (Figure 13–13). Large numbers of beams continue to be used in the electrical industry as cross-members on utility poles, and construction timbers are driven into the ground as pilings upon which bridges, docks, and buildings are constructed. Structural timbers are being replaced for many of these uses, however, by laminated wood products, metal, and reinforced concrete.

The value of some woods as structural materials can be enhanced by treating them with chemicals to resist decay. This is usually done by injecting chemical compounds directly into the wood. This process extends the useful life of wood construction materials for many years. Some woods, such as redwood, are naturally resistant to decay. This quality makes this wood particularly valuable for construction purposes.

HARDWOODS

Hardwoods are generally considered to be superior-quality wood for many purposes in comparison with softwoods. High-quality hardwoods are used to construct hardwood floors in homes and gymnasiums where durability is a high priority (Figure 13–14). Most of the world's finest furniture is constructed of high-quality hardwoods that combine beauty with durability. The market for antique

Figure 13–13 One of the uses of structural cuts of wood is railroad ties to which train rails are fastened in the construction of railroads.

hardwood furniture bears out the durability claims that are made for hardwoods. In addition to being used to construct doors and wood furniture, hardwoods are also used in the frames of chairs and couches that are stuffed and covered with leather and other materials. Hardwoods are used for many purposes where

Figure 13–14 Hardwoods have the exceptional qualities of beauty and durability that are required to make attractive and lasting floors.

durable construction material is required. A growing demand for hardwoods is for the construction of durable pallets for industrial use.

Hardwoods such as black walnut and a few other species are so valuable that great care is taken to avoid waste. Some of the high-value hardwoods are peeled or sliced to make high-quality veneers for furniture. Some of the best hardwood logs are sawed into thick pieces of high-quality wood. Each of these is called a **flitch.** Flitches are usually sliced into high-quality veneers for use in the construction of expensive furniture. Hardwood veneers are made from large diameter trees (greater than 16" dbh) that are free of knots, decay, and other defects.

LAMINATED AND RECONSTITUTED PRODUCTS

Wood is a versatile building material that can be shaped to create structural materials. A group of products of this kind that has found acceptance by the construction industry includes laminated structural beams (Figure 13–15). Laminated materials consist of layers of individual pieces that overlap, all of which are smaller than the finished product. Laminated structural beams are constructed of multiple layers of wood that have been glued together to form long, uniform structural supports for the roof of a large building. Laminated beams are often used when long, rounded roof spans are desired, such as in gymnasiums where there are no internal supporting walls. Individual pieces of

Figure 13–15 Wooden structural materials can be glued together to increase strength and to create structural beams that are longer than the original materials. These products are called laminated structural beams.

lumber are bent to the desired curvature of the beam, and they are held in place with clamps while the glue sets up between the layers. Once the glue has cured, these massive beams will hold their shape.

Plywoods and Veneers

Veneer consists of thin sheets of wood, ¼" or less in thickness, that are peeled off the surface of a log using a specialized lathe. The process begins by cutting the logs to proper lengths. A log that has been cut to a standard length to fit the lathe is called a **block.** Blocks are prepared for the lathe by heating them with hot water or steam. Next, the blocks are peeled, forming a long, thin ribbon that is clipped to the proper size. Veneer is then cured in dryers to remove moisture.

Some veneers are sold to the furniture industry where they are bonded to inexpensive wood in the manufacture of furniture. Veneer is also used to make crates and packing containers. Most veneers are used to make plywood (Figure 13–16). **Plywood** is a laminated wood product made of several sheets of veneer that are joined together by adhesives. The orientation of the grain of the wood in each layer is alternated to align the long wood fibers across one another. This practice is called **cross-banding.** It gives added strength to the plywood sheet and minimizes contraction and expansion of the material. Each succeeding layer is joined to the others with special adhesives. Each different layer in plywood is called a *ply*, hence the name *plywood*. With the glue in place, the sets of

Plywood Construction

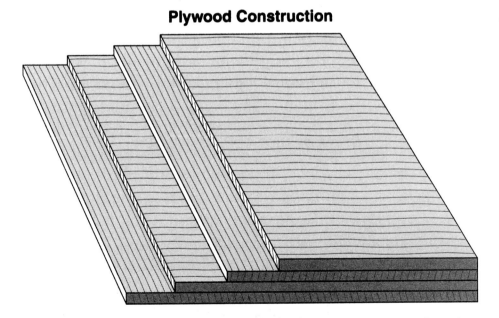

Figure 13–16 Plywood is a laminated wood product in which several layers of wood veneer are glued together in such a way that the long fibers are aligned across one another to increase the strength.

plies are introduced in the hot presses. When the boards emerge, they are trimmed and sanded.

One of the advantages that plywood construction materials offer is that small trees can be converted to building materials with large, standard dimensions. Plywood also can make use of lower-grade materials (such as particleboard in the center of the board) than is evident on the outer layers of the board. The more expensive grades of plywood use high-quality material throughout the board. Much of this material is used to manufacture furniture or cabinets. The main uses of construction grade plywoods include house siding, paneling, and sheathing materials.

Reconstituted Boards

It has been estimated that only 40–70% of the clean wood that enters a sawmill actually ends up as saleable lumber. The rest of the wood was burned or wasted until recent years. Today, most of the material that is trimmed from lumber products is used for reconstituted wood products, or it is reduced to wood chips by a machine called a hog for use as pulpwood or biomass. The hog and the chip-n-saw are wood-chipping machines that are used to reduce large slabs or other wasted wood materials to wood chips. A reconstituted board contains wood material bonded together by an adhesive. The size of the wood particles ranges from fairly large in particleboard to medium or small chips in fiberboard and hardboard.

Particleboard contains a high percentage of wood shavings in the central core of the board with layers of wood flakes on either side of the core and fine sawdust near the surfaces (Figure 13–17). These materials along with an adhesive are pressed together between two steel sheets until the glue has bonded. Some cross-banding occurs, but not to the extent that it is found in plywood. Particleboard has replaced much of the plywood that was formerly used as sheathing in the construction of buildings.

Cross-section of Particle-board

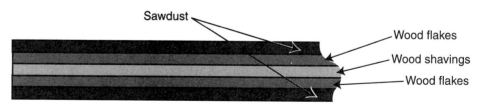

Figure 13–17 Particleboard is a wood product that contains wood shavings in the center with layers of wood flakes on either side of the core and sawdust near the surfaces.

Figure 13–18 A combination of hardboard, particleboard, and lumber is used in the construction of most houses.

Fiberboard is made of wood fibers that become cross-banded in the panel due to the random arrangement of the fibers in the mat from which the board is formed. Fiberboard is a rather loose arrangement of wood fibers in comparison with other reconstituted boards. The density range of this product is .16 to .5 grams/cubic centimeter. Fiberboard is used as acoustical tiles to deaden sound between rooms and in large rooms like auditoriums and gymnasiums. Fiberboard that is used as sheathing on buildings is treated with asphalt to make it more durable and water repellent.

Hardboard is made of similar materials to fiberboard, but the boards are press-bonded between heated steel plates, resulting in a density range for this product of .5 to 1.3 grams/cubic centimeter. Hardboard is probably better known as Masonite, which is a brand name under which it is marketed. Among the uses for which hardboard is suited are siding, panels, tiles, and pegboard.

Approximately 60% of lumber and reconstituted wood products is used in the building construction industry, and the other 40% is used to make furniture or other wood products (Figure 13–18). Wood panels account for approximately 30–35% of all of the lumber products that are used. Lumber products of all kinds are processed in high-tech mills that are fitted with modern machines. These mills are designed to produce all of the different kinds of wood construction materials that are available in the modern construction industry (Figure 13–19).

Figure 13–19 Modern lumber mills are designed to make a variety of construction and specialty products from wood.

MARKETING OF LOGS AND WOOD PRODUCTS

Marketing is one of the most important business activities associated with the timber industry. Wood products must be promoted to potential customers in order for businesses to be profitable. Most wood products are sold through large building supply stores that carry all kinds of products for building construction. Some wood products are marketed through local lumber supply stores. Another fast-growing market has emerged for lumber products such as prefabricated roof trusses, joists, and laminated beams. Other similar markets exist for doors and windows made from wood.

Domestic markets in the United States account for only part of the growth of markets for wood products. International trade has grown for all types of wood products, ranging from unprocessed logs to paper and other specialty wood products. Logs and wood chips are regularly shipped to international markets for processing. In some instances, they are processed on large factory ships that lie in international waters just off the coasts of North America. It is common for many of these products to be returned to the United States where they are sold in competition with domestic products.

Markets for wood products face strong competition from products that are made from materials other than wood. For example, asphalt and slate shingles have replaced wood shingles in the construction of many homes. Aluminum and vinyl siding has taken over a large part of the siding market due to its durability and long-lasting enamel paint surface. In each instance, market share for wood products has been lost. Marketing is an important key to success in creating and maintaining demand for wood products.

A segment of the industry that has grown in recent years is the market for prefabricated houses. These homes are readily available and usually are less expensive than homes that are constructed from the ground up. This market is an example of what can be done to increase demand for a product through aggressive marketing activities. In many respects, the prefabricated home market is similar to the market for cars. New models are created and promoted each year. The customer can buy a standard model or select options that add to the price of the house. This is a good example of the importance of marketing in the wood products industry.

PROFILE ON FOREST SAFETY

Log Handling

Logs are heavy and awkward to handle, and they are hazardous to those who transport and process them. Workers are sometimes injured at the landings by logs that roll down slopes or off piles. Logs tend to shift positions as they are limbed or loaded, crushing unwary workers. Steep slopes are hazardous to harvest because once the limbs are removed, logs tend to roll on sloping terrain. Despite the fact that most logs are handled by large machines equipped with grappling devices, they are sometimes accidentally dropped or placed in unstable positions. Some sawmill operations require logs to be cut into proper lengths for lumber before they are processed. This can be a dangerous job, because piled logs tend to shift in the piles as the cuts are made. A worker with a chain saw in hand is not very mobile, and a moving log can easily crush both the worker and the saw. Piles should be pulled apart by machines before workers begin to cut them to standard lengths.

Logs and green cuts of lumber move rapidly through processing mills on chains or rollers. They build up momentum that can seriously injure a worker if he or she gets hands, fingers, or any part of the body between a stationary object and the moving log or lumber. Mill workers should become highly conscious of where they are positioned with respect to logs and lumber products. No chances should be taken, and safety precautions should be exercised at all times.

CAREER OPTION

Lumber Grader

A lumber grader is responsible for grading lumber products according to the standards of quality. He or she inspects the lumber products as they move across the conveyor system or in the lumber stacks after processing is complete. The lumber grader looks for defects such as decay, splits, milling defects, knots, and stains, and then assigns a grade to the product. The lumber is measured to ensure that standard dimensions are met, and it is marked according to its grade. The grader also may be responsible for removing wood that detracts from the grade of a package of lumber, and for keeping a tally according to the grade and board footage of the lumber. Training and work experience in a lumber mill prepare a worker for this forest industry career.

LOOKING BACK

Wood is an excellent material for construction purposes. Hardwood features that help distinguish hardwoods from softwoods include the presence of pores in hardwoods, but not in softwoods. Softwoods contain resin ducts, and hardwoods do not. Wood characteristics that favor its use in construction include good tensile strength, aesthetic beauty, insulating qualities, flexibility, and ease of cutting to desired shapes. Wood characteristics that detract from its usefulness include the tendency of wood to decay, the inability to fuse wood, and the tendency of wood to resist molding to shapes (except for reconstituted wood). Wood is processed in high-tech lumber mills into standard dimensions of lumber, timbers, and reconstituted wood products. High-quality hardwoods and softwoods are processed into veneer that is used in plywood panels and in the furniture industry.

QUESTIONS FOR DISCUSSION AND REVIEW

Essay Questions

1. What differences exist between processed woods of the hardwood and softwood varieties?

2. What properties of wood make it suitable as a material for construction purposes? What properties detract from its usefulness?

3. Describe the general process by which logs are processed into lumber.

4. List the different cuts of processed wood according to standard lumber and timber dimensions.

5. List the duties that are performed by a person who chooses a career as a lumber grader.

6. What force causes green lumber to warp when it is not properly stacked as it cures?

7. What are some ways that structural wood products are used?

8. What are some major uses for which hardwoods are best adapted?

9. Explain what veneer is and how it is obtained from logs.

10. What is the process by which plywood is manufactured?

11. How are reconstituted boards such as particleboard, fiberboard, and hardboard made?

12. How are plywood, particleboard, fiberboard, and hardboard different from one another?

Multiple-Choice Questions

1. Which of the following is a distinguishing visual feature of hardwoods?
   A. pores
   B. resin ducts
   C. density
   D. odor

2. Which of the following is a distinguishing feature of a softwood?
   A. pores
   B. color
   C. resin ducts
   D. heartwood

3. The strength of wood along the long axis of a tree is called:
   A. ductile strength
   B. density
   C. toughness
   D. tensile strength

4. A saw that makes two or more cuts at the same time is called a:
   A. gang saw
   B. band saw
   C. head rig
   D. circular saw

5. A log from which only the slabs have been removed is called a:
   A. timber
   B. cant
   C. beam
   D. block

6. A timber with a cross-section measurement of 8" × 8" or greater is called a:
   A. cant
   B. block
   C. beam
   D. flitch

7. A source of power used in lumber mills that is provided by compressed air is called:
   A. hydraulic power
   B. pneumatic power
   C. wind power
   D. steam power

8. A condition that occurs as green lumber dries unevenly, resulting in distorted shapes in the wood, is known as:
   A. lamination
   B. asymmetry
   C. symmetry
   D. warp

9. A thin sheet of wood that is peeled from a block is called:
   A. veneer
   B. plywood
   C. particleboard
   D. reconstituted wood

10. The practice of aligning long wood particles and fibers across one another in the manufacture of reconstituted wood is called:
    A. articulation
    B. cross-banding
    C. integration
    D. cross-bonding

11. Another name by which hardboard is well known is:
    A. particleboard
    B. waferboard
    C. Masonite
    D. fiberboard

12. A reconstituted wood product in which wood shavings are a major component is called:
    A. waferboard
    B. fiberboard
    C. hardboard
    D. particleboard

LEARNING ACTIVITIES

1. Visit a home building supply warehouse, either individually or as a class, and report on the dimensions, grades, and prices of wood products that were available for home construction. Assign students to small work groups, and assign them some log dimensions (debarked). Have each group calculate the highest possible retail value that could be obtained from the log based on the lumber dimensions and prices that were observed at the warehouse.

2. Obtain some samples of reconstituted wood products, and allow the students to examine them. Assign students to small work groups, and have each group do an in-depth written and oral report on the types of materials that were used to make the product. Have each group explain the probable process by which the product was manufactured.

Chapter 14

Specialty Forest Products

Terms to Know

cellulose
monomer
monosaccharide
polymer
polysaccharide
hemicellulose
lignin
mechanical pulping
semichemical pulping
disk refiner
chemical pulping
hydrapulper
dissolving process
cellulose xanthate
saccharification
fermentation
ethanol
gasohol
destructive distillation
carbonization process
thermochemical liquefaction
syngas
extractive
naval stores
wood naval stores
gum naval stores
oleoresin
sulfate naval stores
British thermal unit (Btu)

Objectives

After completing this chapter, you should be able to

* distinguish between a monosaccharide and a polysaccharide

* explain the significance of cellulose to the fiber and paper industries

* describe the different methods that are used to convert wood fiber to pulp

* suggest reasons why the mechanical pulping method is widely used in the paper and fiber industry

* distinguish the differences between the bleaching and brightening processes

* explain why different grades of wood pulp are sometimes blended together

* define the process by which wood is converted to ethanol

* identify products that are obtained by destructive distillation of wood

* name the different types of products that are extracted from wood using solvents, and explain the processes by which they are obtained

* discuss the importance of biomass as a fuel for generating electrical power

351

Wood is only one of many forest products that are significant to the economy and the way of life we have come to expect in North America. Some forest products contribute in major ways to the economy, but trees are also important sources of substances that we hear little about. Raw materials for the medical, textile, and apparel industries come from forests. Foods, spices, and extracts for flavoring and coloring foods are obtained from trees. The forests of North America have proven to be sources for many products besides wood.

Half of the total annual harvest of timber is used for something besides lumber. The largest consumer of wood for specialty products is the wood fiber industry. One-fourth of all harvested wood is converted into paper or cardboard products. Another 25% of the timber harvest is burned for energy production or used for chemical production.

FIBER AND PAPER INDUSTRIES

The key component of plant cell walls is **cellulose.** It is the most abundant raw material in the manufacturing process that yields paper and cardboard products (Figure 14–1). The percentage of cellulose is slightly higher in hardwoods

Figure 14–1 Cardboard products are produced from wood pulp that has not been bleached. Large amounts of cardboard are used each year as packaging material for manufactured products. *(Photo courtesy of Marilyn Parker)*

(44%) than in softwoods (42%), but the length of wood fibers is greatest in the softwoods, making them more desirable for making paper. A cellulose molecule consists of a long chain of sugar molecules known as **monomers** or **monosaccharides.** A long chain of monomers formed by a plant as it grows is called a **polymer** or **polysaccharide.** Polysaccharides called **hemicelluloses** are also found in wood. They make up about 27–29% of the material found in wood. The other major component of wood (25%) is **lignin.** This is the material that binds wood components together.

Wood must be processed into wood chips before it can be used to make wood products. This is sometimes done at the processing plant, but much of the wood is now processed in the forest. Large tractor-trailers, specially designed to haul wood chips, are used to transport the wood chips from the forest to the mill. Rail cars are also used for this purpose. Chip size is important because the length of the wood fibers must be preserved during the pulping process (Figure 14–2). Wood fibers that are too short do not yield strong paper.

Pulping Processes

Wood chips are converted to pulp using several different methods. One of these methods is **mechanical pulping.** This is a method that is used to make paper that is newsprint quality. A mechanical pulping system uses a stone grinder or a disk refiner to separate the wood fibers as it grinds up the pulpwood bolts or wood chips (Figure 14–3). The fiber that is produced is mixed with water, forming

Figure 14–2 The production of paper products requires large amounts of chipped wood as a raw material. These wood chips are the ideal size for papermaking. *(Photo courtesy of Potlatch Corporation)*

Disk Refiner

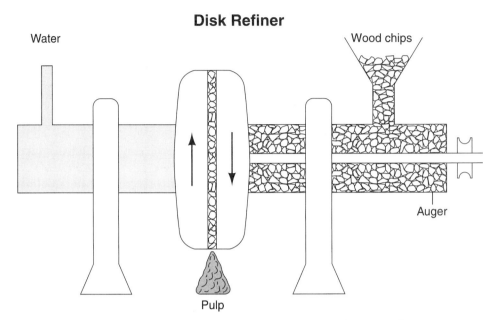

Figure 14–3 One mechanical pulping method introduces wood chips and water between large disks that are rotating in opposite directions.

a slurry that is converted to wood pulp. This system causes damage to individual wood fibers, making it necessary to add higher-grade pulp to mechanical pulps to increase the strength of the paper. Most commercial newsprint consists of approximately 70% mechanical pulp and 30% chemical pulp. The big advantage that is gained from using mechanical pulping methods is that up to 95% of the volume of wood can be converted to paper.

It is becoming more common for wood chips to be pretreated with chemicals and/or steam as part of the mechanical pulping process. This results in wood fiber that makes strong paper because it is not seriously damaged. Newsprint can be made with this mechanical pulping method that is strong enough for use without adding high-quality chemical pulp to the mixture. Mechanical pulping methods preserve the hemicellulose and lignin components of wood in the pulp, resulting in high yields of paper products.

Semichemical pulping is a process for producing wood pulp that is used mostly in the production of cardboard boxes (Figure 14–4). Wood is exposed to a mild chemical treatment to partially separate the fibers before it is processed through a disk refiner. A **disk refiner** is a machine that separates wood fibers between two mechanical disks as they rotate. When this method is used, the most common chemical treatment is a combination of sodium sulfite and sodium carbonate. This process yields paper products equal to 70–85% of the volume of the wood that is used as raw material. The strength of the paper product that is produced by this process is higher than that produced strictly from mechanical pulp.

Figure 14–4 A major industry has emerged within the wood products industry that constructs cardboard boxes from huge rolls of cardboard products purchased from pulp and paper mills. *(Photo courtesy of Marilyn Parker)*

Paper products with high strength requirements are produced using a method called **chemical pulping.** This pulping method uses chemicals to dissolve the lignin component of wood. This is done by placing the wood and chemicals in a large container called a digester. Heat and pressure are applied to the mixture, causing a chemical reaction that combines lignin with water. In this form, the lignin component of wood is removed along with the hemicelluloses. The wood fibers that remain are undamaged by the process, and they form strong paper that is of high quality. Most chemical pulping processes yield only 40–55% of the total wood volume as paper products (Figure 14–5). Several different chemical pulping processes have been developed. The Kraft process, developed in Germany in 1884, is the most used chemical pulping process in the United States.

In addition to the wood pulp that is obtained directly from wood, some pulp is recovered from recycled paper. Approximately 25% of used paper is recycled in the United States (Figure 14–6). Recycled paper is reduced to pulp in a machine called a **hydrapulper.** The recycled paper is mechanically processed in water to separate the wood fibers. Ink is removed by treating the pulp with sodium hydroxide, and the recycled pulp is screened to remove fine materials that reduce the strength of the paper. After the pulp has been cleaned, it is bleached. Among the uses of recycled paper are newsprint, tissues, and paper towels.

Paper Yields

Pulping Process	Percent Paper Yield	Paper Strength
Mechanical pulping	Up to 95%	Low
Semichemical pulping	70–85%	Moderate
Chemical pulping	40–55%	High

Figure 14–5 Pulping methods vary in the percentage of raw product that is converted into paper. High yields are associated with low quality. Most paper is manufactured by mixing pulp from different processes for the purpose of maintaining high yields of acceptable quality.

Figure 14–6 Approximately 25% of the paper products used in the United States each year is recycled to make new paper products. *(Photo courtesy of Marilyn Parker)*

Bleaching

Some pulping processes produce light-colored pulp, but many of them produce pulp that is the color of cardboard. Wood pulp that is to be used to produce paper must be bleached or brightened to remove colored pigments. Lignin that remains in the paper pulp is often dark in color. Bleaching removes the lignin from the pulp. Brightening is a process that leaves the lignin in the pulp, but it is modified to form a compound that is lighter in color.

The bleaching process usually involves a series of processes to obtain whiteness. It is an expensive process that requires large investments in equipment, and maintenance costs for the equipment are high because the chemicals that are used for bleaching are corrosive. The waste materials that are produced by the bleaching process require treatments to prevent pollution of the environment with highly toxic substances.

Paper Products

The production of paper from wood pulp begins with the preparation of paper stock, which involves subjecting the pulp to a refining or beating process by which the pulp fibers are further separated, crushed, and cut to uniform sizes. This is done with machines such as beaters and refiners that use abrasive action to enhance the fibers. This process improves the potential for strong chemical and ionic bonds to develop in the finished paper, holding the fibers together. Most paper products are blends of different kinds of paper pulp. Blends of pulp are obtained by mixing bales of dried pulp of the desired quality with liquefied pulp and reprocessing the mixture.

Paper is made by spreading the liquid pulp on the moving screen of a Fourdrinier paper machine. A smooth and uniform mat of liquid paper is formed on the screen by the machine. Excess water is drained off the paper that forms during this process. The continuous paper ribbon is pressed and dried by heated rollers and hot drum dryers [Figure 14–7(a)]. Surface coatings are applied to the better grades of paper that are to be used for printing or writing, and the paper is rolled under pressure to ensure smoothness of the paper surfaces. The paper is then rolled up in huge rolls as the dried paper comes off the end of the machine in a continuous ribbon [Figure 14–7(b)].

The final step in the production of paper products is to cut the paper into standard sizes or to form the paper into useful shapes, such as paper plates, cups, boxes, and so on. This is followed by packaging the paper products for shipment.

CHEMICAL PRODUCTS

One of the processes that is used to make wood pulp is the **dissolving process.** This process is used to dissolve cellulose into a viscous liquid called **cellulose xanthate.** This material is used to produce products like rayon, photographic films, and cellophane (Figure 14–8). Rayon is made by extruding a solution of cellulose xanthate through tiny holes and spinning the fibers together to produce thread and cloth. It can also be cast in thin layers to form cellophane or converted to an acetate product to produce photographic film. Other uses of cellulose include explosives such as dynamite or nitroglycerin. Cellulose is also the source of cellulose ethers used in the production of lacquers, adhesives, latex paints, pharmaceuticals, and cosmetics.

Through a process called **saccharification,** the polysaccharides that make up the cellulose in wood are converted to form simple sugars. This can be done commercially by treating wood chips with acids. **Fermentation** of these sugars produces **ethanol,** which is a fuel-grade alcohol. All of the simple sugars obtained from wood can be fermented to ethanol by using a combination of

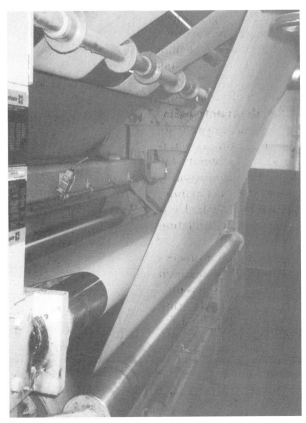

Figure 14–7(a) Paper is manufactured by spreading wet paper pulp on a moving screen where it is dried to form a continuous ribbon of paper. *(Photo courtesy of Marilyn Parker)*

Figure 14–7(b) This paper is packaged in huge rolls that are used later to make many kinds of paper products. *(Photo courtesy of Potlatch Corporation)*

Chemical Products from Wood Sources

rayon	lacquers
photographic film	adhesives
cellophane	latex paints
explosives	pharmaceuticals
cosmetics	ethanol
charcoal	oil
syngas	tannin
turpentine	rosin
naval stores	food flavors

Figure 14–8 Many different kinds of products of a chemical nature are obtained from timber products.

yeasts. When enthanol is mixed at the rate of 10% ethanol with 90% gasoline, a fuel called **gasohol** is produced (Figure 14–9). This fuel burns well in the internal combustion engines that provide power for cars and trucks. Another alcohol that is obtained by heating wood is methyl alcohol. It also works well as a fuel when it is blended with gasoline.

A number of gases can be recovered from wood by heating it above its combustion point in the absence of oxygen. This process is called **destructive distillation,** because the wood is reduced to charcoal and the volatile gases are released from the wood. The charcoal that is produced is a wood product that is high in its carbon content (74–81%) due to the concentration of carbon during the distillation process. For this reason, the destructive distillation of wood is also a **carbonization process.** The charcoal that is produced in this manner burns with a very hot, smokeless flame. When it is used as an industrial fuel, it is called coke. Among the end products that are collected by destructive distillation of wood are acetic acid, acetone, methane, and tar.

An industrial oil similar to petroleum can be produced through a process called **thermochemical liquefaction.** Wood chips are heated under high pressure in a hydrogen gas or syngas atmosphere. **Syngas** is a synthetic gas produced from methane and carbon monoxide, both of which are recovered during the carbonization process. Oils produced by this process have potential for future use, but they are not economical at the present time.

Figure 14–9 Ethanol is an alcohol that is obtained from fermenting the cellulose in wood products. When it is mixed at the rate of 10% ethanol with 90% gasoline, a fuel called gasohol is produced.

PROFILE ON FOREST SAFETY

Machinery Safety

Modern sawmills and paper mills are equipped with many large, complex machines, and each machine is designed to perform a specific task. For example, a disk refiner is a machine that is used to separate wood fibers in preparation for manufacturing paper products. This large machine tears wood particles apart between large abrasive wheels that rotate in opposite directions. For their own safety, workers must carefully avoid contact with any of the moving parts on this machine while it is operating. It is capable of crippling or killing any worker who becomes careless.

Another example of a machine that can pose danger to the operator is a gang saw. This machine is equipped with several saw blades that are used to cut lumber into standard dimensions. One or more blades may be operating at any given time. If a board should enter the gang saw at the wrong angle, it can bind the blades and stop them from turning. In this circumstance, the operator of the saw must learn that the only safe way to remove the wood from the vicinity of the blades is to turn the power off before reaching into the blades to remove the wood. The operator should personally turn the switch to the off position before working to free the saw. It is also very important that only the operator of a machine is allowed to operate the power switch. Imagine the hazards that a worker would face if someone else turned the machine on while it was being repaired.

These are only two of the many examples of the machines and safety issues that are encountered during the production of products from wood. Each product benefits from the use of machinery, and each machine presents new concerns for human safety.

Extracted Wood Products

Some valuable wood products can be extracted from wood using solvents. The products are called **extractives.** Water soluble extractives include a class of chemicals called tannins. These chemicals are used to process animal hides into leather. These chemicals are extracted from the bark and heartwood of some varieties of trees by dissolving the water-soluble component in water that has been heated to 80–120° C. Some of the tannins are also used in the production of adhesives.

Extractives that are obtained from wood using organic solvents include fatty acids, turpentine, and rosin. These products are called **naval stores** due to their historic use in caulking and sealing wooden ships to prevent them from leaking.

They form rich deposits in the heartwood of pine trees. **Wood naval stores** are obtained from the chipped or shredded wood from pine stumps and logs by dissolving the extractives using organic solvents. **Gum naval stores** are obtained by injuring pine trees and collecting a mixture called **oleoresin** that flows from the wounds. The oleoresins are composed of oils, turpentine, and rosin, and the different products are separated by distilling them. **Sulfate naval stores** are obtained as by-products from the Kraft pulping process.

Turpentine and rosin are important wood products. Turpentine is used for many things including the production of a synthetic pine oil that is used to make cough syrup. It is also the source of flavors and fragrances such as spearmint, lilac, peppermint, menthol, lemon, and others. Rosins are used as a sizing agent for paper products to reduce their absorption of water. They are also used to manufacture some types of adhesives.

BIOMASS (ENERGY)

Biomass was defined in Chapter 9 as all of the wood that is available in the stem, leaves, branches, and roots of a tree. This chapter will consider biomass as a fuel product. Wood is an efficient source of renewable energy, and biomass production is an efficient use of wood. This is because nothing is wasted when wood is used for biomass production. The entire tree can be used to produce heat. The measurement of heat from a fuel such as wood is measured in **British thermal units (Btu's).** Most biomass fuels are used to produce steam which is used to operate generators in the production of electrical energy.

Much of the wood that is used in biomass production grows in plantations where biomass is the intended crop (Figure 14–10). In addition to wood that is

Figure 14–10 Biomass is obtained from tree plantations and waste materials that are recovered from lumber mills and manufacturing plants. Large amounts of biomass are used as fuel by electrical generating plants.

produced for this purpose, wood that is recovered from the waste materials in lumber mills is also called biomass. The forest industry has become much more efficient in the use of wood by-products in recent years, and much of the wood that was once wasted is now used in reconstituted wood products or as biomass fuel in electrical generating plants.

Trees that are harvested for use as biomass are chipped or shredded. New varieties of fast-growing trees have made it feasible to produce competitively priced electricity with the heat that is generated from burning biomass products. Hybrid varieties of trees that have been developed for their rapid growth are capable of reducing the length of time that is required to produce a biomass crop. Research has demonstrated that high-density plantings are capable of high yields of biomass products. Biomass production has great promise as a commercial crop.

Wood of all varieties tends to be very similar in its actual chemical makeup. A pound of hardwood will produce approximately the same amount of heat as a pound of softwood (approximately 8,300 Btu). A pound of biomass has been

CAREER OPTION

Environmental Quality Technician

A person who works as an environmental quality technician in the paper and pulp industry is responsible for monitoring all phases of processing to identify areas within the processing plant where abnormal amounts of pollutants enter the water stream or escape into the atmosphere. He or she collects samples of water and air on a routine basis, submits samples for laboratory testing, and interprets the results of the water and air quality tests (Figure 14–11). Environmental quality technicians work with process engineers to devise ways to monitor waste water outputs and smoke-stack emissions. It is the responsibility of the water quality technician to ensure that government standards are met for water quality before the water is discharged from the plant. A college/university degree is usually required for employment in this career.

Figure 14–11 A paper mill uses a lot of water in the manufacturing process. The people who test the waste water and control the water treatment processes are water quality technicians and engineers. It is their job to make sure the water is clean before it is returned to rivers and lakes.

demonstrated to produce about the same amount of heat as wood when both products are dry. Differences do exist in woods of different varieties, but the differences are in their densities. The volume of a pound of hardwood is less than the volume of a pound of softwood. On a per pound basis, biomass products, regardless of variety, are very similar in the amount of energy they produce.

Biomass as an energy crop is destined to play an increasing role in power production because it is renewable and the yields are high. In a modern world where efficiency of production is valued and the demand for electricity is rising, biomass can be expected to emerge as an important source of energy.

LOOKING BACK

Half of the annual timber harvest is used for something besides building materials. The cellulose that is contained in the cells of trees is valuable for paper products. It is processed by reducing wood to pulp from which paper products are made (Figure 14–12). The cellulose found in pulp is also converted into other products such as rayon, photographic films, lacquers, adhesives, cosmetics, explosives, cellophane, and many other products. Wood is also converted to charcoal through a carbonization process. By-products of this process include acetic acid, acetone, methane, and tar. An industrial oil is obtained from wood using the process called thermochemical liquefaction. Naval stores and tannins are extracted from wood by dissolving these materials in solvents. Biomass is produced by chipping entire trees including stems, branches, roots, and foliage. Electrical energy is generated by burning biomass obtained from plantation biomass crops and sawmill wastes.

Figure 14–12 A pulp mill is a massive enterprise that reduces wood fiber to pulp and then reconstitutes it to make paper and paper products.

QUESTIONS FOR DISCUSSION AND REVIEW

Essay Questions

1. What is the relationship between monosaccharides and polysaccharides?

2. Why is cellulose important to the fiber and paper industries?

3. What is the purpose of adding chemical pulp to mechanical pulp in the manufacture of newsprint?

4. Name the different pulping processes, and compare the methods that are used by each of them.

5. Explain the bleaching and brightening processes, and describe the differences between them.

6. Describe the process by which wood is converted to ethanol.

7. What are the end products of the process called destructive distillation?

8. What are some extracted products, and how are they obtained from wood?

9. List some commercial products that are obtained from naval stores.

10. What is biomass, and how is it used commercially?

Multiple-Choice Questions

1. Another name for a simple sugar that forms in long chains to make cellulose is:
   A. monomer
   B. lignin
   C. hemicellulose
   D. polysaccharide

2. A pulping process in which wood fibers are separated from each other by grinding or abrasion is called:
   A. chemical pulping
   B. hydrapulping
   C. mechanical pulping
   D. semichemical pulping

3. A pulping process in which wood fibers are separated by dissolving the lignin that cements them together is called:
   A. hydrapulping
   B. chemical pulping
   C. mechanical pulping
   D. disk refining

4. A machine that is used to reduce recycled paper to pulp is the:
   A. Fourdrinier
   B. stone grinder
   C. disk refiner
   D. hydrapulper

5. A process that changes the lignin in paper pulp to a compound that is lighter in color is known as:
   A. bleaching
   B. brightening
   C. coloring
   D. blending

6. Which of the following products does *not* come from cellulose xanthate?
   A. naval stores
   B. photographic film
   C. cellophane
   D. rayon

7. A fuel that is composed entirely of the product that is obtained by converting cellulose to simple sugars and fermenting them to form alcohol is called:
   A. methanol
   B. charcoal
   C. gasohol
   D. ethanol

8. Oil can be produced from wood using a process called:
   A. destructive distillation
   B. saccharification
   C. thermochemical liquefaction
   D. fermentation

9. Oleoresin is a product that is obtained by collecting the sap from trees. Which of the following materials contains oleoresin?
   A. wood naval stores
   B. gum naval stores
   C. sulfate naval stores
   D. food grade margarine

10. Which of the following materials produces the hottest flame when it is burned?
    A. hardwood
    B. softwood
    C. charcoal
    D. biomass

LEARNING ACTIVITIES

1. Obtain microscopes with which to observe the structure of paper and paper products. Point out the overlapping structure of the fibers, and explain to the students that chemical and ionic bonds also attract and hold the wood fibers together.

2. Collect as many products as you can find that are obtained from wood. Assign pairs of students to make and display posters illustrating how one of the products was manufactured. Give each group of students an opportunity to discuss their product with the class. Keep the collection of products together to be used in future classes.

Chapter 15

Plantation Products and Practices

Terms to Know

plantation forest
monoculture
containerized seedlings
cuttings
nursery stock
bare-root stock
pruning

Objectives

After completing this chapter, you should be able to

* define *monoculture* as it is related to forestry

* explain the cultural practices that are involved in the production of containerized seedlings

* describe how cuttings are used to produce trees for transplanting

* list some silviculture practices that are used in the management of a Christmas tree plantation

* distinguish between bare-root stock and containerized seedlings

* explain the beneficial effects of pruning on lumber quality

* evaluate the production of biomass as a plantation crop in contrast with biomass production in most forest environments

A **plantation forest** is created when trees are established in an area by planting seeds or seedlings. In most instances, the trees that are planted in a plantation are of a single variety. A population of trees consisting of a single variety is called a **monoculture.** A monoculture is not restricted to plantation forests. A monoculture also can occur naturally in a forest when a single species of trees such as Douglas-fir or lodgepole pine becomes dominant. Plantation forests have become widely established because it is possible to reduce the rotation age and to increase total forest production using intensive silviculture practices that favor monoculture forest populations.

Management of a forest plantation is very much like the cultivation of crops on a farm (Figure 15–1). Silviculture has been defined as the cultivation of trees, and the management of a plantation forest is often referred to as tree farming. It should be noted that there are many different types of tree plantations. Some of these produce a tree crop every year in the form of seedlings. Some tree plantations raise trees for ornamental purposes in landscapes for homes and businesses. In these instances, a crop may require several years to mature.

A Christmas tree plantation produces trees that require several years to mature and produce the crop (Figure 15–2). Biomass is produced on plantations, often requiring twelve to twenty years to attain maximum economic yields. Pulpwood plantations may require twenty to forty years to produce a single crop,

Figure 15–1 A tree plantation is very much like a row-crop farm. The only real difference is that the crop is trees instead of a traditional farm crop like corn or beans. Silviculture is the cultivation of trees.

Figure 15–2 A Christmas tree plantation has a relatively short rotation time of a few years in comparison with tree plantations that produce pulpwood or lumber.

and timber plantations often have rotations of fifty to seventy years. All of these tree crops are products of tree and forest plantations.

SEEDLINGS

Seedlings are young trees that are generated from seeds. They are raised under controlled conditions to ensure high survival and growth rates. Some of them are produced in beds located in field nurseries (Figure 15–3). Seeds are planted directly into the soil in narrow rows with close spacing of the seeds in each row. This method of planting helps to control weeds and makes it possible to produce a large number of seedlings in a very small area.

Containerized seedlings are produced in greenhouses in small containers of soil (Figure 15–4). The roots and soil are left intact when the seedlings are transplanted in forest locations. Seedlings that are produced under these conditions are more expensive than those raised from seed in nursery plantings. This is due to the cost of containers, soil medium, buildings, and labor.

Seedlings vary in the length of time that is required from germination to transplanting. Some species of trees take much longer than others to reach a seedling size that can be successfully transplanted. The use of seedlings over direct seeding has advantages such as uniformity in the spacing of trees and in the age of the stand. Less seed is needed for seedling production than for direct seeding of a forest site. This is because of the high seed losses that occur due to birds and rodents when seed is broadcast on soil surfaces.

Figure 15–3 Tree seedlings that are produced in plantation nurseries grow in densely populated beds where the seeds have been planted close together in narrow rows.

Figure 15–4 Containerized seedlings grow in individual packets of soil to minimize the damage to roots when they are transplanted into the forest. *(Photo courtesy of Potlatch Corporation)*

A very large industry has developed in North America for the production of tree seedlings. Many new hybrid varieties of trees with specific characteristics have been developed, and these are actively marketed for commercial tree production on plantations and for ornamental purposes in parks, city streets, and homes. Seedling sales are promoted in trade journals and industry trade shows. Large numbers of seedlings are produced each year by government forest agencies and by private forest industry facilities. They are used to replant thousands of acres of harvested forest lands each year.

CUTTINGS

Some types of trees can be regenerated from cuttings. This is a type of asexual reproduction that was discussed in Chapter 4. Fresh branches or twigs are cut from trees during the dormant season and buried in the debris on the forest floor to accomplish reforestation. An adaptation of this method is to generate new tree growth by rooting the tips of branches in soil medium in a greenhouse or cold frame. These vegetative tree parts are called **cuttings,** and large numbers of trees are generated from these materials.

Some types of trees such as poplars can be generated under damp climatic conditions from vegetative cuttings that have been pushed down into the moist soil (Figure 15–5). Other trees, such as some of the conifers, are generated in large numbers from small cuttings from the tips of the branches. Under greenhouse conditions, these plant materials can be rooted in large numbers. Once the cuttings have rooted, they can be transplanted to plantation sites in the same manner as seedlings for the purpose of growing the trees to larger sizes or populating the plantation for tree crop production.

NURSERY STOCK

Approximately 1.6 billion forest tree seedlings are produced in the United States each year. A significant number of these are produced in field nurseries by federal and state forestry agencies and by the private sector of the forest industry (Figure 15–6). Many of these seedlings are used to regenerate public and private forest lands, and some of them are used to produce trees that are used for decorative purposes in public parks or for landscaping purposes. Although these types of plantings do not conform to the traditional image of forestry, they contribute to the urban forestry movement that is sweeping through the cities and towns of North America.

Commercial nurseries produce trees of many different sizes ranging from seedlings to mature trees (Figure 15–7). All of these plants are considered to be **nursery stock,** and most of these plants are exposed to intensive management practices. Among these practices are weed control, pruning, fertilization, insect control, irrigation, and transplanting. Commercial nurseries produce many trees that are hybrid tree varieties. The hybrids have been developed to express resistance to specific problems or to exhibit aesthetic qualities that make them desirable for decorative purposes.

Figure 15–5 Vegetative cuttings of some tree varieties such as poplars will grow under damp climatic conditions when their stems are pushed down into moist soil.

CHRISTMAS TREES

The production of Christmas trees on plantations has replaced much of the harvesting of trees in evergreen forests as young stands of trees were thinned. Most of the fresh Christmas trees that are sold each year are raised in Christmas tree plantations because the plantation trees are generally higher in quality than wild

Figure 15–6 Many state and federal agencies produce forest seedlings in government-owned field nurseries.

Figure 15–7 Commercial nurseries produce trees of many species and stages of maturity.

trees (Figure 15–8). A number of evergreen varieties are acceptable for Christmas trees, but the Scotch pine variety has become quite popular for this purpose.

Quality in Christmas trees is measured by the color of the needles, density of the branches, and fullness of the foliage. All of these characteristics can be controlled by using silviculture practices. Needle color can be affected by applying fertilizers to the soil in proper amounts. The density of branches is increased by cutting back the central leader of the tree. This causes the tree to produce additional branches from the stem. Pruning the ends of the branches with a shearing knife increases the production of foliage, causing the tree to have a full appearance.

Christmas tree plantations that are located near urban centers sometimes sell their trees to customers who come to the farms to select them. Trees are also cut and shipped from tree farms to retailers in cities and towns a few weeks before Christmas (Figure 15–9). Christmas tree plantation managers usually produce only as many trees as they think they can sell each year, and they plant a new crop each year to replace the trees that were harvested.

BIOMASS

Biomass production was discussed as a specialty crop in Chapter 14 of this textbook. It will be discussed in this chapter as a tree crop that is well adapted for high-yield plantation farming. In recent years, it has been demonstrated that biomass production can be increased dramatically by producing biomass crops under intense management systems on plantations (Figure 15–10). Some of the hybrid varieties of trees, such as hybrid poplars, grow at extremely rapid rates. It

Figure 15–8 Many Christmas tree farms offer the customer an opportunity to bring the family to the farm to choose and cut their own tree, or to buy a live tree that can be planted in the yard when the Christmas season is over.

Figure 15–9 Large numbers of evergreen trees are grown under plantation conditions to supply the Christmas tree lots that are set up in urban areas a few weeks before the Christmas season.

Figure 15–10 Biomass crops that are produced under intensive management conditions are capable of producing high yields of wood for commercial uses.

is evident that plantings with high populations of high-yielding trees are capable of yielding massive amounts of biomass material over relatively short periods of time. The fact that biomass has found a market niche in the production of electricity has contributed to its success as a crop.

Perhaps the most significant aspect of biomass production is in the potential of this forest product to replace fossil fuels as an energy source in the production of electricity. Whole-tree chipping converts entire trees to useful fuel that can be used to replace coal, diesel, and natural gas in the production of large amounts of electricity. Efficient production of biomass is having the effect of reducing our dependency on nonrenewable fuels. The production of many different types of fuels from biomass is now possible, and research is continuing to refine and enhance these processes. Biomass may well be our most important energy source of the future.

CULTURAL PRACTICES

Cultural practices that are used to manage plantation forests include all of the practices that were discussed in Chapter 8. A review of the silviculture practices that were discussed there would be appropriate as part of this discussion on plantation management practices. In addition to those already mentioned, some additional practices are common in plantation forests that are too expensive to be profitable in most commercial forests.

Transplanting

Most of the trees that are transplanted to plantations or forests are in the seedling stage of development. If they have been raised in nurseries under field conditions, they are prepared for transplanting by removing them from the soil. Seedlings that have been produced in this manner are called **bare-root stock** (Figure 15–11). Bare-root stock is often removed from the nursery in the late fall and stored in cellars or refrigerated storage areas until the seedlings are planted in the spring.

Seedlings must be transplanted with great care to increase their chances of survival. Before planting, the trees should be stored at cool temperatures, but they must not be allowed to freeze. The optimal storage temperature range is 32–35° F. The storage temperature can be increased to 45° F during the last two or three days preceding transplanting. During the planting process, the roots must be protected from exposure to the wind or the sun to prevent the fine roots from drying out. Serious damage is sustained by seedlings when the roots become too dry.

Soil condition is important when seedlings are transplanted to permanent sites. The soil should consist of fine, moist particles that can be packed firmly around the seedling roots. Seedbed preparation is important if seedlings are to survive. Seedlings should be planted slightly deeper than they were growing in the nursery. The roots must retain their conical shape during transplanting. Any twisting, bending, bunching, or shortening of the root system must be avoided. (Figure 15–12). Roots must be placed in the hole pointing downward, and they

Figure 15–11 Seedlings that are produced in field nurseries are usually harvested by removing their roots from the soil and storing them in a refrigerated area to simulate or induce dormancy. *(Photo courtesy of Boise National Forest)*

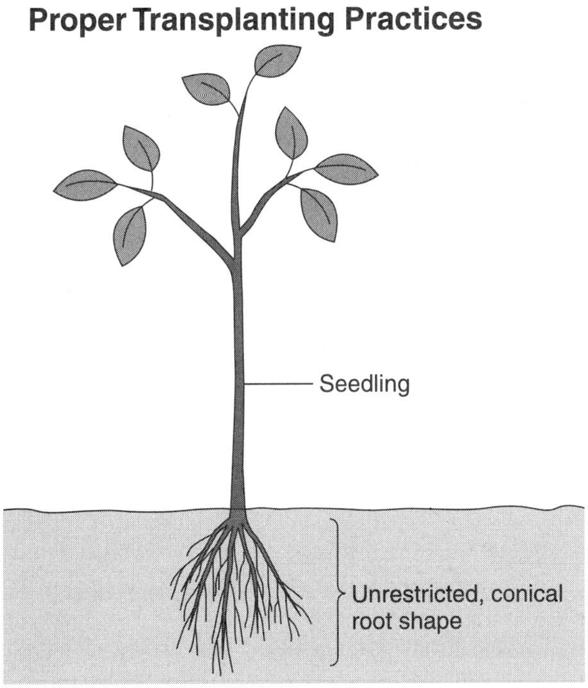

Proper Transplanting Practices

Seedling

Unrestricted, conical root shape

Figure 15–12 It is important to the survival of the seedling to maintain a conical root shape as the young tree is transplanted into the soil.

should not contact the hard edge of the hole. Fine, moist soil should be packed firmly, but not too tightly, on all sides of the roots.

Seedlings that are planted in plantations often benefit from supplemental water and weed control. Many of the seedlings that are transplanted to plantation locations are planted by mechanical planters in prepared seedbeds. Workers riding on the planter place the seedlings in the furrow that is opened by the machine, and packing wheels compress the soil on either side of the seedling to ensure good contact between the soil and the roots. Spacing of the seedlings depends on the tree variety and the intended use of the trees. Biomass plantations provide the highest yields when row spacings and seedling placements within the rows are close together, resulting in a high-density stand of trees. A lower-density stand is desirable for pulpwood production, and trees that are cultivated for lumber production must be thinned extensively to provide proper spacing (Figure 15–13).

Seedlings planted in forest locations must depend on rain for moisture, but some steps can be taken to improve their access to the moisture that is available. Planting the seedling in the bottom of a scrape or gouged area allows moisture from the surrounding area to flow to the bottom of the scrape where the seedling is located (Figure 15–14). Site preparation of this kind is fairly common in forest plantings. Seedling survival rates of 75–80% are considered to be normal for seedlings that have been properly transplanted. It may be necessary, however, to replant in the following year when seedling mortality is high or where large areas have experienced heavy seedling losses.

Figure 15–13 Trees that are raised for lumber production must be thinned to allow them adequate space to grow and mature properly. Failure to thin the crop results in greatly reduced timber yields.

Figure 15–14 One method of ensuring that a transplanted seedling gets as much moisture as possible is to plant it in the bottom of a scrape or gouged area in the soil surface.

Pruning

Removing unwanted branches to change the shape or growth pattern of a tree is called **pruning** (Figure 15–15). It is done to improve the quality of a tree for a specific purpose. For example, pruning the lower branches from the stem of a tree increases the quality of the lumber because fewer knots are evident in the wood. On the other hand, a Christmas tree will generate many more branches if the central leader of the tree is cut each year. The shape of the tree is modified by trimming the ends of the branches with a sharp shearing knife.

Trees that are managed under plantation conditions are pruned during the early stages of growth to control the growth patterns of individual trees. Removing all of the lower branches at the sapling stage of growth will often add significantly to the value of trees when they are mature (Figure 15–16). Pruning is a valuable cultural practice for trees that are raised on plantations, but this practice should be restricted to dominant trees of those species that are used for finishing work and cabinets. One such species is white pine. Pruning enables the tree to produce a knot-free shell of clear wood that is high in value.

Pruning is expensive, and it is seldom economical to prune to greater heights than 16'. The first pruning should not occur until the tree can be pruned to the height of the first 8' log. This pruning will result in a uniform shell of clear wood up to the length of a standard cut of lumber. Pruning should be restricted to the dominant trees that can be expected to remain in the stand after it is thinned for the last time.

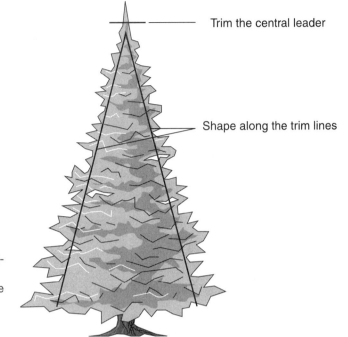

Trim the central leader

Shape along the trim lines

Figure 15–15 Pruning is necessary to produce a high-quality Christmas tree. It is done by swinging a sharp pruning machete or knife downward across the foliage of the tree and forming a conical shape. Extensive branching is encouraged by trimming the central leader of the tree.

Pruning Lumber Trees

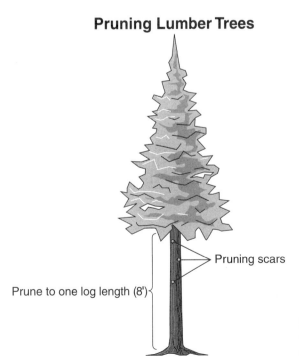

Pruning scars

Prune to one log length (8')

Figure 15–16 Removing the lower branches of trees during the sapling and pole stages of growth will result in higher-quality lumber when the trees are mature.

PROFILE ON FOREST SAFETY

Safe Tractor Operation

Tree farming involves the use of many of the same kinds of machinery that are used in other farming operations. For example, plows and disks are frequently used to prepare seedbeds for planting trees. Tractors provide the power to perform field operations in preparation for planting, and they are used during the transplanting process. They are used to pull the implements that provide mechanical weed control, and they are also used during harvest operations on many of the smaller privately owned forest plantations. Three aspects of tractor safety are especially important to plantation workers. The first is prevention of rollovers, the second is safe use of the power take-off capability, and the third is prevention of damage or injuries caused by pulling logs through standing timber.

Tractors are usually operated at modest speeds in the fields, so a rollover that occurs from side to side is rare. The more common rollover occurs when the front wheels come off the ground and the tractor tips over backward. This kind of rollover is usually due to an implement that is hitched too high to the tractor. For example, a plow or subsoil tillage implement that is hitched to the upper link of the three-point hitch will sometimes pull the tractor over backward if the implement encounters a large rock or heavy root. An alert operator can prevent the rollover from occurring by immediately cutting the power or pushing on the clutch pedal to disengage the power.

The power take-off (PTO) is a source of power to a variety of implements. A driveshaft is connected to a rotating shaft at the rear of the tractor, and power is transferred from the tractor to an implement such as a tiller or a rotary mower. The dangerous aspect of a PTO power source is in the tremendous turning power of the driveshaft. Loose clothing that becomes caught on the shaft will wrap around it, pulling a person into a very dangerous situation. If the power is not turned off in time, it is likely that the victim of the accident will be severely mangled or even killed.

A tractor that is used in the forest environment should be equipped with a safety cab and roll bar to prevent injuries to the driver due to falling trees. A log that is being pulled through standing timber will occasionally become caught on a tree, causing the tree to fall toward the tractor and its driver. Care should also be taken to avoid hitting standing timber, and to hitch the logs to the standard tractor hitch to prevent a roll over from occurring. Caution should also be used to avoid sharp turns that would cause the log chain to catch on the rear wheel of the tractor. This dangerous situation can pull the log right up on the rear wheel, endangering the safety of the person who is operating the tractor. The tractor should be stopped immediately if this happens. The driver should check implements or logs carefully as sharp turns are made.

CAREER OPTION

Silviculturist

A career as a silviculturist requires an understanding of the science and culture of trees. A person engaged in this career should have a college education plan that includes the study of natural sciences along with a strong component in business, forestry, and soil science. He or she will need to be knowledgeable in farming methods including the use of machinery for preparing seedbeds, planting, fertilizing, and harvesting. It will be important to study the latest research and to seek information about improved tree varieties. A thorough understanding of forest needs, from seedlings to mature trees, will contribute to success in this occupation.

LOOKING BACK

Plantation forests usually consist of monocultures that are established for a specific purpose. Plantations are established for the commercial production of seedlings, Christmas trees, biomass, wood pulp, and lumber. Some of these purposes overlap, and two or even more of these products are sometimes obtained as a crop matures. Other products, such as Christmas trees and seedlings, are fairly restrictive in their uses. Plantation populations of seedlings are usually established from seeds, but a few varieties reproduce from vegetative plant parts. Both containerized seedlings and bare-root stock require proper care at transplanting to ensure that viable populations are established. Some plantation crops benefit from pruning, shearing, or other silviculture practices, and plantation trees are usually subjected to much more intense management practices than wild trees. One of the most promising plantation tree crops is biomass. Biomass research is establishing this product as an important energy source for both the present and the future.

QUESTIONS FOR DISCUSSION AND REVIEW

Essay Questions

1. Explain why plantation forest plantings tend to be monocultures.
2. What steps are involved in the production of containerized seedlings?
3. How are cuttings used to produce young trees for transplanting?
4. What are two silviculture practices that are used in the production of Christmas trees to cause the branches and foliage of the tree to increase in density?
5. Explain how pruning the lower branches from the stem of a tree usually results in improved lumber quality.
6. Distinguish between containerized seedlings and bare-root stock.
7. How is a biomass planting under plantation conditions superior to biomass production in a natural forest environment?

Multiple-Choice Questions

1. A forest planting consisting of a single variety of tree is called a:
   A. plantation
   B. monoculture
   C. nursery
   D. silviculture
2. A young tree that is generated from seed in a container filled with soil is a:
   A. sapling
   B. cutting
   C. containerized seedling
   D. twig
3. A young tree that is generated by vegetative reproduction is called a:
   A. sapling
   B. cutting
   C. containerized seedling
   D. twig

4. The shape and density of the branches and foliage of a Christmas tree are improved by which of the following practices?
   A. shearing
   B. cutting
   C. flocking
   D. culturing
5. Lumber quality can be improved by eliminating knots from the main stem of a growing tree using a cultural practice called:
   A. cutting
   B. culturing
   C. shearing
   D. pruning
6. A young tree that has had its roots removed from the soil in preparation for planting is known as:
   A. a cutting
   B. bare-root stock
   C. a sapling
   D. a containerized seedling
7. A dense plantation planting of fast-growing trees for energy production is called:
   A. biomass
   B. prune production
   C. pulpwood
   D. jungle
8. Nursery stock consists of:
   A. young cattle or sheep that are grazed in nurseries to control weeds
   B. trees that are maintained in nurseries from which rootstock is obtained for grafting
   C. a product obtained from tree sap that is used as a soup base for human consumption
   D. all of the trees and other plant materials that are maintained for sale by nurseries

LEARNING ACTIVITIES

1. Obtain some nursery supplies and select some seeds for trees that are adapted to your area. Instruct the students on the proper planting procedures, and have each student plant a small tray of seeds. Have the students identify their trays, and make each student responsible for caring for his or her own plants as they sprout and grow. Keep the plantings damp and maintain them in a warm place as they germinate. Once they have emerged, place the plantings in a greenhouse or near a window to allow exposure to light. Allow the students to take their seedlings home, or find a planting location near the school when the trees are big enough to be transplanted outside.

2. Take a field trip or assign students to visit a local nursery. Make a list of the trees that are available in landscape and shade tree varieties. Ask the customer service representative to explain how these trees should be transplanted for best results.

NEW DIRECTIONS AND TECHNOLOGIES IN FORESTRY

Chapter

16

Terms to Know

urban forestry
arboriculture
hardiness
zone map
growth habit
tensiometer
integrated pest management (IPM)
Plant Health Care (PHC)

Urban Forestry

Objectives

After completing this chapter, you should be able to

* appraise career opportunities in the emerging field of urban forestry

* define the roles of trees in urban settings

* identify factors that should be considered in selecting trees for urban uses

* explain how a zone map should be used to guide tree selection

* name three basic functions of soil

* describe the relationship between soil characteristics and root development in trees

* evaluate the use of a tensiometer as a water management tool

* explain why it is important to prune trees

* describe a systematic approach to diagnosing problems in trees

* explain how cables and other hardware items are used to stabilize and repair damaged trees

* analyze the differences between the Plant Health Care (PHC) system for managing trees and traditional methods of management

Forestry is not always practiced in the woods. An emerging branch of forestry that occurs in or near population centers is called **urban forestry.** It is also known as **arboriculture,** which is the scientific care of shrubs and trees in cities and towns (Figure 16–1). Trees are becoming more and more important to people, and caring for them in large population centers presents a whole new set of problems and challenges. In comparison with forest management that is practiced in remote or isolated environments, urban forestry is never practiced in isolation. It is highly visible to the public.

LANDSCAPE AND SHADE TREES

Humans place value on trees for many reasons. Trees are valued for their beauty, and they are often included in landscape designs (Figure 16–2). They add value to homes and commercial properties by contributing to their visual attraction. They also provide shade that cools the environment on hot days. Still another function of trees is to provide protection from wind. Many homes and commercial properties are protected by trees that have been planted in windbreaks for this purpose (Figure 16–3). Urban forestry includes the care and management of trees that are used for all of these purposes.

Figure 16–1 Arboriculture or urban forestry is a relatively new branch of forestry that deals with care of trees in urban areas.

Figure 16–2 Humans place value on trees for the beauty that they lend to a landscape, and many urban areas have abundant populations of trees.

Figure 16–3 Trees offer an effective form of protection from wind when they are planted close together on the upwind side of the property they are intended to protect.

Most people enjoy outdoor environments, and city parks are designed to create outdoor environments in populated areas. Such environments reduce some of the stresses that afflict people in their fast-paced lives. Trees are key components of parks, and urban foresters are employed by many city park systems. They are responsible for the health and care of the trees and other woody plants that grow in the parks and on the streets of the city (Figure 16–4).

Some varieties of trees are more useful for urban purposes than others, and the trees that are used for these purposes in one area may not be adapted for use in other areas. Trees also differ in the amount of stress that they can tolerate. Trees that grow in urban areas are subjected to conditions such as hard-packed soil, paved surfaces, and environmental pollutants in the air and water supplies (Figure 16–5). Landscape and shade trees must be tolerant of these conditions if they are to survive in cities and towns.

Many of the trees that are used in landscapes and along the streets of urban communities are hybrid tree varieties that have been developed to exhibit wide ranges of tolerance for urban environmental conditions. For example, some trees can tolerate paved surfaces that cover much of the root zone of the tree, and others will die under such conditions. Some trees can grow near roads where soil has been compacted to form roadbeds, but many varieties of trees will be stunted by the inability of their root systems to penetrate the soil (Figure 16–6).

Figure 16–4 Large numbers of trees are planted around homes, along streets, and in city parks.

Figure 16–5 Trees in urban areas encounter many problems that are not found under wild conditions such as hard-packed soil, polluted water and air, and paved surfaces.

Figure 16–6 Some trees are tolerant of packed soils and paved surfaces, but many kinds of trees will become stunted or die when they encounter such conditions.

Figure 16–7 Trees should be planted only in those climatic zones where they are capable of survival. Some trees, such as these palm trees, are restricted to warm climates.

TREE SELECTION

An important principle in tree selection is to match tree varieties to environments that resemble those to which they have become adapted in nature. Some trees are able to exist only in very restricted environments (Figure 16–7). **Hardiness** is a measurement of the tolerance of a particular kind of tree or other plant to restrictive factors such as climate, altitude, temperature, and availability of moisture. Most tree nurseries rate the hardiness of the trees that they sell. They do this to guide the choices of the customer in the tree and plant varieties that they select. The tendency of a tree to die when it is subjected to different intensities and durations of cold temperatures is recorded on a **zone map.** The zone map illustrates the severity of climatic conditions that can be expected in the different areas that are represented (Figure 16–8).

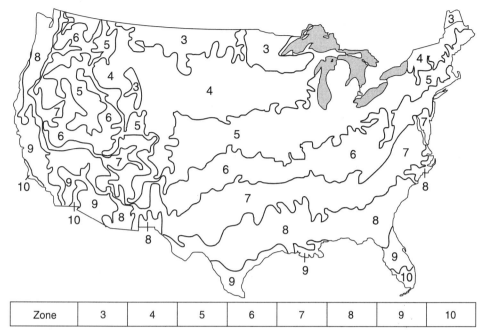

Zone	3	4	5	6	7	8	9	10

Figure 16–8 A hardiness map illustrates the climatic zones according to the severity of the climate. The hardiness of trees and other plants should be matched to the climatic zones to which they are best adapted. Low numbers indicate very restrictive, cold zones, while a "10" is a tropical zone.

A single factor in the environment that is restrictive to a tree will cause it to become unhealthy (Figure 16–9). An example of this is the stunted growth that occurs to wild trees that are growing at timberline. Even the most hardy trees do not grow above this altitude. Many varieties of trees are restricted in the same way at lower altitudes. All of the other restricting factors have similar effects on trees and other plants. The most restricting factor in the environment will determine the ability of a tree to survive (Figure 16–10).

Tree selection should favor trees that are adapted to the conditions found in the community and to the restrictions that are imposed by the urban environment. Among these restrictions is acid precipitation that is known to be caused by high levels of exhaust gases in the atmosphere. Other factors include restrictions such as saturated soils or compacted soils.

Trees should also be selected for their functional uses. For example, fast-growing trees should be included in a planting that is intended as a windbreak. These tree varieties will make the windbreak functional long before the late-maturing trees have much impact. Most of these trees will also be removed by the time late-maturing trees begin to dominate the planting.

Figure 16–9 Injuries or restrictive factors in the environment of a tree can cause stunting, loss of health, and even death.

Limiting Factors

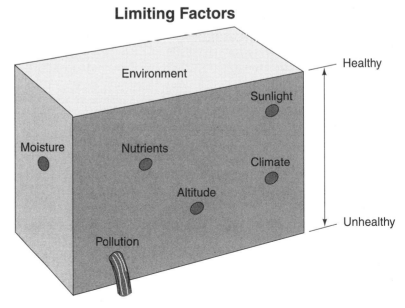

Figure 16–10 Just as a hole in a tank of water restricts the water level, the most restrictive factor in the environment determines the ability of a tree to survive.

The mature size of a tree, along with its **growth habit** or shape, should be considered when trees are selected. Trees should fit in the planting area and complement the size of the house, building, or other feature around which a landscape is designed. For example, the mature size of a tree should not exceed 30–40' in height for ranch-style homes, and 60–80' trees are usually considered to be the appropriate size for two-story homes (Figure 16–11). Many woody plants and trees have distinctive forms that allow landscape architects to choose plants that complement particular architectural styles. The trees listed in Figure 16–12 represent a sample of the tree varieties that are used in urban parks and landscapes.

Rooting patterns of trees should be considered when large trees are desired. Restrictive zones in the soil profile can be expected to modify the rooting patterns of deep-rooted trees (Figure 16–13). Large trees with shallow root systems are much more likely to be blown down by wind than large trees with deep, widely dispersed roots. Windfalls are dangerous to people, buildings, power lines, and other property. Shallow roots also cause other problems such as raising the soil around the base of the tree. Expanding roots can break concrete paving, and they frequently break and invade sewer pipes and underground utility lines (Figure 16–14). For these reasons, placement and rooting patterns are important factors to consider when large trees are desired.

Wood strength is important in trees that are planted in urban areas. Trees in parks and yards sometimes sustain damage from people climbing in them. Serious damage also occurs in trees that lack wood strength from heavy winds or

Figure 16–11 The mature size of a tree should be considered when it is planted in a home landscape. The height of the house should determine the mature size of the tree.

Common Shade and Landscape Trees

Tree	Zone	Height	Tree	Zone	Height
Hybrid Willow	3–8	35–40'	Hybrid Elm	3–10	55–60'
Weeping Willow	2–9	40–50'	Seedless Ash	3–9	40–50'
Crimson Maple	4–8	40–50'	Mountain Ash	3–7	20–25'
Norway Maple	3–7	45–50'	Purple Ash	3–9	40–60'
Harlequin Maple	4–8	40–50'	Russian Olive	2–8	25–30'
Sugar Maple	3–8	60–75'	Locust	3–8	40–50'
Scarlet Maple	3–9	50–60'	Pin Oak	4–8	60–70'
White Birch	2–7	40–50'	Red Oak	4–9	40–60'
Red-leaf W. Birch	3–7	30–35'	Hackberry	2–9	50–60'
Bolleana Poplar	3–8	30–45'	American Sycamore	4–8	60–80'
Lombardy Poplar	3–9	45–50'	Redmond Linden	2–8	40–60'
Hybrid Poplar	3–8	45–50'	Quaking Aspen	1–8	30–40'
Cottonwood	2–9	65–75'	Dawn Redwood	4–8	60–80'

Figure 16–12 Some hybrid and native tree varieties are adapted for use in urban environments.

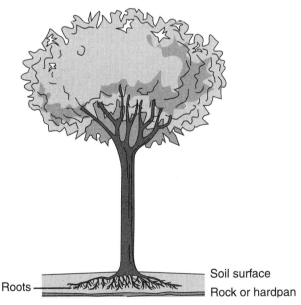

Roots —
Soil surface
Rock or hardpan

Figure 16–13 A restrictive layer of rock or hardpan beneath the soil surface restricts the downward growth of tree roots. This weakens the tree, making it susceptible to wind damage.

Figure 16–14 The placement of a tree in relation to sewer lines, sidewalks, and foundations should be considered at planting time to avoid damage to these structures by the roots of the tree when it matures.

from accumulations of ice on the branches. Slow-growing hardwood trees tend to be stronger than fast-growing species of trees.

The need for water should be considered in selecting trees. Some trees use a large volume of water while others can survive with very little water. The needs of most trees fall somewhere between these extremes. The availability of supplemental water is important to the success of a planting. The need for supplemental water or other kinds of care may affect the amount of labor that is required to maintain healthy trees located in city streets, parks, golf courses, and yards (Figure 16–15). Many trees are able to obtain sufficient water from the soil once their root systems develop, but they may require a lot of individual care until they are well established.

Other features about a particular kind of tree should also be considered before it is used in an urban setting. The density of its leaves may cast such heavy shade that it kills the lawn grass. Some trees are quite messy. Sap may fall freely from some trees during the spring when the leaves are forming. Others have messy or foul-smelling fruits that fall in yards and on streets. Still others lose their leaves over a long period, and they create a lot of labor to keep the area clean and tidy.

Figure 16–15 Trees are among the most prominent features on a golf course. Some golfers consider a golf course with young trees to be inferior to a similar course that has trees in a more advanced stage of maturity.

PLANTING SITES

Forest soils were discussed in detail in Chapter 5, but the planting sites that were described in earlier chapters of this textbook apply to seedbed preparation for plantation or forest plantings of trees. This chapter will discuss planting sites for trees such as city parks, paved areas, yards, planters, and interior landscapes (Figure 16–16). These planting sites are common in most urban areas, and they require different management practices than those that are found in traditional forest plantings.

The principles of soil science that were discussed in Chapter 5 are in effect in urban forestry, but they are complicated by the tendency that people have to disturb the soil during the construction of cities and towns. Soil is packed to prevent roads and foundations from cracking. The yards of most houses located in subdivisions no longer have a normal soil profile. Instead, the soil profile often consists of buried concrete, bricks, trash, and gravel mixed with the original soil. In most cases, a shallow layer of topsoil is placed on the surface to allow for planting lawns and landscaping yards. Such soil problems sometimes interfere with the establishment of healthy trees.

Figure 16–16 It is not uncommon to see a tree growing in a large container in an area where most or all of the surface is paved.

Three basic functions are performed by soil (Figure 16–17). The first function is to provide support for the plant through the network of roots that spreads in the soil. Soil provides an anchor that keeps the tree upright. The second function of soil is to act as a storage reservoir for air and water close to the roots of plants. The third function of soil is to provide minerals and organic nutrients to the plant. Water is attracted to soil particles, and it adheres within pores in the soil profile even after excess water has drained away. Minerals and nutrients become dissolved in water, and water becomes a carrier for these materials as they move into plants.

Most trees require soil that is well drained and that has no restrictive layers to interfere with the growth of roots. These conditions are important whether the

Basic Functions of Soil

❊ Support for the tree
❊ Storage reservoir for water and air
❊ Sources of minerals and organic nutrients

Figure 16–17 Soil performs three basic functions upon which a tree depends for survival.

FORESTRY
PROFILE

Arbor Day

Every year as the spring season arrives, the individual states in the United States celebrate a holiday to call attention to the importance of trees. Many communities, schools, and individuals celebrate Arbor Day by planting trees in their neighborhoods and yards (Figure 16–18). The date of this holiday changes from year to year in most states. The idea to celebrate Arbor Day was established in Nebraska in 1872 by Julius Sterling Morton.

Figure 16–18 Arbor Day is celebrated on behalf of trees, and citizens are encouraged to plant a tree on that day. Was it Arbor Day when the tree in this photograph was planted by President Harrison?

tree is planted in a park or yard, near a city street, in a container, or in an interior landscape. Containers must be large enough to accommodate the roots of the tree. The walls of a container function much like a hard restrictive barrier in the soil. Care must be taken to select a tree that has a small enough root pattern to be adaptable to a container. Containers should be selected that are large enough to minimize restrictions on the tree.

A coarse layer of soil that is located beneath a fine-textured soil often acts as a root barrier because water will not move into the coarse layer from the fine-textured layer until the upper layer becomes saturated. Roots tend to remain in the soil zone where moisture is plentiful, and they do not readily penetrate the coarse soil layer. When it is apparent that this condition exists, it is usually wise to thoroughly mix the two layers of soil using a trencher or other mechanical device.

CULTURAL PRACTICES

Cultural practices consist of management techniques that are used to provide care for plants and animals. Cultural practices that are used in urban forestry or arboriculture include those that enhance or improve the growth and development of woody plants and trees. Among these practices are water management, pruning, diagnosing and repairing problems, and providing preventive maintenance to plants.

Water Management

Water management requires an awareness of soil conditions and an understanding of the needs of plants. Experienced managers often learn to estimate soil moisture conditions by feeling the soil. When sufficient moisture is available in the soil to form a ball when it is rolled between the hands, yet crumble when it is rubbed, the moisture level is usually ideal to support plant life (Figure 16–19). This rule does not hold true with sandy soils, however, because they crumble even when they are saturated with water.

A **tensiometer** is an instrument that measures the amount of moisture in soil. These instruments are most useful in monitoring the availability of moisture in soil. The ceramic tip of the tensiometer is placed at an appropriate depth in the root zone. As the soil at that depth dries out, water is lost from the instrument, and vacuum pressure is created. The dial of the tensiometer provides a reading that indicates when irrigation is needed.

Enough water should be applied at each irrigation to replace the water that has been lost. Deep penetration of irrigation water will encourage deep rooting of trees. Shallow rooting is sometimes caused by applying only light irrigations. Irrigation systems should be designed to deliver adequate amounts of water without the soil becoming waterlogged. Drip irrigation systems can be adjusted using one or more tensiometers to determine the correct flow settings to achieve deep irrigation.

Figure 16–19 Young trees require an adequate supply of soil moisture if they are to survive transplanting. The correct amount of moisture is usually present when you can form the soil into a ball and it tends to crumble in your hand.

Pruning

Pruning is the practice of removing branches from a tree (Figure 16–20). This should be done in late winter for most species, but dead or hazardous branches

Figure 16–20 Most trees in landscape settings require pruning to remove unwanted branches and to develop a desirable shape when the tree is mature.

can be removed at any time. The health of the tree can be protected by removing weak or diseased branches as they are noticed. Some species such as birch and maple trees should be pruned in the summer because they tend to bleed or lose sap when they are pruned in late winter. Transplanted trees and saplings should not be pruned at all, because they need all of their leaves to produce food for the tree while the roots become established.

Pruning should always be done with a purpose in mind. For example, a branch that is infected with a disease should be removed from the tree and burned. In many cases, such trees can be saved if infected branches are removed before the infection moves throughout the entire tree. Any time that disease is thought to be the cause of an unhealthy branch, all of the pruning tools should be disinfected with a mixture of tincture of iodine and rubbing alcohol to prevent the spread of the infection to other trees.

Pruning during the first four to six years of the life of a tree is an important practice that determines the strength and shape of the tree when it is mature. Removal of weak or poorly placed branches is important because the strength of the tree is affected by the placement of its main branches. A strong branch should be identified on each side of the tree that has a wide angle at its point of attachment to the trunk (Figure 16–21). These should be retained, and competing branches should be removed. Branches that compete with the main leader for dominance also should be removed, along with branches that are rubbing on

Figure 16–21 A tree that has been properly pruned during the sapling stage should have strong lower branches with wide angles of attachment where the branches join the trunk of the tree.

other branches. It is important to remember that trees should not be topped because it will destroy the form of the tree.

One purpose of pruning is to thin the branches of the tree to allow light to penetrate the interior of the canopy. Thinning also decreases wind resistance in the tree and makes it less vulnerable to wind damage. As the tree grows, lateral branches should be spaced to give the tree balance. Spacing of the laterals also improves the strength of the branches. The growth of a tree occurs only at the tips of its branches, and a branch will be at the same height when the tree is old that it was when the tree was young. If a branch seems too low for the purpose for which the tree was planted, it should eventually be removed.

The pruning cut should be made without leaving a branch stub, but it must not be made too deep. The branch collar is a thickened ridge that surrounds the branch where it attaches to the trunk. The collar should be left intact because it forms protective tissue to cover the wound that is created by the cut. In time, it will seal the cut to prevent the entry of insects or disease organisms. A special effort should be made when pruning large branches to avoid tearing the bark. This can be done by making three cuts on the branch as illustrated (Figure 16–22).

Diagnosis of Tree Problems

The problems that affect trees are of the two basic types discussed in Chapter 6, biotic and abiotic. Biotic agents are living organisms, and abiotic agents are nonliving causes of plant problems. Abiotic agents account for the majority of

Pruning a Large Branch

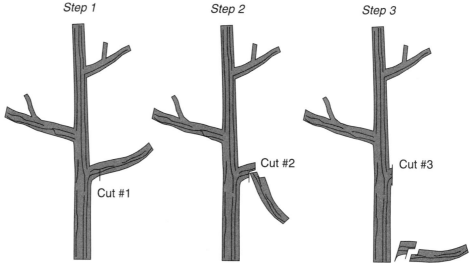

Figure 16–22 Three cuts, made in the proper sequence, will remove a large branch without causing the bark of the tree to tear beneath the last cut.

problems in trees in urban locations. Abiotic problems often have symptoms that are easily confused with biotic problems. A diagnosis should not be attempted on the basis of a single symptom.

A systematic approach to the diagnosis of tree problems includes a thorough study of the condition of the site, the site history, and the tree itself. Most urban problems are not the result of insect damage or disease, so it is important to look for the real cause of the problem. The following questions should help uncover the nature of the tree problem:

* ❋ What species of trees are affected? Some insects and diseases affect only specific kinds of trees.

* ❋ What damage patterns exist? If all species are affected equally, consider pollution or weather damage as possible causes. If all of the affected trees are of the same species, a biotic agent may be to blame.

* ❋ What site conditions appear likely to contribute to the problem? Perhaps a restrictive soil layer exists, or the trees are not getting enough water.

* ❋ What changes have occurred at the site? Consider everything that has been done at the site, including chemical applications, excavations, new installations, and so on.

* ❋ What visible changes have occurred in the foliage of the tree? Changes in leaf color, visible leaf damage, or damage to the foliage on certain parts of the tree are indicators that can be used to discover the primary cause of the problem.

* ❋ What symptoms are visible on the branches and trunk of the tree? Look for visible signs of insects, lawnmower damage, wire-girdling action, or other unusual damage.

* ❋ Are there any signs of root damage? New construction may have severed some of the root system, or the soil may be waterlogged.

Diagnosis of tree problems requires careful observation and analysis. By seeking answers to these questions, it is likely that the causes of many tree problems can be understood. Through the process of elimination, it is often possible to discover the cause of a problem.

Tree Maintenance and Repair

Maintaining a tree can extend its life and usefulness, especially when it is discovered to be structurally weak or when it has been damaged. By using cables and braces to support the weight of weakened branches, a damaged tree can be stabilized. This usually involves drilling holes through the damaged branch and the trunk of the tree, and fastening the cable snugly between eyebolts that have been inserted into the holes. Large washers are usually used to keep the bolts from pulling through the live wood. Several weak branches may be cabled together to reduce their range of motion. Triangular cabling has the advantage of adding direct support and minimizing twisting of the branches (Figure 16–23).

Triangular Cabling

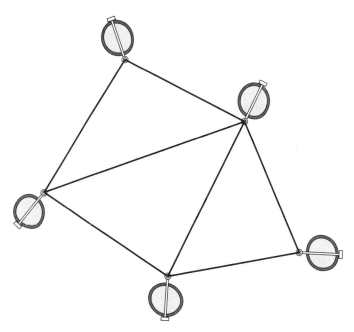

Figure 16–23 Several weak branches may be stabilized by placing eyebolts through each of the branches and connecting them with cables. Triangular cable placement is much more secure than a single cable placed around the perimeter.

Lag screws have been demonstrated to be as effective as bolts in hardwood trees that are free from decay.

Nearly any kind of hardware that can be used for heavy duty wood construction can be used to repair or maintain damaged or weakened trees. Holes should be drilled that are the same size as the bolts that are to be used. Holes for lag screws should be $\frac{1}{16}$" smaller than the diameter of the lag screws, and the lag screws should be turned all the way into the branch at the time the hole is drilled. Failure to do this may allow swelling to bind the lag screw in place, preventing it from turning later.

Weak or split crotches, where two main leaders separate, can be strengthened by installing a threaded rod completely through the trunk of the tree just below the split (Figure 16–24). Large washers should be placed on either end of the rod, and the nuts should be tightened until the damaged split is securely pulled together. This usually should be supplemented with two more rods that pass through the two limbs above the damaged crotch for added support. When more than one bolt is used in the same limb, care must be taken to ensure that they do not pass through the same part of the wood grain.

Braces are sometimes needed to hold two limbs apart. A small length of pipe with a washer at each end can be inserted between the two branches and held

Repair of Damage to a Tree

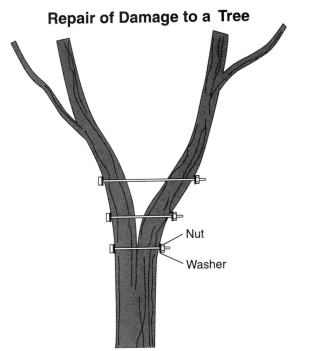

Figure 16–24 A tree that splits at the fork of two main branches can be repaired by drilling holes at strategic points and holding the branches in place with bolts.

Installation of a Spacer

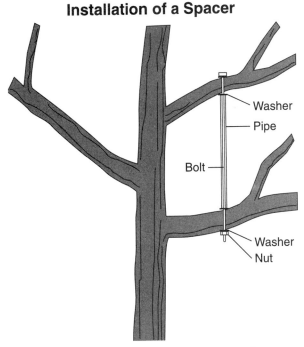

Figure 16–25 Branches that tend to sag together may be held apart with a spacer made from a pipe through which a bolt is passed.

there by a bolt passing through the spacer and both branches (Figure 16–25). Sometimes two branches are even bolted securely together to prevent rubbing.

Trees that have weak or nonexistent upper branches may require props under their existing branches to take pressure off the root systems. Props may be made of wood or metal, but care must be taken to avoid girdling of the branch at the point where the prop touches the branch. It is usually best to stabilize the prop with a bolt that passes through the branch rather than attempting to hold the prop in place with a device that goes around part or all of the branch. The main value of a prop is to take weight and pressure off the root system or off a damaged or weakened tree crotch where two branches meet.

One of the most serious injuries that can occur to a tree is to have its bark girdled around the base of the tree. In such cases, bark or branches can be grafted to the area to restore the flow of nutrients between the branches and the roots of the tree (Figure 16–26). A repair of this nature must be made soon after the injury is sustained, and the cambium layers of the tree and the bark or stem implant must be sealed securely together with nails or binding material while healing takes place. Valuable trees can sometimes be saved when this procedure is used.

Decay is a deadly problem in some trees, but trees have their own way of dealing with it. They tend to form compartments within the tree to isolate the

Bark Graft

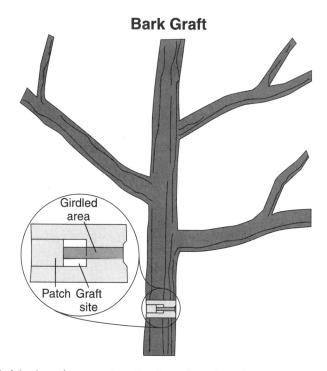

Figure 16–26 A bark graft can restore the flow of nutrients between the leaves and roots when the repair is made soon after the girdling of the tree occurs.

decay. One cultural practice that has been used to deal with decay has been to clean out the decayed material and fill the cavity with concrete. A more recent development has been to fill the cavity with urethane foam. It is generally believed that filling the cavity has little effect on the life or strength of the tree, but it does tend to improve the appearance of the tree.

INTEGRATED PEST MANAGEMENT

Trees that are located in urban sites such as parks, yards, and city streets are subject to the same insects, diseases, parasites, and weeds that afflict trees in plantations and forests. A multipronged attack on harmful insects is the most practical approach to insect control (Figure 16–27). **Integrated pest management (IPM)** is a concept for controlling harmful insects or other pests while providing some protection for organisms that are useful. It involves the use of some chemical pesticides, but also relies on natural insect enemies and other biological control strategies to control harmful insects. When insecticides are the only source of control, they kill both harmful and useful insects. Integrated pest management is an insect control program that does not attempt to kill all of the harmful insects, since insect control of this kind also kills the natural enemies.

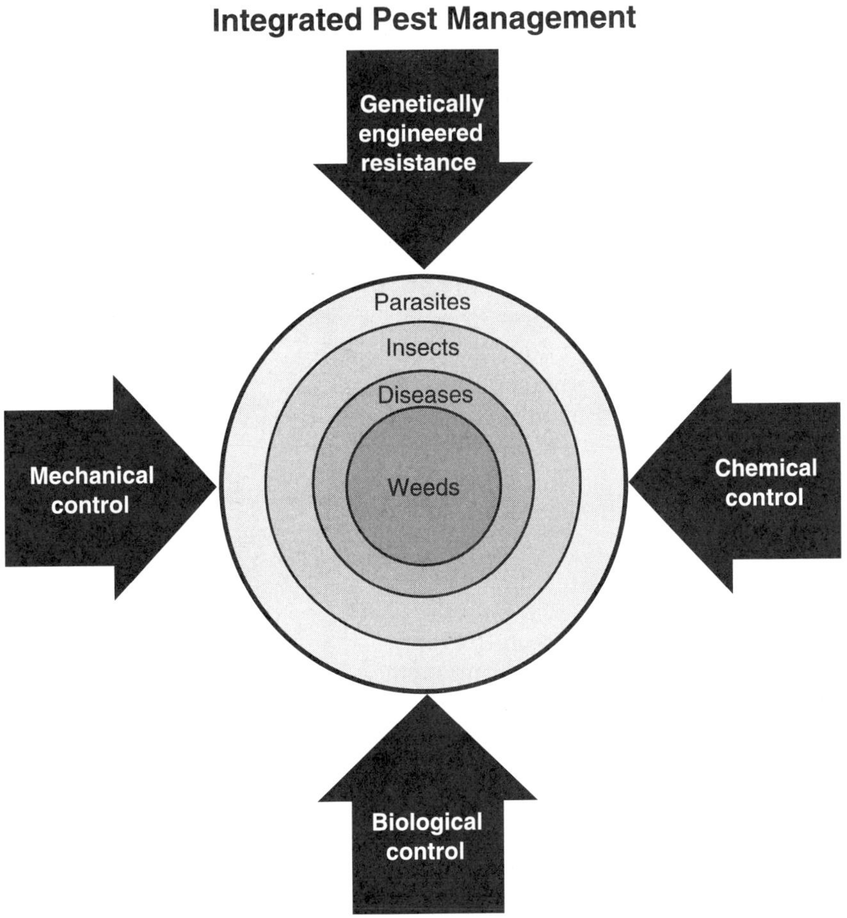

Integrated Pest Management

Figure 16–27 The most effective form of pest management uses a variety of pest control methods. It is the most acceptable form of pest control because it does minimal damage to species for which the control method is not intended.

For natural insect enemies to survive, they must have a small population of harmful insects upon which they can prey.

Integrated pest management is not a new idea. It was widely used prior to the introduction of modern insecticides, and it has emerged in recent years as the best alternative to complete reliance on pesticides. Integrated pest management is an ecosystem approach to controlling insect problems. It takes into account the effects that a particular form of insect control might have on the other living things that are found in the ecosystem.

Two major factors favor the adoption of IPM in urban forestry. They are the resistance of urban residents to the use of chemicals near their homes, and the ability of insects and other pests to develop genetic tolerances for chemicals. An

PROFILE ON FOREST SAFETY

Ladder Safety

One of the most important pieces of equipment to urban forestry is the ladder. Despite the fact that many cities have trucks that are equipped to place workers high above the ground, ladders are still used often in the routine care of trees. Improper use of ladders is a leading cause of injuries to workers who provide tree care. An injury sometimes occurs because the ladder has not been repaired when a repair was needed. Another common problem that occurs when ladders are used is the malfunction of extension ladders, allowing the upper section to slide down over the lower section while someone is on it. This type of accident frequently causes broken bones in the feet and hands of the victim.

Workers should remember that the top three rungs of a ladder are intended to be places to hold on to. They are not intended to be places to stand. A person who stands on the top rungs of a ladder does not have a good place to hold on, and most falls from ladders occur when the user has climbed too high. Falls also occur when ladders have not been placed in stable positions prior to using them. A ladder needs to have solid footing beneath it, and it should be placed against a tree at a wide enough angle to keep it from tipping backward.

Rules for safe use of ladders should include the following:

1. Keep the ladder in good repair.
2. Always place the legs of the ladder on solid footing.
3. Place the ladder at a safe angle to the tree or other object.
4. Have a partner hold the ladder to stabilize it when necessary.
5. Never climb so high on the ladder that you have no rungs to hold.
6. Reduce the size of the container or load that is carried as the height increases.
7. Avoid walking beneath a ladder.
8. Take the ladder down when it is no longer being used.

effective alternative to insect control other than complete reliance on chemical control is needed.

A new approach to maintaining healthy trees that is promoted by the International Society of Arboriculture is **Plant Health Care (PHC)** (Figure 16–28). This approach incorporates IPM, but it focuses on improving the health of

---

Plant Health Care (PHC)

❋ Select proper tree for the site
❋ Tolerate some pests to encourage natural enemies
❋ Monitor the tree regularly for pests and disease
❋ Educate property owners about pest tolerance levels
❋ Apply appropriate cultural practices as needed

Figure 16–28 The Plant Health Care (PHC) approach to tree management focuses on improving the health of the tree instead of on the destruction of harmful pests.

the tree instead of on the destruction of harmful pests. Prevention of health problems is the key to this management system. The key elements of PHC are: (1) preventing problems by selecting a tree that fits the site, and using proper practices during planting, pruning, and caring for the tree; (2) tolerating some pests to encourage populations of their natural enemies and to reduce the need for pesticides; (3) monitoring the health of the tree on a regular basis for diseases and pest damage, and recommending treatments as necessary; (4) educating property owners to determine reasonable pest tolerance levels; and (5) treating problems in an intelligent manner by applying appropriate cultural practices as they are needed.

Another useful technology for controlling harmful insects is becoming available. It is genetic insect resistance. Some plants are naturally resistant to insects. They may give off an odor that insects avoid, or they may contain natural insecticides in their plant juices. Through the use of genetic engineering techniques,

---

CAREER OPTION

Urban Forester

An urban forester is usually employed in a metropolitan area to care for trees and shrubs in urban landscapes. Many cities and towns now employ urban foresters to ensure that their investments in trees and woody plants receive proper care. Urban foresters apply principles of biology in the management of plants and wildlife. They plan and carry out tree plantings as site barriers and for noise abatement. They manage the trees on streets and in city parks, and they develop landscapes and plantings for environmental purposes. University degrees in forestry, horticulture, or the biological sciences provide the education that is required for this emerging profession.

it is now possible to transfer the genes responsible for producing natural insecticides to plants that have no insect resistant traits. The advantage of this technology is that only those insects attempting to eat the plant are killed. Pollinating insects and natural insect enemies are not subjected to insecticides, so they survive to aid in controlling the damaging insect species.

LOOKING BACK

Urban forestry, also known as arboriculture, is the scientific care of trees and woody plants in cities and towns. Urban foresters provide the care for these trees. Trees are valued for their beauty, and they are frequently included in landscapes. They are also valued for shade and for windbreaks. Trees often require individual care in urban sites due to soil and pollution problems and injuries caused by people. Trees should be selected that are adapted to the sites where they will be used. Water management is an important part of arboriculture, especially with young trees that have been transplanted. The diagnosis of tree problems and the methods for solving them are part of urban forestry. Integrated pest management (IPM) is an attempt to use a variety of methods to control pests while reducing the amount of chemicals that are used. Plant Health Care (PHC) is a system of caring for trees that focuses on the health of the plant instead of on the elimination of all pests.

QUESTIONS FOR DISCUSSION AND REVIEW

Essay Questions

1. What career opportunities exist in urban forestry, and what training is required?
2. What roles do trees play in cities and towns?
3. List some factors that should be considered in selecting a tree for planting in a town or city.
4. How is a zone map used in the selection of trees?
5. What are the three basic functions of soil?
6. How does the depth of the soil profile affect the root development of trees?
7. What function does a tensiometer perform that makes it useful as a water management tool?
8. What useful purposes are accomplished by pruning trees?
9. List several important considerations that should be made in diagnosing a problem with a tree.
10. Describe some methods for repairing some of the more common types of damage that occur to trees.
11. How is the PHC system for managing the health of trees superior in some ways to traditional methods of management?

Multiple-Choice Questions

1. A career that involves caring for trees in cities and towns is:
   A. extension forester
   B. regional forester
   C. urban forester
   D. district forester
2. Which of the following tree problems is most associated with urban environments?
   A. compacted soil
   B. disease infestations
   C. insect damage
   D. weather damage
3. Which term is directly related to the hardiness of a tree?
   A. form
   B. zone map
   C. shape
   D. growth pattern
4. The most important factor to tree survival following transplanting is:
   A. soil fertility
   B. pruning
   C. timely water applications
   D. favorable temperature
5. A tensiometer is an instrument that is used to:
   A. calculate the strength of a tree
   B. measure the purity of irrigation water by testing the surface tension of water
   C. estimate the flexibility of tree limbs
   D. measure the water content of soil
6. Most problems with urban trees are the result of:
   A. abiotic agents
   B. insects
   C. biotic agents
   D. disease
7. The practice of cabling a tree is performed for what purpose?
   A. removing a dangerous tree that is damaged or diseased
   B. stabilizing branches of the tree that have been damaged
   C. repairing a tree crotch between two leaders
   D. controlling the direction the tree falls when it is cut down
8. A concept for controlling harmful insects or other pests while providing some protection for organisms that are useful is called:
   A. PHC
   B. GPS
   C. IPM
   D. GIS

9. A management system for trees that focuses on the health of the trees instead of on pests and diseases is called:
   A. PHC
   B. GPS
   C. IPM
   D. GIS

10. An advantage of controlling insects by engineering new tree varieties with genetic insect resistance is that:
    A. the wood of engineered trees is stronger than other wood
    B. this is a mechanical method of insect control
    C. all of the insects become susceptible to the same insecticides
    D. the only insects that are affected are those that attempt to feed on the tree

LEARNING ACTIVITIES

1. Develop an arboretum at or near the school consisting of the varieties of trees and woody shrubs that are adapted to your local area. Make the arboretum available to community groups for educational purposes. This is a long-term school project that will require special care and management.

2. Take a field trip to a public park, and record the different varieties of trees and shrubs that are found there. Observe the trees carefully to detect any problems that may be present, and consider ways to improve the health of the trees and shrubs that are observed. Have the class members write field reports that list the woody plants and trees along with observations about the health and condition of the trees.

Chapter

17

Terms to Know

computer
hardware
program
software
database
model
network
modem
Internet
home page
electronic mail (e-mail)
forest inventory
forest growth model
earth resource satellite
topographic map
stereoplotter
geographic information system (GIS)
global positioning system (GPS)

Computers and Space-age Forest Technologies

Objectives

After completing this chapter, you should be able to

* identify the key components of a computer

* distinguish between hardware and software

* describe the functions of a computer model

* explain what a computer network is and how it can be used in the forest industry

* distinguish between a computer network and the Internet system

* describe ways that computers are used in timber processing

* identify ways that computers are used for financial management purposes

* list some ways that computers are used to gather and organize forest data to create forest inventories

* speculate on some ways that computers may be used in the future to manage forest resources

* identify emerging military and space technologies that are being adapted to forest uses

* explain how new technologies may be used to increase forest productivity and reduce environmental problems in forests

A**computer** is a machine that performs mathematical calculations in a pro-grammed sequence. It is capable of recording and integrating large amounts of information. The forest industry has entered the computer age, and many of the clerical and record-keeping functions that were formerly done by people are now performed by computers. It is important to remember, however, that a computer does not create new information—it only interprets the information that people have entered into the system. A conclusion or interpretation that is provided by a computer is only as accurate as the data on which it is based.

UNDERSTANDING COMPUTERS

Computer **hardware** consists of the mechanical and electronic structures, devices and components that are used to make a computer. This includes the monitor, keyboard, disk drives, modem, printer, and any other similar compo-nents found in a computer system (Figure 17–1). There are many different kinds of computer hardware. Computer hardware has been adapted for use in the field by making the components small and portable enough to be used by timber cruisers and forest managers.

A computer **program** consists of the organized commands and routines that are used by computers to perform calculations. A portable disk, tape, or

Figure 17–1 Computer hardware consists of the monitor, keyboard, disk drives, modem, printer, and other similar devices.

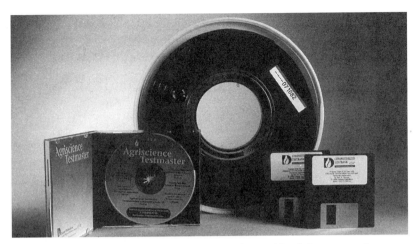

Figure 17–2 Computer software consists of portable disks, magnetic tapes, and compact discs upon which computer programs or original data are stored.

compact disc on which a computer program or original data is stored is called **software** (Figure 17–2). Information that is stored in a computer system is called a **database.** Specialized software has been developed by the forest industry that makes it possible for a computer to perform forestry applications. A computer **model** consists of advanced programming available as a software package that is used to simulate or imitate an actual system. For example, a forest management model should be able to imitate the responses of the entire forest to changes in forest conditions.

Computers have become important office machines in every aspect of business. One of the important functions of computers in the forest industry is for word processing purposes. This includes the preparation of letters, reports, documents, technical manuals, and any other purpose that requires written materials. Computers make it possible for owners of private forests as well as managers of public lands to prepare correspondence and other paperwork in a professional manner.

Among the most valuable uses of computers is the ability to transfer information from one system to another. This is done by linking computers together, allowing them to share information. A computer **network** is a delivery system that links two or more computers together and allows them to have access to the same information (Figure 17–3). A network can be developed within a business or processing operation using electrical cable connections between computers.

Computer networks can also be set up using a device called a **modem** to connect the computer to a telephone line. A modem is a device that allows computer data to be transmitted over telephone lines. The data is changed by the modem to a format that is compatible with the telephone data system. A modem at the other end of the phone connection transforms the data back to its original form.

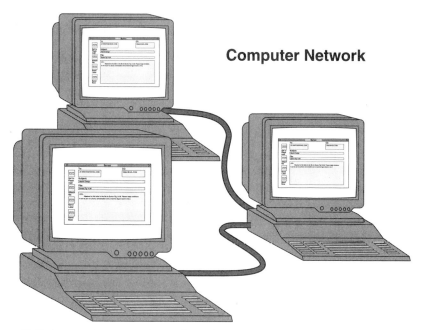

Computer Network

Figure 17–3 A computer network consists of two or more computers linked together to allow them to have access to the same information. They may be connected by wires, cables, telephone lines, or other methods.

The **Internet** is a computer information network that has emerged as the leading computer network in the world. Many smaller networks can be accessed using the Internet system, and much of the important information in the world is available to anyone who cares to view it. The libraries and data-banks of the world are available using an Internet connection. Any person or business that uses the Internet can market its own products on the system by creating a **home page** on the Internet system. The home page is a menu of the information and databases that an individual or organization has made available to others who use the Internet system.

Those who use the Internet or other computer networks can communicate with one another using **electronic mail,** also known as **e-mail** (Figure 17–4). This communication system uses the network connections to transfer messages to a database that can be accessed by the intended user.

MACHINERY OPERATION

The forest industry is highly mechanized, and some of the machines used by the industry are operated by computers. Computer use is evident in a modern sawmill where logs are processed into lumber. Through the use of modern technology, each log is measured. The computer generates a profile of the log and determines how the log will be positioned as it is cut (Figure 17–5). The

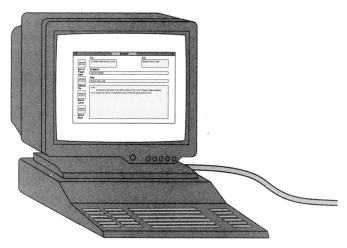

Figure 17–4 Electronic mail is used to send and receive messages to and from persons who are served by the same or a connected computer network.

computer even determines which cuts should be made to reduce the amount of waste and to obtain the highest value from a log.

Two important roles are served by computer linkages to machines. One role is to control the operation of machines through electronic signals. In this instance, a computer is programmed to react to a particular measurement or visual stimulus and select an appropriate procedure for processing a raw product.

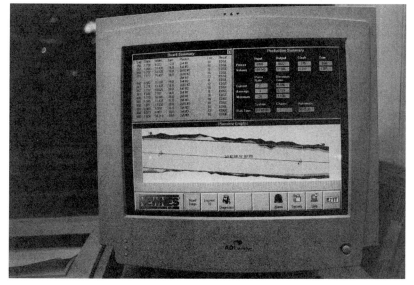

Figure 17–5 Computers are used to select the most efficient milling option for each log. *(Photo courtesy of Potlatch Corporation)*

The second role of a computer linked to a machine is to record data, such as the output of a machine. This kind of record can be used to create a product inventory or to measure the efficiency of a manufacturing process.

Computers are used to control and monitor many of the complex procedures that occur as wood is processed and manufactured into useful forest products. Some processing operations, such as the operation of the drying drums in a pulp mill, occur at such high speeds that computers must be used to control the operation. Lasers and electronic eyes linked to computers are used in the manufacture of paper products to monitor visual activities that occur more rapidly than the human eye can focus. Control rooms with computers and monitors are found throughout modern processing and manufacturing plants where wood is processed and used (Figure 17–6). Many of these computers are linked to the control systems of the machines that do the work.

MACHINERY MANAGEMENT

Machines require routine preventive maintenance to keep them in proper operating condition. When large numbers of machines are in use on a daily basis, it becomes very difficult to keep track of maintenance schedules. Maintenance management becomes even more complex when multiple mechanical systems with different maintenance requirements are combined

Figure 17–6 Modern sawmills are equipped with computer control rooms complete with computers, monitors, lasers, and electronic eyes that measure and monitor events that occur more quickly than the human eye can focus. *(Photo courtesy of Potlatch Corporation)*

on a single machine. Computers are very useful tools in establishing routine procedures such as machinery maintenance (Figure 17-7). Once the maintenance requirements of a machine have been entered into a computer system, maintenance lists can be generated on a daily or weekly basis to ensure that machines are properly serviced.

Many modern tractors and other power units are manufactured with computer components designed into the control panels for the purpose of recording the number of hours the machine has been used. Some of these systems are designed to alert the operator when it is time for maintenance procedures to be performed.

Machinery purchases make up one of the largest expense items in a silviculture budget. New machines are very expensive, and the managers who use heavy machinery and equipment can now use computer programs to help make decisions related to purchasing or leasing. Computer software that is developed for this purpose is capable of making cost-benefit assessments of specific machines in comparison with other machines that are available. Forest industry managers can use software packages to compare machinery options such as purchasing versus leasing of machinery, and new versus used machinery purchases, and to make cost-benefit comparisons between specific machines.

RECORD MANAGEMENT

Record management is a system for staying in control of the many different kinds of documents and papers, such as receipts, vouchers, invoices, and order forms, that are assembled by managers of both privately and publicly owned forests and forest industries. A high level of accountability is required of everyone who

Figure 17-7 A computer is valuable as a tool to track routine maintenance schedules for machinery and equipment.

manages natural resources. This makes it necessary for many different kinds of records to be maintained. Computers help to generate and organize those records that are needed.

Forest management requires a record system that addresses the different aspects of forestry. Records are needed for such purposes as forest inventories, harvest yields, business inventories, product sales, business income and expenses, accounts payable, accounts receivable, cash flows, cash balance sheets, and tax purposes. All of these records can be assembled using computers. Some very good computer software packages are available for these purposes.

FINANCIAL MANAGEMENT

Computer applications for financial management purposes have become widely used in every aspect of business. The forest industry has followed the trend to computerize financial and business records. Many excellent business management programs are being used in businesses that deal in forest products.

Businesses in the timber industry require large amounts of operating and investment capital. Many of these businesses borrow funds from banks and other lending institutions to operate their businesses and to buy the equipment and machinery they need. In addition to net worth statements and balance sheets, most banks require cash flow records that show when the business experiences its greatest demands for cash. Such records also help the lender to determine when income is available to repay loans.

Managers also need complete records to determine which business enterprises are profitable and which are not. For example, it may be advisable for a logging company to sell its trucks and contract with another business to haul timber. This would allow the business to invest more of its resources in equipment to improve harvesting efficiency. Reliable financial records are needed in order to make sound business decisions of this kind.

Tax management is an aspect of business that is possible only when accurate financial records have been assembled. Many different legal strategies can be used to delay or minimize the amount of tax that is due in any given year. This makes cash available to the business for investment and other business purposes. Tax management software helps business managers to understand tax laws and to file accurate tax returns.

FOREST INVENTORY

A **forest inventory** is a record of the estimated volume of timber of the different varieties that are present in the forest at any given time. It is an estimate because it is not possible to record and measure every living tree in North American forests. In Chapter 9, tree measurements were studied, and sampling methods were discussed. Small, portable, battery-powered computers are used by cruisers in the forest to gather and organize forest data. Forest inventories are based on the measurements obtained from samples of the population, and the average values for a few trees are assigned to all of the trees in the population as estimates of the total timber inventory are prepared.

Figure 17–8 Computers are capable of computing complex equations. These computations are necessary for forest measurements and to organize forestry data. *(Photo courtesy of Michael Dzaman)*

Computers are excellent tools for organizing vast amounts of information, and they are also capable of computing complex equations (Figure 17–8). Both of these functions are important in working with forest inventories. Computers are used to convert the forest measurements obtained from a few trees in the different sectors of a forest into estimated timber inventories for the entire forest. They are also used to calculate forest growth rates based on tree measurements.

When a computer is used to combine forest growth rates with forest inventories, the annual increases in forest inventories can be predicted with reasonable accuracy. Many of the calculations that are used to develop forest inventories are now done using computers as tools. A **forest growth model** is a computer program that estimates forest yield by applying all of the factors known to affect forest growth. For example, research has indicated that normal growth of a tree is strongly influenced by these factors: (1) the size or age of the tree, (2) the quality of the site or environment where the tree is growing, (3) the degree to which the tree is affected by competition from other trees, and (4) effects of diseases or insects (Figure 17–9). When these variables are converted to mathematical equations based on research data, a computer program can be created for the purpose of predicting future forest growth and timber yields. Such a program is an example of a forest growth model.

Dependable computer models have been developed for forest conditions in different environments and for different species of trees. The Hubbard Brook computer models and the FOREST model are some of the better known of these. Models have even been developed for mixed-species forests. Models must be tailored to the specific conditions in each forest environment, and they are

---

Factors Affecting Forest Growth

❈ Size or age of the tree
❈ Quality of the site or environment
❈ Competition from other trees and plants
❈ Effects of disease or insects

Figure 17–9 A computer program can integrate the known factors that affect forest growth into forest growth models. Such models are used to predict future forest growth.

validated by testing with real data to check the accuracy of their predictions. In some cases, a computer model is used to make a series of predictions based on differences in conditions from one sector of the forest to another. The estimates from the different forest sectors are then combined to provide an estimated inventory value for the entire forest.

FOREST RESOURCE MANAGEMENT

Every phase of forest management involves paperwork of some kind. The use of computers has made it possible to record, organize, store, and communicate forest management data in an efficient and effective manner (Figure 17–10). Charts are generated, graphs are constructed, data is recorded and organized, financial records are analyzed, inventories are recorded, letters and reports are written, and electronic messages are sent and received using computers.

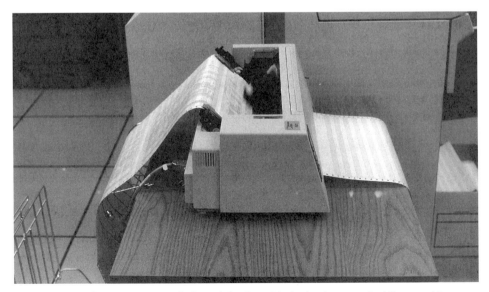

Figure 17–10 Government regulation of the forest industry has created the need to gather extensive data and generate detailed reports. Computers make it possible to do this efficiently.

One of the promising uses for computers as a forest resource management tool is to predict fire behavior. Using research data related to the combustion process and applying mathematical formulas to calculate the rate of spread and fire intensity, a computer program called BEHAVE has been developed. This fire simulation model assembles the data associated with a surface fire, and it predicts the location of the fire perimeter at any given time. As conditions change that affect the fire, the data must be updated. This computer program is proving to be an effective tool in fire management and suppression.

Computer models have been developed for forestry simulations that imitate forest reactions to pollution, harvesting, forest growth, and forest yields. Some of the better known of the computerized timber harvest scheduling models are TREES, ECHO, SIMAC, MAX MILLION II, SORAC, and TIMBER RAM. All of these models are used to determine the amount of timber that can be harvested while still meeting different kinds of forest objectives.

AERIAL PHOTOGRAPHY

Aerial photography has many applications to the forest industry. Many of our forested regions are so vast that it is difficult to visually inspect some forested areas. Aerial photography has been used to perform surveillance of some of these areas. The technique requires flying over the area in question and taking photographs from the air. By aligning the photographs with known landmarks, forest regions can be visually inspected in this manner.

Infrared photographs, recorded on special film, provide information about the forest that is invisible to human eyes. Infrared films of several types are available that are sensitive to different wavelengths of light. The color film records these differences as different colors on film, and the color prints that are produced can provide information on stress levels such as drought or insect damage to the trees (Figure 17–11). The principle by which infrared film works is based on the differences in the amount of sunlight that is reflected from the surfaces of plants and other objects. Healthy plants that are not stressed reflect sunlight at different rates than do stressed plants. This technology makes it possible to identify areas in the forest where a problem exists. These areas can then be inspected by a ground crew to determine the cause and possible responses to the condition.

Aerial photography plays an important role in fire detection. The number of fire lookout towers has declined in recent years, and aerial surveillance has taken on an important role. Once a fire is detected, the progress of the fire can be followed through the use of aerial surveillance and aerial photography. Some of the functions that have been performed by aerial photography are now being replaced with photographs and images obtained from a satellite platform.

REMOTE SENSING

Recent adaptations of space technology to other productive uses are beginning to have an impact in forest management. Special satellites known as **earth resource satellites** have been placed in earth orbits that allow them to pass repeatedly over every part of the surface of the earth. These satellites are

Multispectral Imagery

Figure 17-11 Infrared photography, using special film that is sensitive to different wavelengths of light, is used to detect forest problems such as stress to trees due to drought conditions or insect infestations.

equipped with sensors of various types that allow them to make and transmit different types of data to ground bases. The United States has developed a series of satellites of this kind known as the Landsat satellites. The French government was joined by Sweden and Belgium to create and launch a similar satellite system known as SPOT.

One of the technologies available using earth resource satellites involves taking aerial photographs or other types of images from vantage points high above the ground. This was done from airplanes until satellites became available, and with satellite platforms in space, photographs and other images of forests and land areas can be readily obtained (Figure 17–12). Several different kinds of images can be obtained using this technology. For example, photogrammetry is a system that is used to obtain forest measurements from photographs. In order to make accurate measurements of the forest, photographs must be taken from two separate vantage points. This is now possible by using photographs taken from two different satellite tracks as the satellites make successive passes over the area.

Photographs from space are used to create maps with photogrammetry. This process was discussed in Chapter 9 as a way to obtain forest measurements. Until the earth resource satellites were in use, some areas of the world had never been photographed, and accurate maps of those regions did not exist. The vast rain forests of Brazil were among some of the unmapped areas.

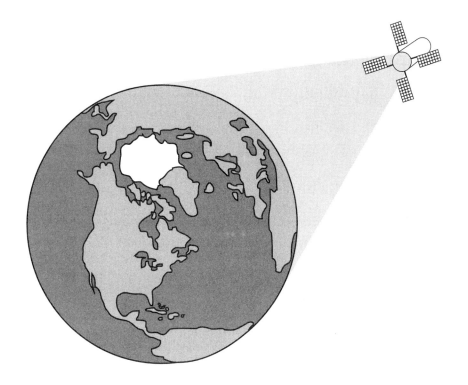

Figure 17–12 Satellites placed in orbits that cross over every segment of the earth during its rotation are capable of photographing the entire surface of the planet.

Remote sensing technologies are now used in other types of mapping, such as maps that show elevation. This kind of map is a **topographic map.** It is created by analyzing photographs taken from satellites using an instrument called a **stereoplotter.** This kind of map is important in the management of forested areas because slope is a variable that affects the production of the forest.

GEOGRAPHIC INFORMATION SYSTEM

The **geographic information system,** also known as **GIS,** combines data obtained by remote sensing technologies with computer mapping technologies to evaluate forest needs (Figure 17–13). Each forest map is divided into sectors, and information such as soil type, harvest yields, forest type, and insect damage is recorded as a management factor in each sector to which it applies. The GIS makes it possible to independently consider the forest management needs of each sector of the forest. It is also capable of monitoring the health of the forest in each sector by using remote sensing technologies.

The strength of the GIS is in the individual treatments that each forest sector can be given, particularly in plantation systems. This system makes it possible to analyze several different production factors at the same time. For example,

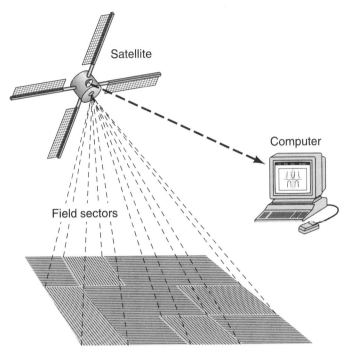

Figure 17–13 The geographic information system (GIS) combines satellite technology with computer mapping technologies to evaluate changing needs in forest environments.

suppose that a particular sector on the forest map shows the area to be susceptible to soil erosion because a heavy clay subsoil prevents precipitation from being absorbed into the soil. A topographic profile of the sector indicates that much of the sector is located on a slope greater than 4%. A map of forest type obtained from an earth resource satellite shows a mixed-age forest to be the predominant forest type. Aerial photographs show the pathway of a destructive fire that damaged the soil in a steep region in the sector. The GIS is capable of combining all of this data to show that this sector is at high risk for serious soil erosion and that management steps should be taken to prevent this problem. A logical conclusion for managing this sector would be that forest managers should consider a selection harvest plan for this forest sector and avoid clear-cutting the area.

The GIS is able to gather specific information about each site from a number of different sources. Management practices can then be recommended for each forest sector that have been shown through research to be effective in dealing with specific forest conditions.

GLOBAL POSITIONING SYSTEM

A geographic information system must be integrated with technology that can accurately identify exact field locations from which the field data were gathered. An example of such a system is the satellite technology known as the **global**

Figure 17–14 A small GPS receiver, carried by a forest worker into the woods, can be used to pinpoint exact locations using the global positioning system.

positioning system (GPS). When this system is integrated with GIS, it is capable of locating the exact location in a forest from which field data were gathered.

The GPS consists of twenty-four satellites that orbit the earth two times each day at altitudes of approximately 12,000 miles. These are military satellites that are being adapted to peacetime uses. To use the system, forest workers carry GPS receivers with them as they work (Figure 17–14). The system can then be used to locate the exact positions of the GPS receivers in the forest. As forest data are entered into handheld computers by cruisers and other forest workers, the GPS can identify the exact locations from which the data were generated.

New technologies such as GPS, GIS, and remote sensing are expensive to implement, but the degree of environmental accountability that is required of

---

PROFILE ON FOREST SAFETY

Do You Know Where You Are?

A handheld GPS receiver is a useful tool to people who work in vast forest regions. Readings taken from the receiver can be correlated with forest maps with a high degree of accuracy. They can identify exact locations within a few meters. The GPS receiver is a useful tool to anyone who spends time outdoors, especially in heavily forested regions where the terrain is hilly. The user simply compares the coordinates obtained from the receiver to the coordinates on a special map of the region. Global positioning system receivers are now available in many sporting goods stores, making this technology available to everyone who spends time in outdoor environments. Hikers, hunters, fishermen, skiers, snowmobile enthusiasts, and many others can now join forest workers in enjoying the security of knowing where they are in the forest.

CAREER OPTION

Technology Specialist

A person who understands the operation of computers, lasers, satellites, and electronic control systems can find a career in nearly any sector of the forest industry. Modern technologies are used in many different aspects of forestry and silviculture (Figure 17–15). Job preparation for this field is available through technical colleges, community colleges, universities, and private industrial training organizations. This career option requires continual education and training experiences to stay current with new and emerging technologies. A person who enters this career must be proficient in setting up, maintaining, diagnosing, and repairing high-tech equipment. Earnings are above average for people who are proficient in the understanding and use of modern technologies.

Figure 17–15 One of the new careers that has emerged in the technology age is Technology Specialist. There are many different specialties in this field such as computers, lasers, satellites, and electronic control systems.

forest managers in both the public and private sectors may one day require that the managers use these modern tools.

LOOKING BACK

The computer is a management tool that is useful in assembling and analyzing information. Its use in the forest industry has expanded rapidly during recent years as computer hardware and software have been adapted to fit forest industry needs. Computers are widely used for cruising forests, modeling forest growth, and creating inventories of forest resources. Computers are used to predict fire behavior and to model the effects of different harvest practices. They are used in timber processing to control machines, schedule routine maintenance, track inventories, and record business transactions. Computers are expected to become even more important as management tools in the future.

Aerial photography and remote sensing technologies using aircraft and satellite systems offer some new approaches to forest management. The geographic information system (GIS) and the global positioning system (GPS) in combination with computer mapping have the potential for managing individual sectors of a forest according to their greatest needs. This approach to forest management is likely to simultaneously reduce environmental problems in the forest and increase forest productivity.

QUESTIONS FOR DISCUSSION AND REVIEW

Essay Questions

1. What are the main components of a computer?

2. How is computer hardware different from computer software?

3. What is a computer model, and how is it used in forestry?

4. Identify some ways that computer networks can be used in the forest industry.

5. In what ways is a computer network different from the Internet system?

6. How are computers used in the forest industry for financial management? Processing? Forest inventories?

7. Speculate on some new ways that computers may be used to manage forest resources in the future.

8. Describe some ways that military and space technologies are being adapted to new uses in forestry.

9. How may new technologies be used to increase the production of forests?

10. What are some new technologies that contribute to solutions for environmental problems in forests?

Multiple-Choice Questions

1. A portable disk, tape, or compact disc on which computer programs or original data are stored is called:
   A. modem
   B. hardware
   C. software
   D. home page

2. A connection between two or more computers that allows each computer to have access to the same information is a:
   A. modem
   B. network
   C. database
   D. program

3. A computer network that provides user access to the information contained in the databases of many smaller computer networks throughout the world is known as:
   A. Internet
   B. modem
   C. database
   D. software

4. Which of the following devices is capable of providing a visual stimulus to a computer?
   A. laser
   B. hard drive
   C. modem
   D. compact disc

5. Which of the following criteria is *not* considered in a computer-generated forest growth model?
   A. machine costs
   B. competition with other trees
   C. size or age of trees
   D. quality of the growth site or environment

6. The name of a computer program that is used to predict the behavior of a forest fire is:
   A. TREES
   B. SIMAC
   C. ECHO
   D. BEHAVE

7. Which of the following computer programs is used to determine the amount of timber that can be harvested based on forest management objectives:
   A. SORAC
   B. GROWTH
   C. SPOT
   D. BEHAVE

8. An American satellite system of the type known as earth resource satellites is called:
   A. SPOT
   B. SIMAC
   C. Landsat
   D. MAX MILLION II

9. A remote sensing technology that is used to map such information as soil and forest types is known as:
   A. GPS
   B. GIS
   C. SPOT
   D. BEHAVE

10. A satellite system that is used to identify exact locations in a forest is known as:
    A. GPS
    B. BEHAVE
    C. GIS
    D. SIMAC

LEARNING ACTIVITIES

1. Plan a field trip to a government agency field office that is responsible for management of forest resources. Request demonstrations of computer technology to include the use of cruising computers and some computer models that are used by the agency. Obtain the names of some of the computer software packages and forest modeling materials that are used.

2. Assign the class to work in groups of two or three students to design a poster showing ways that computers are now used or may be used in the future to manage forest resources.

APPENDIX I
North American Trees and Bushes

Family and Common Names	**Scientific Name**
Ginkgo Family, *Ginkgoaceae:*	
Ginkgo (Maidenhair-tree)	*Ginkgo biloba*
Yew Family, *Taxaceae:*	
Florida Yew	*Taxus floridana*
Pacific Yew (Western Yew)	*Taxus brevifolia*
California Torreya (California-nutmeg)	*Torreya californica*
Florida Torreya (Stinking-cedar)	*Torreya taxifolia*
Pine Family, *Pinaceae:*	
Balsam Fir (Canada Balsam, Eastern Fir)	*Abies balsamea*
Bristlecone Fir (Santa Lucia Fir, Silver Fir)	*Abies bracteata*
California Red Fir (Red Fir, Silvertip)	*Abies magnifica*
Fraser Fir (Balsam, She-balsam)	*Abies fraseri*
Grand Fir (Lowland White Fir, Lowland Fir)	*Abies grandis*
Noble Fir (Red Fir, White Fir)	*Abies procera*
Pacific Silver Fir (Amabilis Fir, Cascades Fir)	*Abies amabilis*
Subalpine Fir (Alpine Fir, Rocky Mountain Fir)	*Abies lasiocarpa*
White Fir (Silver Fir, Concolor Fir)	*Abies concolor*
Atlas Cedar (Atlas Mountain Cedar)	*Cedrus atlantica*
Cedar-of-Lebanon	*Cedrus libani*
Deodar Cedar (Deodar, California Christmas-tree)	*Cedrus deodara*
European Larch	*Larix decidua*
Subalpine Larch (Alpine Larch, Tamarack)	*Larix lyallii*
Tamarack (Hackmatack, Eastern Larch)	*Larix laricina*
Western Larch (Hackmatack, Western Tamarack)	*Larix occidentalis*
Black Spruce (Bog Spruce, Swamp Spruce)	*Picea mariana*
Blue Spruce (Colorado Spruce, Silver Spruce)	*Picea pungens*
Brewer Spruce (Weeping Spruce)	*Picea brewerana*
Engelmann Spruce	*Picea engelmannii*

Family and Common Names	Scientific Name
Norway Spruce	*Picea abies*
Red Spruce (Eastern Spruce, Yellow Spruce)	*Picea rubens*
Sitka Spruce (Coast Spruce, Tideland Spruce)	*Picea sitchensis*
White Spruce (Canadian Spruce, Skunk Spruce)	*Picea glanca*
Apache Pine (Arizona Longleaf Pine)	*Pinus engelmannii*
Austrian Pine (European Black Pine)	*Pinus nigra*
Bishop Pine (Santa Cruz Island Pine, Prickle-cone Pine)	*Pinus muricata*
Bristlecone Pine (Foxtail Pine, Hickory Pine)	*Pinus aristata*
Chihuahua Pine (Yellow Pine)	*Pinus leiophylla*
Coulter Pine (Bigcone Pine, Pitch Pine)	*Pinus coulteri*
Digger Pine (Bull Pine, Gray Pine)	*Pinus sabiniana*
Eastern White Pine (White Pine, Northern White Pine)	*Pinus strobus*
Jack Pine (Scrub Pine, Gray Pine)	*Pinus banksiana*
Jeffrey Pine (Western Yellow Pine, Bull Pine)	*Pinus jeffreyi*
Knobcone Pine	*Pinus attenuata*
Limber Pine (White Pine, Rocky Mountain White Pine)	*Pinus flexilis*
Loblolly Pine (Oldfield Pine, North Carolina Pine)	*Pinus taeda*
Longleaf Pine (Longleaf Yellow Pine, Southern Yellow Pine)	*Pinus palustris*
Monterey Pine (Insignis Pine)	*Pinus radiata*
Parry Pinyon (Four-needle Pinyon, Nut Pine)	*Pinus quadrifolia*
Pinyon (Two-leaf Pinyon, Colorado Pinyon)	*Pinus edulis*
Ponderosa Pine (Western Yellow Pine, Blackjack Pine)	*Pinus ponderosa*
Red Pine (Norway Pine)	*Pinus resinosa*
Sand Pine (Scrub Pine, Spruce Pine)	*Pinus clausa*
Scotch Pine (Scots Pine)	*Pinus sylvestris*
Shortleaf Pine (Shortstraw Pine, Southern Yellow Pine)	*Pinus echinata*
Singleleaf Pinyon (Nut Pine, Singleleaf Pinyon Pine)	*Pinus monophylla*
Slash Pine (Yellow Slash Pine, Swamp Pine)	*Pinus elliottii*
Southwestern White Pine (Mexican or Border White Pine)	*Pinus strobiformis*
Spruce Pine (Cedar Pine, Walter Pine)	*Pinus glabra*
Sugar Pine	*Pinus lambertiana*
Table Mountain Pine (Hickory Pine, Mountain Pine)	*Pinus pungens*
Torrey Pine (Del Mar Pine, Soledad Pine)	*Pinus torreyana*
Virginia Pine (Scrub Pine, Jersey Pine)	*Pinus virginiana*
Washoe Pine	*Pinus washoensis*
Western White Pine (Mountain White Pine, Idaho White Pine)	*Pinus monticola*
Whitebark Pine (Scrub Pine, White Pine)	*Pinus albicaulis*
Bigcone Douglas-fir (Bigcone-spruce)	*Pseudotsuga macrocarpa*
Douglas-fir (Douglas-spruce, Oregon-pine)	*Pseudotsuga menziesii*
Carolina Hemlock	*Tsuga caroliniana*
Eastern Hemlock (Canada Hemlock, Hemlock Spruce)	*Tsuga canadensis*

Family and Common Names	**Scientific Name**
Mountain Hemlock (Black Hemlock, Alpine Hemlock)	*Tsuga mertensiana*
Western Hemlock (Pacific Hemlock, West Coast Hemlock)	*Tsuga heterophylla*

Redwood Family, *Taxodiaceae*:

Cryptomeria (Japanese-cedar, Sugi)	*Cryptomeria japonica*
Giant Sequoia (Sierra Redwood, Bigtree)	*Sequoiadendron giganteum*
Redwood (Coast Redwood, California Redwood)	*Sequoia sempervirens*
Baldcypress (Cypress, Swamp-cypress)	*Taxodium distichum*
Montezuma Baldcypress (Mexican-cypress, Sabino)	*Taxodium mucronatum*

Cypress Family, *Cupressaceae*:

Alaska-cedar (Alaska Yellow-cedar, Nootka Cypress)	*Chamaecyparis nootkatensis*
Atlantic White-cedar (Southern White-cedar, Swamp-cedar)	*Chamaecyparis thyoides*
Port-Orford-cedar (Oregon-cedar, Lawson Cypress)	*Chamaecyparis lawsoniana*
Sawara False-cypress (Sawara Cypress, Retinospora)	*Chamaecyparis pisifera*
Arizona Cypress	*Cupressus arizonica*
Baker Cypress (Siskiyou Cypress, Modoc Cypress)	*Cupressus bakeri*
Gowen Cypress	*Cupressus goveniana*
MacNab Cypress	*Cupressus macnabiana*
Monterey Cypress	*Cupressus macrocarpa*
Sargent Cypress	*Cupressus sargentii*
Alligator Juniper	*Juniperus deppeana*
Ashe Juniper (Rock-cedar, Post-cedar)	*Juniperus ashei*
California Juniper	*Juniperus californica*
Common Juniper (Dwarf Juniper)	*Juniperus communis*
Drooping Juniper (Weeping Juniper, Mexican Drooping Juniper)	*Juniperus flaccida*
Eastern Redcedar (Red Juniper)	*Juniperus virginiana*
Oneseed Juniper	*Juniperus monosperma*
Pinchot Juniper (Redberry Juniper)	*Juniperus pinchotti*
Redberry Juniper	*Juniperus erythrocarpa*
Rocky Mountain Juniper (Rocky Mountain Redcedar, River Juniper)	*Juniperus scopulorum*
Southern Redcedar (Sand-cedar)	*Juniperus silicicola*
Utah Juniper	*Juniperus osteosperma*
Western Juniper (Sierra Juniper)	*Juniperus occidentalis*
Northern White-cedar (Eastern White-cedar, Eastern Arborvitae)	*Thuja occidentalis*
Oriental Arborvitae (Chinese Arborvitae)	*Thuja orientalis*

Family and Common Names	Scientific Name
Western Redcedar (Giant Arborvitae, Canoe-cedar)	*Thuja plicata*

Palm Family, *Palmae*:

Cabbage Palmetto (Carolina Palmetto, Cabbage-palm)	*Sabal palmetto*
Canary Island Date (Canary Island Date-palm)	*Phoenix canariensis*
California Washingtonia (California-palm, Fanpalm)	*Washingtonia filifera*
Mexican Washingtonia (Mexican Fanpalm)	*Washingtonia robusta*

Lily Family, *Liliaceae*:

Aloe Yucca (Spanish-bayonet, Spanish-dagger)	*Yucca aloifolia*
Faxon Yucca (Spanish-bayonet, Spanish-dagger)	*Yucca faxoniana*
Joshua-tree (Tree Yucca, Yucca-palm)	*Yucca brevifolia*
Mohave Yucca (Spanish-dagger)	*Yucca schidigera*
Moundlily Yucca (Spanish-bayonet, Spanish-dagger)	*Yucca gloriosa*
Schott Yucca (Spanish-bayonet, Spanish-dagger)	*Yucca schottii*
Soaptree Yucca (Soapweed, Palmilla)	*Yucca elata*
Torrey Yucca (Spanish-bayonet, Spanish-dagger)	*Yucca torreyi*
Giant Dracaena (Green Dracaena, Cracaena-palm)	*Cordyline australis*

Casuarina Family, *Casuarinaceae*:

River-oak Casuarina (Cunningham Casuarina, Beefwood)	*Casuarina cunninghamiana*

Willow Family, *Salicaceae*:

Balsam Poplar (Tacamahac, Balm)	*Populus balsamifera*
Bigtooth Aspen (Largetooth Aspen, Poplar)	*Populus grandidentata*
Black Cottonwood (Western Balsam Poplar, California Poplar)	*Populus trichocarpa*
Eastern Cottonwood (Carolina Poplar, Southern Cottonwood)	*Populus deltoides*
Fremont Cottonwood (Rio Grande Cottonwood, Meseta Cottonwood)	*Populus fremontii*
Lombardy Poplar	*Populus nigra*
Narrowleaf Cottonwood (Mountain Cottonwood, Black Cottonwood)	*Populus angustifolia*
Quaking Aspen (Trembling Aspen, Golden Aspen)	*Populus tremuloides*
Swamp Cottonwood (Swamp Poplar, Black Cottonwood)	*Populus heterophylla*
White Poplar (Silver Poplar)	*Populus alba*
Arroyo Willow (White Willow)	*Salix lasiolepis*
Balsam Willow (Bog Willow)	*Salix pyrifolia*
Basket Willow (Osier, Silky Osier)	*Salix viminalis*

Family and Common Names	**Scientific Name**
Bebb Willow (Beak Willow, Diamond Willow)	*Salix bebbiana*
Black Willow (Swamp Willow, Goodding Willow)	*Salix nigra*
Bonpland Willow (Red Willow, Polished Willow)	*Salix bonplandiana*
Coastal Plain Willow (Southern Willow, Ward Willow)	*Salix caroliniana*
Crack Willow (Brittle Willow, Snap Willow)	*Salix fragilis*
Feltleaf Willow	*Salix alaxensis*
Florida Willow	*Salix floridana*
Hinds Willow (Sandbar Willow, Valley Willow)	*Salix hindsiana*
Hooker Willow (Bigleaf Willow, Coast Willow)	*Salix hookerana*
Littletree Willow	*Salix arbusculoides*
Mackenzie Willow	*Salix mackenzieana*
Northwest Willow (Velvet Willow, Soft-leaf Willow)	*Salix sessilifolia*
Pacific Willow (Western Black Willow, Yellow Willow)	*Salix lasiandra*
Peachleaf Willow (Peach Willow, Almond Willow)	*Salix amygdaloides*
Pussy Willow	*Salix discolor*
River Willow (Sandbar Willow)	*Salix fluviatilis*
Sandbar Willow (Coyote Willow, Narrowleaf Willow)	*Salix exigua*
Scouler Willow (Black Willow, Fire Willow)	*Salix scoulerana*
Sitka Willow (Silka Willow, Coulter Willow)	*Salix sitchensis*
Tracy Willow	*Salix tracyi*
Weeping Willow (Babylon Weeping Willow)	*Salix babylonica*
White Willow (European White Willow)	*Salix alba*
Yewleaf Willow (Yew Willow)	*Salix taxifolia*

Bayberry (Waxmyrtle) Family, *Myricaceae:*

Odorless Bayberry (Odorless Waxmyrtle, Waxmyrtle)	*Myrica inodora*
Pacific Bayberry (Pacific Waxmyrtle, Western Waxmyrtle)	*Myrica californica*
Southern Bayberry (Candle-berry, Southern Waxmyrtle)	*Myrica cerifera*

Corkwood Family, *Leitneriaceae:*

Corkwood	*Leitneria floridana*

Walnut Family, *Juglandaceae:*

Bitternut Hickory (Bitternut, Pignut)	*Carya cordiformis*
Black Hickory (Buckley Hickory, Pignut Hickory)	*Carya texana*
Mockernut Hickory (White Hickory, Mockernut)	*Carya tomentosa*
Nutmeg Hickory (Swamp Hickory, Bitter Water Hickory)	*Carya myristiciformis*
Pecan (Sweet Pecan)	*Carya illinoensis*
Pignut Hickory (Pignut, Smoothbark Hickory)	*Carya glabra*
Sand Hickory (Pale Hickory, Pignut Hickory)	*Carya pallida*
Scrub Hickory (Florida Hickory)	*Carya floridana*
Shagbark Hickory (Scalybark Hickory, Shellbark Hickory)	*Carya ovata*

Family and Common Names   Scientific Name

Shellbark Hickory (Big Shagbark Hickory, Kingnut)   *Carya laciniosa*
Water Hickory (Swamp Hickory, Bitter Pecan)   *Carya aquatica*

Arizona Walnut (Arizona Black Walnut, Nogal)   *Juglans major*
Black Walnut (Eastern Black Walnut, American Walnut)   *Juglans nigra*
Butternut (White Walnut, Oilnut)   *Juglans cinerea*
Little Walnut (Texas Walnut)   *Juglans microcarpa*
Northern California Walnut (Hinds Walnut)   *Juglans hindsii*
Southern California Walnut (California
    Walnut, California Black Walnut)   *Juglans californica*

Birch Family, *Betulaceae:*

Arizona Alder (New Mexican Alder, Mexican Alder)   *Alnus oblongifolia*
European Alder (Black Alder, European Black Alder)   *Alnus glutinosa*
Hazel Alder (Common Alder, Tag Alder)   *Alnus serrulata*
Mountain Alder (Thinleaf Alder, River Alder)   *Alnus tenuifolia*
Red Alder (Oregon Alder, Western Alder)   *Alnus rubra*
Seaside Alder   *Alnus maritima*
Sitka Alder (Mountain Alder, Wavyleaf Alder)   *Alnus sinuata*
Speckled Alder (Tag Alder, Gray Alder)   *Alnus rugosa*
White Alder (Tag Alder, Gray Alder)   *Alnus rhombifolia*

European White Birch (European Birch, European Weeping Birch) *Betula pendula*
Gray Birch (White Birch, Wire Birch)   *Betula populifolia*
Paper Birch (Canoe Birch, White Birch)   *Betula papyrifera*
River Birch (Red Birch, Black Birch)   *Betula nigra*
Sweet Birch (Black Birch, Cherry Birch)   *Betula lenta*
Virginia Roundleaf Birch (Ashe Birch, Virginia Birch) *Betula uber*
Water Birch (Red Birch, Black Birch)   *Betula occidentalis*
Yellow Birch (Gray Birch, Silver Birch)   *Betula alleghaniensis*

American Hornbeam (Blue-beech, Water-beech)   *Carpinus caroliniana*

Chisos Hophornbeam (Big Bend Hophornbeam)   *Ostrya chisosensis*
Eastern Hophornbeam (American Hophornbeam, Ironwood)   *Ostrya virginiana*
Knowlton Hophornbeam (Western Hophornbeam, Ironwood)   *Ostrya knowltonii*

Beech Family, *Fagaceae:*

Allegheny Chinkapin   *Castanea pumila*
American Chestnut (Chestnut)   *Castanea dentata*
Florida Chinkapin (Trailing Chinkapin)   *Castanea alnifolia*
Ozark Chinkapin (Ozark Chestnut)   *Castanea ozarkensis*

Family and Common Names	Scientific Name
Giant Chinkapin (Golden Chinkapin, Goldenleaf Chestnut)	*Castanopsis chrysophylla*
American Beech (Beech)	*Fagus grandifolia*
European Beech	*Fagus sylvatica*
Tanoak (Tanbark-oak)	*Lithocarpus densiflorus*
Arizona White Oak (Arizona Oak)	*Quercus arizonica*
Arkansas Oak (Water Oak)	*Quercus arkansana*
Bear Oak (Scrub Oak)	*Quercus ilicifolia*
Black Oak (Yellow Oak, Quercitron Oak)	*Quercus velutina*
Bluejack Oak (Sandjack, Upland Willow Oak)	*Quercus incana*
Blue Oak (Mountain White Oak, Iron Oak)	*Quercus douglassi*
Bur Oak (Blue Oak, Mossycup Oak)	*Quercus macrocarpa*
California Black Oak (Black Oak, Kellogg Oak)	*Quercus kelloggii*
California Scrub Oak (Scrub Oak)	*Quercus dumosa*
Canyon Live Oak (Canyon Oak, Goldcup Oak)	*Quercus chrysolepis*
Chapman Oak (Chapman White Oak, Scrub Oak)	*Quercus chapmanii*
Chestnut Oak (Rock Chestnut Oak, Rock Oak)	*Quercus prinus*
Chinkapin Oak (Chestnut Oak, Rock Oak)	*Quercus muehlenbergii*
Coast Live Oak (California Live Oak, Encina)	*Quercus agrifolia*
Dunn Oak (Palmer Oak)	*Quercus dunnii*
Durand Oak (Bluff Oak, White Oak)	*Quercus durandii*
Emory Oak (Black Oak, Blackjack Oak)	*Quercus emoryi*
Englemann Oak (Evergreen White Oak, Mesa Oak)	*Quercus engelmannii*
English Oak	*Quercus robur*
Gambel Oak (Rocky Mountain White Oak, Utah White Oak)	*Quercus gambelii*
Georgia Oak	*Quercus georgiana*
Gray Oak (Scrub Oak, Shin Oak)	*Quercus grisea*
Interior Live Oak (Highland Live Oak, Sierra Live Oak)	*Quercus wislizeni*
Island Live Oak (Island Oak)	*Quercus tomentella*
Lacey Oak (Rock Oak, Smoky Oak)	*Quercus glaucoides*
Laurel Oak (Darlington Oak, Diamond-leaf Oak)	*Quercus laurifolia*
Live Oak (Virginia Live Oak)	*Quercus virginiana*
McDonald Oak (Island Scrub Oak)	*Quercus macdonaldii*
Mexican Blue Oak	*Quercus oblongifolia*
Mohr Oak (Shin Oak, Scrub Oak)	*Quercus mohriana*
Myrtle Oak (Scrub Oak)	*Quercus myrtifolia*
Netleaf Oak	*Quercus rugosa*
Northern Pin Oak (Black Oak, Jack Oak)	*Quercus ellipsoidalis*
Northern Red Oak (Red Oak, Gray Oak)	*Quercus rubra*
Nuttall Oak (Red Oak, Pin Oak)	*Quercus nuttallii*
Oglethorpe Oak	*Quercus oglethorpensis*
Oregon White Oak (Garry Oak, Oregon Oak)	*Quercus garryana*

Family and Common Names	Scientific Name
Overcup Oak (Swamp Post Oak, Water White Oak)	*Quercus lyrata*
Pin Oak (Swamp Oak, Spanish Oak)	*Quercus palustris*
Post Oak (Iron Oak)	*Quercus stellata*
Sandpaper Oak (Scrub Oak, Shin Oak)	*Quercus pungens*
Scarlet Oak (Red Oak, Black Oak)	*Quercus coccinea*
Shingle Oak (Laurel Oak)	*Quercus imbricaria*
Shumard Oak (Spotted Oak, Swamp Oak)	*Quercus shumardii*
Silverleaf Oak (White-leaf Oak)	*Quercus hypoleucoides*
Southern Red Oak (Spanish Oak, Swamp Red Oak)	*Quercus falcata*
Swamp Chestnut Oak (Basket Oak, Cow Oak)	*Quercus michauxii*
Swamp White Oak	*Quercus bicolor*
Toumey Oak	*Quercus toumeyi*
Turbinella Oak (Shrub Live Oak, Scrub Oak)	*Quercus turbinella*
Turkey Oak (Catesby Oak, Scrub Oak)	*Quercus laevis*
Valley Oak (Valley White Oak, California White Oak)	*Quercus lobata*
Water Oak (Spotted Oak, Possum Oak)	*Quercus nigra*
White Oak (Stave Oak)	*Quercus alba*
Willow Oak (Pin Oak, Peach Oak)	*Quercus phellos*

Elm Family, *Ulmaceae:*

Georgia Hackberry (Dwarf Hackberry, Upland Hackberry)	*Celtis tenuifolia*
Hackberry (Sugarberry, Nettletree)	*Celtis occidentalis*
Lindheimer Hackberry (Palo Blanco)	*Celtis lindheimeri*
Netleaf Hackberry (Western Hackberry, Sugarberry)	*Celtis reticulata*
Sugarberry (Sugar Hackberry, Hackberry)	*Celtis laevigata*
Water Elm (Planertree)	*Planera aquatica*
American Elm (White Elm, Soft Elm)	*Ulmus americana*
Cedar Elm (Basket Elm, Southern Rock Elm)	*Ulmus crassifolia*
Chinese Elm	*Ulmus parvifolia*
English Elm	*Ulmus procera*
Rock Elm (Cork Elm)	*Ulmus thomassi*
September Elm (Red Elm)	*Ulmus serotina*
Siberian Elm (Asiatic Elm, Dwarf Elm)	*Ulmus pumila*
Slippery Elm (Red Elm, Soft Elm)	*Ulmus rubra*
Winged Elm (Cork Elm, Wahoo)	*Ulmus alata*
Japanese Zelkova	*Zelkova serrata*

Mulberry Family, *Moraceae:*

Paper-mulberry	*Broussonetia papyrifera*

Family and Common Names	**Scientific Name**
Osage-orange (Bodark, Hedge-apple)	*Maclura pomifera*
Red Mulberry (Moral)	*Morus rubra*
Texas Mulberry (Mexican Mulberry, Mountain Mulberry)	*Morus microphylla*
White Mulberry (Silkworm Mulberry, Russian Mulberry)	*Morus alba*

Protea Family, *Proteaceae*:

Silk-oak (Silky-oak)	*Grevillea robusta*

Magnolia Family, *Magnoliaceae*:

Florida Anise-tree (Stinkbush, Star Anise)	*Illicium floridanum*
Yellow Anise-tree (Small-flower Anise-tree, Star Anise)	*Illicium parviflorum*
Yellow-poplar (Tuliptree, Tulip-poplar)	*Liriodendron tulipifera*
Ashe Magnolia (Sandhill Magnolia)	*Magnolia ashei*
Bigleaf Magnolia (Silverleaf Magnolia, Umbrella-tree)	*Magnolia macrophylla*
Cucumbertree (Cucumber Magnolia)	*Magnolia acuminata*
Fraser Magnolia (Mountain Magnolia, Umbrella-tree)	*Magnolia fraseri*
Pyramid Magnolia (Southern Cucumbertree, Mountain Magnolia)	*Magnolia pyramidata*
Saucer Magnolia (Chinese Magnolia)	*Magnolia Xsoulangiana*
Southern Magnolia (Evergreen Magnolia, Bull-bay)	*Magnolia grandiflora*
Sweetbay (Swampbay, Swamp Magnolia)	*Magnolia virginiana*
Umbrella Magnolia (Umbrella-tree, Elkwood)	*Magnolia tripetala*

Annona (Custard Apple) Family, *Annonaceae*:

Pawpaw (Pawpaw-apple, False-banana)	*Asimina triloba*

Laurel Family, *Lauraceae*:

Camphor-tree	*Cinnamomum camphora*
California-laurel (Oregon-myrtle, Pepperwood)	*Umbellularia californica*
Redbay (Shorebay)	*Persea borbonia*
Sassafras	*Sassafras albidum*

Witch-Hazel Family, *Hamamelidaceae*:

Witch-hazel (Southern Witch-hazel)	*Hamamelis virginiana*
Sweetgum (Redgum, Sapgum)	*Liquidambar styraciflua*

Family and Common Names	Scientific Name

Sycamore Family, *Platanaceae*:

Arizona Sycamore (Arizona Planetree, Alamo)	*Platanus wrightii*
California Sycamore (Western Sycamore, Aliso)	*Platanus racemosa*
London Planetree	*Platanus xacerifolia*
Sycamore (American Sycamore, American Planetree)	*Platanus occidentalis*

Rose Family, *Rosaceae*:

Downy Serviceberry (Shadbush, Juneberry, Shadblow)	*Amelanchier arborea*
Roundleaf Serviceberry (Roundleaf Juneberry, Shore Shadbush)	*Amelanchier sanguinea*
Western Serviceberry (Saskatoon, Western Shadbush)	*Amelanchier alnifolia*
Birchleaf Cercocarpus (Birchleaf Mountain-mahogany, Hardtack)	*Cercocarpus betuloides*
Catalina Cercocarpus (Bigleaf Mountain-mahogany)	*Cercocarpus traskiae*
Curlleaf Cercocarpus (Curlleaf Mountain-mahogany)	*Cercocarpus ledifolius*
Hairy Cercocarpus (Hairy Mountain-mahogany)	*Cercocarpus breviflorus*
Cliffrose (Quinine-bush, Stansbury Cliffrose)	*Cowania mexicana*
Barberry Hawthorn (Bigtree Hawthorn, Barberrylear Hawthorn)	*Crataegus berberifolia*
Beautiful Hawthorn	*Crataegus pulcherrima*
Biltmore Hawthorn (Thicket Hawthorn, Allegheny Hawthorn)	*Crataegus intricata*
Black Hawthorn (Douglas Hawthorn, River Hawthorn)	*Crataegus douglasii*
Blueberry Hawthorn (Big Haw)	*Crataegus brachyacantha*
Brainerd Hawthorn	*Crataegus brainerdii*
Broadleaf Hawthorn (Apple-leaf Hawthorn)	*Crataegus dilatata*
Cerro Hawthorn	*Crataegus erythropoda*
Cockspur Hawthorn (Hog-apple, Newcastle-thorn)	*Crataegus crus-galli*
Columbia Hawthorn	*Crataegus columbiana*
Dotted Hawthorn (Large-fruit Thorn)	*Crataegus punctata*
Downy Hawthorn	*Crataegus mollis*
Fanleaf Hawthorn	*Crataegus flabellata*
Fireberry Hawthorn (Roundleaf Hawthorn, Golden-fruit Hawthorn)	*Crataegus chrysocarpa*
Fleshy Hawthorn (Long-spine Hawthorn, Succulent Hawthorn)	*Crataegus succulenta*
Frosted Hawthorn (Waxy-fruit Thorn)	*Crataegus pruinosa*
Green Hawthorn (Southern Hawthorn)	*Crataegus viridis*
Gregg Hawthorn	*Crataegus greggiana*
Harbison Hawthorn	*Crataegus harbisonii*
Kansas Hawthorn	*Crataegus coccinioides*
Littlehip Hawthorn (Small-fruit Hawthorn, Pasture Hawthorn)	*Crataegus spathulata*
May Hawthorn (Apple Hawthorn, Shining Hawthorn)	*Crataegus aestivalis*
Oneflower Hawthorn (Dwarf Hawthorn)	*Crataegus uniflora*
Oneseed Hawthorn (Single-seed Hawthorn, English Hawthorn)	*Crataegus monogyna*
Parsley Hawthorn	*Crataegus marshallii*

Family and Common Names	**Scientific Name**
Pear Hawthorn	*Crataegus calpodendron*
Pensacola Hawthorn (Weeping Hawthorn, Sandhill Hawthorn)	*Crataegus lacrimata*
Reverchon Hawthorn	*Crataegus reverchonii*
Riverflat Hawthorn (May Hawthorn, Apple Haw)	*Crataegus opaca*
Scarlet Hawthorn	*Crataegus coccinea*
Texas Hawthorn	*Crataegus texana*
Threeflower Hawthorn	*Crataegus triflora*
Tracy Hawthorn (Mountain Hawthorn)	*Crataegus tracyi*
Washington Hawthorn (Washington-thorn)	*Crataegus phaenopyrum*
Willow Hawthorn	*Crataegus saligna*
Yellow Hawthorn (Summer Haw)	*Crataegus flava*
Toyon (Christmas-berry, California-holly)	*Heteromeles arbutifolia*
Lyontree (Lyonothamnus, Catalina-ironwood)	*Lyonothamnus floribundus*
Apple (Common Apple, Wild Apple)	*Malus sylvestris*
Oregon Crab Apple (Pacific Crab Apple, Western Crab Apple)	*Malus fusca*
Prairie Crab Apple (Iowa Crab)	*Malus ioensis*
Southern Crab Apple (Narrow-leaf Crab Apple)	*Malus angustifolia*
Sweet Crab Apple (Garland-tree)	*Malus coronaria*
Allegheny Plum (Allegheny Sloe, Sloe Plum)	*Prunus alleghaniensis*
American Plum (Red Plum, River Plum)	*Prunus americana*
Bitter Cherry (Quinine Cherry, Wild Cherry)	*Prunus emarginata*
Black Cherry (Wild Cherry, Rum Cherry)	*Prunus serotina*
Canada Plum (Horse Plum, Red Plum)	*Prunus nigra*
Catalina Cherry	*Prunus lyonii*
Carolina Laurelcherry (Cherry-laurel, Carolina Cherry)	*Prunus caroliniana*
Chickasaw Plum (Sand Plum)	*Prunus angustifolia*
Common Chokecherry (Eastern Chokecherry, Western Chokecherry)	*Prunus virginiana*
Flatwoods Plum (Hog Plum, Black Sloe)	*Prunus umbellata*
Garden Plum (Damson Plum, Bullace Plum)	*Prunus domestica*
Hollyleaf Cherry (Evergreen Cherry, Islay)	*Prunus ilicifolia*
Hortulan Plum (Miner Plum)	*Prunus hortulana*
Klamath Plum (Sierra Plum, Pacific Plum)	*Prunus subcordata*
Mahaleb Cherry (Mahaleb, Perfumed Cherry)	*Prunus mahaleb*
Mexican Plum (Inch Plum, Bigtree Plum)	*Prunus mexicana*
Peach	*Prunus persica*
Pin Cherry (Fire Cherry, Bird Cherry)	*Prunus pensylvanica*
Sour Cherry (Pie Cherry, Morello Cherry)	*Prunus cerasus*
Sweet Cherry (Mazzard, Mazzard Cherry)	*Prunus avium*
Wildgoose Plum (Munson Plum)	*Prunus munsoniana*

Family and Common Names	Scientific Name
Pear (Common Pear)	*Pyrus communis*
American Mountain-ash (American Rowan-tree, Roundwood)	*Sorbus americana*
European Mountain-ash (Rowan-tree)	*Sorbus aucuparia*
Greene Mountain-ash (Western Mountain-ash)	*Sorbus scopulina*
Showy Mountain-ash	*Sorbus decora*
Sitka Mountain-ash (Western Mountain-ash)	*Sorbus sitchensis*
Torrey Vauquelinia (Arizona-rosewood)	*Vauquelinia californica*

Legume Family, *Leguminosae:*

Green Wattle (Green-wattle Acacia)	*Acacia decurrens*
Gregg Catclaw (Devilsclaw, Catclaw Acacia)	*Acacia greggii*
Huisache (Sweet Acacia, Cassie)	*Acacia farnesiana*
Roemer Catclaw (Roemer Acacia)	*Acacia roemeriana*
Wright Catclaw (Texas Catclaw, Wright Acacia)	*Acacia wrightii*
Silktree (Mimosa-tree, Powderpuff-tree)	*Albizia julibrissin*
Carob (St. Johns-bread, Algarroba)	*Ceratonia siliqua*
Blue Paloverde	*Cercidium floridum*
Yellow Paloverde (Foothill Paloverde, Littleleaf Paloverde)	*Cercidium microphyllum*
California Redbud (Western Redbud, Judas-tree)	*Cercis occidentalis*
Eastern Redbud (Judas-tree)	*Cercis canadensis*
Yellowwood (Virgilia)	*Cladrastis kentukea*
Smokethorn (Smoketree, Indigobush)	*Dalea spinosa*
Southeastern Coralbean (Cherokee-bean, Red-cardinal)	*Erythrina herbacea*
Kidneywood	*Eysenhardtia polystachya*
Honeylocust (Sweet-locust, Thorny-locust)	*Gleditsia triacanthos*
Waterlocust	*Gleditsia aquatica*
Kentucky Coffeetree	*Gymnocladus dioicus*
Littleleaf Leucaena (Littleleaf Leadtree, Wahoo-tree)	*Leucaena retusa*
Tesota (Arizona Ironwood, Desert Ironwood)	*Olneya tesota*

Family and Common Names	Scientific Name
Jerusalem-thorn (Horsebean, Mexican Paloverde)	*Parkinsonia aculeata*
Honey Mesquite	*Prosopis glandulosa*
Screwbean Mesquite (Screwbean, Tornillo)	*Prosopis pubescens*
Velvet Mesquite	*Prosopis velutina*
Black Locust (Yellow Locust, Locust)	*Robinia pseudoacacia*
Clammy Locust	*Robinia viscosa*
New Mexico Locust (New Mexican Locust, Southwestern Locust)	*Robinia neomexicana*
Chinese Scholartree (Japanese Pagodatree)	*Sophora japonica*
Mescalbean (Texas Mountain-laurel, Frijolillo)	*Sophora secundiflora*
Texas Sophora (Eves-necklace, Coralbean)	*Sophora affinis*

Caltrop Family, *Zygophyllaceae:*

Texas Lignumvitae (Texas Porliera, Soapbush)	*Guaiacum angustifolium*

Rue (Citrus) Family, *Rutaceae:*

Sour Orange (Seville Orange)	*Citrus aurantium*
Common Hoptree (Wafer-ash)	*Ptelea trifoliata*
Common Prickly-ash (Toothache-tree, Northern Prickly-ash)	*Zanthoxylum americanum*
Hercules-club (Toothache-tree, Tingle-tongue)	*Zanthoxylum clava-herculis*
Lime Prickly-ash (Wild-lime)	*Zanthoxylum fagara*

Quassia Family, *Simaroubaceae:*

Ailanthus (Tree of Heaven)	*Ailanthus altissima*

Bursera Family, *Byrseraceae:*

Elephant-tree (Elephant Bursera, Small-leaf Elephant-tree)	*Bursera microphylla*

Mahogany Family, *Meliaceae:*

Chinaberry (Chinatree, Pride of India)	*Melia azedarach*

Spurge Family, *Euphorbiaceae:*

Tallowtree (Chinese Tallowtree)	*Sapium sebiferum*

Cashew Family, *Anacardiaceae:*

American Smoketree (Yellowwood, Chittamwood)	*Cotinus obovatus*

Family and Common Names	Scientific Name
Texas Pistache (Wild Pistachio, American Pistachio)	*Pistacia texana*
Laurel Sumac	*Rhus laurina*
Lemonade Sumac (Lemonade-berry, Mahogany Sumac)	*Rhus integrifolia*
Prairie Sumac (Prairie Flameleaf Sumac, Texas Sumac)	*Rhus lanceolata*
Shining Sumac (Dwarf Sumac, Winged Sumac)	*Rhus copallina*
Smooth Sumac (Scarlet Sumac, Common Sumac)	*Rhus glabra*
Staghorn Sumac (Velvet Sumac)	*Rhus typhina*
Sugar Sumac (Sugarbush, Chaparral Sumac)	*Rhus ovata*
Poison-sumac (Poison-dogwood, Poison-elder)	*Toxicodendron vernix*
Peppertree (California Peppertree, Peru Peppertree)	*Schinus molle*

Cyrilla Family, *Cyrillaceae:*

Buckwheat-tree (Titi, Black Titi)	*Cliftonia monophylla*
Swamp Cyrilla (Leatherwood, Titi)	*Cyrilla racemiflora*

Holly Family, *Aquifoliaceae:*

American Holly (Holly, White Holly)	*Ilex opaca*
Carolina Holly	*Ilex ambigua*
Dahoon (Dahoon Holly, Christmas-berry)	*Ilex cassine*
English Holly (European Holly)	*Ilex aquifolium*
Mountain Winterberry (Mountain Holly)	*Ilex montana*
Myrtle Dahoon (Myrtle-leaf Holly)	*Ilex myrtifolia*
Possumhaw (Winterberry, Swamp Holly)	*Ilex decidua*
Sarvis Holly (Serviceberry Holly)	*Ilex amelanchier*
Yaupon (Cassena, Christmas-berry)	*Ilex vomitoria*

Bittersweet Family, *Celastraceae:*

Canotia (Crucifixion-thorn)	*Canotia holacantha*
Eastern Burningbush (Eastern Wahoo, Euonymus)	*Euonymus atropurpureus*

Bladdernut Family, *Staphyleaceae:*

American Bladdernut	*Staphylea trifolia*

Maple Family, *Aceraceae:*

Bigleaf Maple (Broadleaf Maple, Oregon Maple)	*Acer macrophyllum*
Black Maple (Hard Maple, Rock Maple)	*Acer nigrum*
Boxelder (Ashleaf Maple, Manitoba Maple)	*Acer negundo*

Family and Common Names   Scientific Name

Canyon Maple (Bigtooth Maple, Sugar Maple) — *Acer grandidentatum*
Chalk Maple (White-bark Maple) — *Acer leucoderme*
Florida Maple (Southern Sugar Maple, Hammock Maple) — *Acer barbatum*
Mountain Maple (Moose Maple) — *Acer spicatum*
Norway Maple — *Acer platanoides*
Planetree Maple (Sycamore Maple) — *Acer pseudoplatanus*
Red Maple (Scarlet Maple, Swamp Maple) — *Acer rubrum*
Rocky Mountain Maple (Dwarf Maple, Mountain Maple) — *Acer glabrum*
Silver Maple (Soft Maple, White Maple) — *Acer saccharinum*
Striped Maple (Moosewood) — *Acer pensylvanicum*
Sugar Maple (Hard Maple, Rock Maple) — *Acer saccharum*
Vine Maple — *Acer circinatum*

Buckeye (Horsechestnut) Family, *Hippocastanaceae:*

California Buckeye — *Aesculus californica*
Horsechestnut — *Aesculus hippocastanum*
Ohio Buckeye (Fetid Buckeye, American Horse-chestnut) — *Aesculus glabra*
Painted Buckeye (Dwarf Buckeye, Georgia Buckeye) — *Aesculus sylvatica*
Red Buckeye (Scarlet Buckeye, Firecracker-plant) — *Aesculus pavia*
Yellow Buckeye (Sweet Buckeye, Big Buckeye) — *Aesculus octandra*

Soapberry Family, *Sapindaceae:*

Western Soapberry (Wild Chinatree, Jaboncillo) — *Sapindus drummondii*

Mexican-buckeye (Texas-buckeye, Spanish-buckeye) — *Ungnadia speciosa*

Buckthorn Family, *Rhamnaceae:*

Blueblossom (Blue-myrtle, Bluebrush) — *Ceanothus thyrsiflorus*
Feltleaf Ceanothus (Catalina Ceanothus, Island-myrtle) — *Ceanothus arboreus*
Greenbark Ceanothus (California-lilac, Redheart) — *Ceanothus spinosus*

Bitter Condalia — *Condalia globosa*
Bluewood (Capul Negro, Brasil) — *Condalia hookeri*

Carolina Buckthorn (Indian-cherry, Yellowwood) — *Rhamnus caroliniana*
Cascara Buckthorn (Cascara Sagrada, Chittam) — *Rhamnus purshiana*
European Buckthorn (European Waythorn, Rhineberry) — *Rhamnus cathartica*
Glossy Buckthorn (Alder Buckthorn) — *Rhamnus frangula*

Basswood (Linden) Family, *Tiliaceae:*

American Basswood (American Linden, Bee-tree) — *Tilia americana*
Carolina Basswood (Linn, Bee-tree, Linden) — *Tilia caroliniana*

Family and Common Names	**Scientific Name**
European Linden (Common Linden)	*Tilia xeuropaea*
White Basswood (Linden, Bee-tree)	*Tilia heterophylla*

Sterculia Family, *Sterculiaceae:*

Chinese Parasoltree (Bottletree, Japanese Varnish-tree)	*Firmiana simplex*
California Fremontia (Flannelbush, California Slippery-elm)	*Fremontodendron californicum*

Tea Family, *Theaceae:*

Common Camellia (Japanese Camellia)	*Camellia japonica*
Franklinia (Franklin-tree)	*Franklinia alatamaha*
Loblolly-bay (Gordonia, Bay)	*Gordonia lasianthus*
Mountain Stewartia (Mountain-camellia, Angle-fruit Stewartia)	*Stewartia ovata*
Virginia Stewartia (Silky-camellia, Round-fruit Stewartia)	*Stewartia malacodendron*

Tamarisk Family, *Tamaricaceae:*

Athel Tamarisk (Athel, Evergreen Tamarisk)	*Tamarix aphylla*
Tamarisk (Saltcedar, Five-stamen Tamarisk)	*Tamarix chinensis*

Cactus Family, *Cactaceae:*

Saguaro (Giant Cactus)	*Cereus giganteus*
Jumping Cholla (Cholla)	*Opuntia fulgida*

Elaeagnus Family, *Elaeagnaceae:*

Russian-olive (Oleaster)	*Elaeaagnus angustifolia*

Ginseng Family, *Araliaceae:*

Devils-walkingstick (Hercules-club, Prickly-ash)	*Aralia spinosa*

Loosestrife Family, *Lythraceae:*

Crapemyrtle	*Lagerstroemia indica*

Myrtle Family, *Myrtaceae:*

Bluegum Eucalyptus (Bluegum, Tasmanian Bluegum)	*Eucalyptus globulus*
Red-ironbark Eucalyptus (Red-ironbark, Mugga)	*Eucalyptus sideroxylon*

Family and Common Names	**Scientific Name**

Dogwood Family, *Cornaceae:*

Alternate-leaf Dogwood (Pagoda Dogwood, Blue Dogwood)	*Cornus alternifolia*
Flowering Dogwood (Eastern Flowering Dogwood, Dogwood)	*Cornus florida*
Pacific Dogwood (Flowering Dogwood, Mountain Dogwood)	*Cornus nuttallii*
Red-osier Dogwood (Kinnikinnik, Red Dogwood)	*Cornus stolonifera*
Roughleaf Dogwood	*Cornus drummondii*
Wavyleaf Silktassel (Tasseltree, Quininebush)	*Garrya elliptica*
Black Tupelo (Blackgum, Pepperidge)	*Nyssa sylvatica*
Ogeechee Tupelo (Ogeechee-lime, Sour Tupelo)	*Nyssa ogeche*
Water Tupelo (Tupelo-gum, Cotton-gum)	*Nyssa aquatica*

Heath Family, *Ericaceae:*

Arizona Madrone (Madrono)	*Arbutus arizonica*
Pacific Madrone (Madrone, Madrono)	*Arbutus menziesii*
Texas Madrone (Texas Madrono)	*Arbutus texana*
Common Manzanita (Parry Manzanita)	*Arctostaphylos manzanita*
Elliottia (Southern-plume)	*Elliottia racemosa*
Mountain-laurel (Calico-bush, Ivybush)	*Kalmia latifolia*
Tree Lyonia (Staggerbush, Titi)	*Lyonia ferruginea*
Sourwood (Sorrel-tree, Lily-of-the-valley-tree)	*Oxydendrum arboreum*
Catawba Rhododendron (Mountain-rosebay, Purple-laurel)	*Rhododendron catawbiense*
Rosebay Rhododendron (Rosebay, Great-laurel)	*Rhododendron maximum*
Tree Sparkleberry (Farkleberry, Tree-huckleberry)	*Vaccinium arboreum*

Sapodilla (or Sapote) Family, *Sapotaceae:*

Buckthorn Bumelia (Buckthorn, Smooth Bumelia)	*Bumelia lycioides*
Gum Bumelia (Woolly Buckthorn, Chittamwood)	*Bumelia lanuginosa*
Tough Bumelia (Narrowleaf Bumelia, Tough Buckthorn)	*Bumelia tenax*

Ebony Family, *Ebenaceae:*

Common Persimmon (Simmon, Possumwood)	*Diospyros virginiana*
Texas Persimmon (Black Persimmon, Mexican Persimmon)	*Diospyros texana*

Family and Common Names	Scientific Name
Snowbell (Storax) Family, *Styracaceae:*	
Carolina Silverbell (Snowdrop-tree, Opossum-wood)	*Halesia carolina*
Little Silverbell (Florida Silverbell)	*Halesia parviflora*
Two-wing Silverbell (Snowdrop-tree)	*Halesia diptera*
Bigleaf Snowbell (Snowbell, Storax)	*Styrax grandifolius*
Sweetleaf Family, *Symplocaceae:*	
Sweetleaf (Horse-sugar, Yellowwood)	*Symplocos tinctoria*
Olive Family, *Oleaceae:*	
Fringetree (Old-man's-beard)	*Chionanthus virginicus*
Desert-olive Forestiera (Desert-olive, Wild-olive)	*Forestiera phillyreoides*
Florida-privet (Florida Forestiera, Wild-olive)	*Forestiera segregata*
Swamp-privet (Common Adelia, Texas Forestiera)	*Forestiera acuminata*
Berlandier Ash (Mexican Ash)	*Fraxinus berlandierana*
Black Ash (Basket Ash, Hoop Ash)	*Fraxinus nigra*
Blue Ash	*Fraxinus quadrangulata*
Carolina Ash (Water Ash, Pop Ash)	*Fraxinus caroliniana*
Chihuahua Ash	*Fraxinus papillosa*
Fragrant Ash (Flowering Ash)	*Fraxinus cuspidata*
Goodding Ash	*Fraxinus gooddingii*
Green Ash (Swamp Ash, Water Ash)	*Fraxinus pennsylvanica*
Gregg Ash (Littleleaf Ash, Dogleg Ash)	*Fraxinus greggii*
Lowell Ash	*Fraxinus lowellii*
Oregon Ash	*Fraxinus latifolia*
Pumpkin Ash (Red Ash)	*Fraxinus profunda*
Singleleaf Ash (Dwarf Ash)	*Fraxinus anomala*
Texas Ash	*Fraxinus texensis*
Two-petal Ash (Flowering Ash, Foothill Ash)	*Fraxinus dipetala*
Velvet Ash (Arizona Ash, Desert Ash)	*Fraxinus velutina*
White Ash	*Fraxinus americana*
Chinese Privet	*Ligustrum sinense*
Olive (Common Olive, European Olive)	*Olea europaea*
Devilwood (Wild-olive)	*Osmanthus americanus*

Family and Common Names	**Scientific Name**

Borage Family, *Boraginaceae:*

Anacua (Sugarberry, Knockaway) — *Ehretia anacua*

Verbena Family, *Verbenaceae:*

Black-mangrove (Blackwood, Mangle) — *Avicennia germinans*

Figwort Family, *Scrophulariaceae:*

Royal Paulownia (Princess-tree, Empress-tree) — *Paulownia tomentosa*

Bignonia Family, *Bignoniaceae:*

Northern Catalpa (Hardy Catalpa, Indian-bean, Cigartree) — *Catalpa speciosa*
Southern Catalpa (Catawba, Indian-bean) — *Catalpa bignonioides*

Desert-willow (Desert-catalpa) — *Chilopsis linearis*

Madder Family, *Rubiaceae:*

Buttonbush (Honey-balls, Globe-flowers) — *Cephalanthus occidentalis*
Pinckneya (Fever-tree, Georgia-bark) — *Pinckneya pubens*

Honeysuckle Family, *Caprifoliaceae:*

American Elder (Elderberry, Common Elder) — *Sambucus canandensis*
Blue Elder (Blue Elderberry, Blueberry Elder) — *Sambucus cerulea*
Mexican Elder (Arizona Elder, Desert Elderberry) — *Sambucus mexicana*
Pacific Red Elder (Coast Red Elder, Red Elderberry) — *Sambucus callicarpa*

Arrowwood (Southern Arrowwood, Arrowwood Viburnum) — *Viburnum dentatum*
Blackhaw (Stagbush, Sweethaw) — *Viburnum prunifolium*
Nannyberry (Blackhaw, Sheepberry) — *Viburnum lentago*
Possumhaw Viburnum (Possumhaw, Swamphaw) — *Viburnum nudum*
Rusty Blackhaw (Bluehaw, Rusty Nannyberry) — *Viburnum rufidulum*
Walter Viburnum (Blackhaw, Small-leaf Viburnum) — *Viburnum obovatum*

Composite Family, *Compositae:*

Eastern Baccharis (Saltbrush, Groundsel-tree) — *Baccharis halimifolia*

APPENDIX II
Insects of Forests and Ornamental Trees

Cone and Seed Insects: Insects that invade the reproductive structures of conifers.

Insect Types	Examples
Coleopterous Insects:	
Cone Beetles	Ponderosa Pine Cone Beetle
	Red Pine Cone Beetle
	White Pine Cone Beetle
Dipterous Insects:	
Cone Maggots	Fir Cone Maggot
	Spruce Cone Maggot
Hemipterous Insects:	
Coreid Bugs	Southern Pine Seed Bug
	Western Conifer Seed Bug
Stink Bugs	Shield-backed Pine Seed Bug
Hymenopterous Insects:	
Sawflies	Conifer Sawflies
	Xyelid Sawflies
Torymids	Douglas-fir Seed Chalcid
Lepidopterous Insects:	
Coneworms	Southern Pine Coneworm
	Webbing Coneworm
Cone Moths	Douglas-fir Cone Moth
Cone Borers	Shortleaf Pine Cone Borer
	Red Pine Cone Borer

Seed Worms	Eastern Pine Seedworm
	Longleaf Pine Seedworm
	Ponderosa Pine Seedworm
	Spruce Seed Moth

Defoliators: Insects that eat the leaves and needles of trees.

Insect Types	**Examples**
Lepidopterous Insects:	
Case and Bag Makers	Larch Casebearer
	Bagworm
Tent Makers and Webworms	Western and Eastern Tent Caterpillars
	Fall Webworm
Leaf Rollers, Leaf Folders, Leaf and Needle Tiers	Oak Leaf Roller
	Spruce Budworms
	Pine Tube Moth
Leaf and Needle Miners	Lodgepole Needleminer
	Aspen Leafminer
Loopers	Fall and Spring Cankerworms
	Elm Spanworm
Naked Caterpillars	Redhumped Oakworm
	Variable Oakleaf Caterpillar
Hairy Caterpillars	Gypsy Moth
	Douglas-fir Tussock Moth
Spiny Caterpillars	Luna Moth
	Cecropia Moth
	Sphinx Moth
Slug Caterpillars	Saddleback Caterpillar
	Hag Moth
Hymenopterous Insects:	
Cimbicid Sawflies	Elm Sawfly
Diprionid Sawflies	European Spruce Sawfly
	Redheaded Pine Sawfly

		Swaine Jack Pine Sawfly
		Introduced Pine Sawfly
		European Pine Sawfly
		Hemlock Sawfly
		Balsam Fir Sawfly
Tenthredinid Sawflies		Mountain Ash Sawfly
		Larch Sawfly
		Yellowheaded Spruce Sawfly
Slug Sawflies		Pin Oak Sawfly
		Pear Sawfly
Leaf Miners		Birch Leafminer
		Leafmining Wasps
Leafcutters		Leafcutting Bees
		Leafcutting Ants

Coleopterous Insects:

Leaf Beetles		Cottonwood Leaf Beetle
		Alder Flea Beetle
		Elm Leaf Beetle
		Pine Colaspis
		Imported Willow Leaf Beetle
Scarab Beetles		May Beetles
		Rose Chafer
		Japanese Beetle
Weevils		Asiatic Oak Weevil

Orthopterous Insects:

Grasshoppers		Shorthorned Grasshoppers
		Longhorned Grasshoppers
		Katydids
		Mormon Cricket
Walkingsticks		Walkingsticks

Gall Makers: Insects that inject substances into plants that cause abnormal growth.

Insect Types	**Examples**
Hymenopterous Insects:	
Gall Wasps	Jumping Oak Gall Wasp
	Mealy Oak Gall Wasp
Sawflies	Euura Sawflies
Coleopterous Insects:	
Borers	Maple Gall Borer
	Poplar Gall Saperda
Dipterous Insects:	
Gall Midges	Balsam Gall Midge
	Birch Midge
	Honeylocust Pod Gall Midge
	Maple Leaf Spot Gall Midge
Homopterous Insects:	
Adelgids	Balsam Woolly Adelgid
	Cooley Spruce Gall Adelgid
	Eastern Spruce Gall Adelgid
	Pine Leaf Adelgid
Aphids	Poplar Vagabond Aphid
	Sugar Beet Root Aphid
	Wooly Apple Aphid
Psyllids	Budgall Psyllid
	Hackberry Blister Gall Maker
	Hackberry Nipplegall Maker
	Petiolegall Psyllid
Lepidopterous Insects:	
Borers	Boxelder Twig Borer
Arachnids:	
Mites	Maple Bladdergall Mite
	Maple Spindlegall Mite

Leaf and Needle Miners: Insects that feed on the tissue located inside a leaf or needle.

Insect Types	Examples
Lepidopterous Insects:	
Leafminers	Arborvitae Leafminer
	Cypress Tipminer
	Aspen Blotchminer
	Aspen Leafminer
	Gregarious Oak Leafminer
	Solitary Oak Leafminer
Needleminers	Lodgepole Needleminer
	Northern Lodgepole Needleminer
Hymenopterous Insects:	
Leafminer	Birch Leafminer
Coleopterous Insects:	
Leafminers	Basswood Leafminer
	Locust Leafminer
Dipterous Insects:	
Leafminers	Holly Leafminer
	Native Holly Leafminer

Phloem and Wood-boring Insects: Insects that use the inner bark and woody tissues of trees for food and habitat.

Insect Types	Examples
Lepidopterous Insects:	
Carpenterworms	Carpenterworm
	Leopard Moth
	Little Carpenterworm
Coleopterous Insects:	
Bark Beetles	Black Turpentine Beetle
	Douglas-fir Beetle
	Fir Engraver
	Mountain Pine Beetle
	Red Turpentine Beetle

Smaller European Elm Bark Beetle
Southern Pine Beetle
Spruce Beetle
Western Pine Beetle

Ips Engraver Beetles

California Fivespined Ips
Eastern Fivespined Ips
Pine Engraver
Sixspined Engraver
Small Southern Pine Engraver

Ambrosia Beetles

Columbian Timber Beetle
Striped Ambrosia Beetle

Powderpost Beetles

Anobiid Beetles
Bostrichid Beetle
Deathwatch Beetle
False Powderpost Beetle
Furniture Beetle
Lyctus Beetle
Old House Borer
True Powderpost Beetle

Roundheaded Borers

Cottonwood Borer
Firtree Borer
Locust Borer
Northern Spruce Borer
Poplar Borer
Red Oak Borer
Southern Pine Sawyer

Flatheaded Borers

Bronze Birch Borer
California Flatheaded Borer
Turpentine Borer
Western Cedar Borer

Hymenopterous Insects:

Ants

Black Carpenter Ant
Florida Carpenter Ant

Bees

Carpenter Bees

Horntails

Blue Horntail
Pigeon Tremex

Lepidopterous Insects:

   Phloem Borers   Clearwing Borers

Dipterous Insects:

   Phloem Borers   Leaf Mining Flies

Isopterous Insects:

   Termites   Aridland Subterranean Termite
Eastern Subterranean Termite
Formosan Subterranean Termite
Nevada Dampwood Termite
Pacific Dampwood Termite
Powderpost Termites
Southeastern Drywood Termite
Western Drywood Termite
Western Subterranean Termite

Root, Shoot, Twig, and Terminal Insects: Insects that feed on tips, terminal branches, lateral branches, stems, root collars, and roots.

Insect Types	**Examples**

Lepidopterous Insects:

   Moths   European Pine Shoot Moth
Nantucket Pine Tip Moth
Pitch Pine Tip Moth
Southwestern Pine Tip Moth
Subtropical Pine Tip Moth
Western Pine Tip Moth

   Borers   Cottonwood Twig Borer
Western Pine Shoot Borer

Coleopterous Insects:

   Weevils   Deodar Weevil
Northern Pine Weevil
Pales Weevil
Pine Reproduction Weevils
Pine Root Collar Weevil
Pitch-eating Weevil
Terminal Weevils

	Warren's Collar Weevil
	White Pine Weevil
Borers	Twig Girdler
	Twig Pruner

Sucking Mites and Insects: Insects that have piercing and sucking mouth parts adapted to suck sap.

Insect Types	**Examples**
Homopterous Insects:	
Cicadas	Dog-day Cicada
	Periodical Cicada
Spittlebugs	Pine Spittlebug
	Saratoga Spittlebug
	Western Pine Spittlebug
Treehoppers	Buffalo Treehopper
	Oak Treehopper
Leafhoppers	Whitebanded Elm Leafhopper
Psyllids	Budgall Psyllid
	Hackberry Psyllids
	Petiolegall Psyllid
Aphids	Balsam Twig Aphid
	Spruce Aphid
	White Pine Aphid
Adelgids	Gall Adelgid
	Wooly Adelgid
Whiteflies	Azalea Whitefly
	Crown Whitefly
	Rhododendron Whitefly
Scales	Armored Scales
	Beech Scale
	Cottony Maple Scale
	European Elm Scale
	European Fruit Lecanium
	Fletcher Scale
	Lecanium Scales

Magnolia Scale
Oystershell Scale
Pine Needle Scale
Pine Tortoise Scale
Pinon Needle Scale
Pit Scales
Red Pine Scale
Tuliptree Scale

Hemipterous Insects:

Plant Bugs

Fourlined Plant Bug
Tarnished Plant Bug

Lace Bugs

Azalea Lace Bug

Boxelder Bugs

Boxelder Bug
Western Boxelder Bug

Arachnids:

Spider Mites

Spruce Spider Mite

APPENDIX III
Schools Offering Forestry/Natural Resource Programs

State or Territory	**School** (FT = Forest Technology)
Alabama	Alabama Agricultural and Mechanical University
	Alabama Southern Community College (Monroeville) (FT Only)
	Alabama Southern Community College (Thomasville)
	Auburn University (FT)
	Bevill State Community College
	Chattahoochee Valley State Community College
	Gadsden State Community College
	Lawson State Community College
	Lurleen B. Wallace State Junior College (FT Only)
	Northeast Alabama State Community College
	Northwest-Shoals Community College
	Reid State Technical College (FT Only)
	Samford University
	Troy State University
Alaska	Alaska Pacific University
	Sheldon Jackson College (FT)
	University of Alaska Anchorage
	University of Alaska Anchorage, Kenai Peninsula College (FT Only)
	University of Alaska Fairbanks
Arizona	Arizona State University
	Eastern Arizona College
	Gateway Community College
	Glendale Community College
	Northern Arizona University
	Prescott College, Adult Degree Program
	Prescott College, Resident Degree Program
	Rio Salado Community College
	University of Arizona
Arkansas	University of Arkansas at Monticello

State or Territory	**School** (FT = Forest Technology)
California	American River College (FT)
	Antelope Valley College
	Bakersfield College
	Butte College
	California Polytechnic State University, San Luis Obispo
	California State University, Bakersfield
	California State University, Chico
	California State University, Fullerton
	Cerritos College
	Citrus College
	City College of San Francisco
	College of the Canyons
	College of the Desert
	College of the Redwoods
	College of the Siskiyous
	Columbia College (FT)
	El Camino College (FT Only)
	Feather River Community College District (FT Only)
	Fresno City College
	Fullerton College
	Hartnell College (FT Only)
	Humboldt State University
	Imperial Valley College
	Kings River Community College (FT Only)
	Los Angeles Pierce College
	Los Angeles Trade-Technical College
	Merritt College
	Modesto Junior College (FT)
	Monterey Institute of International Studies
	Mt. San Antonio College (FT Only)
	National University
	Palomar College
	Palo Verde College
	Pasadena City College (FT Only)
	Rancho Santiago College
	Sacramento City College
	San Diego Mesa College
	San Francisco State University
	San Joaquin Delta College
	San Jose State University
	Santa Barbara City College
	Santa Rosa Junior College (FT Only)
	Shasta College
	Sierra College (FT)

State or Territory	**School** (FT = Forest Technology)
	University of California, Berkeley (FT)
	University of California, Davis
	University of California, Irvine
	University of California, Los Angeles
	University of California, Santa Barbara
	University of California, Santa Cruz
	University of San Francisco
	University of Southern California
	Ventura College
	West Coast University
Colorado	Colorado Mountain College, Timberline Campus
	Colorado Northwestern Community College
	Colorado School of Mines
	Colorado State University
	Metropolitan State College of Denver
	The Naropa Institute
	Northeastern Junior College
	Red Rocks Community College
	Trinidad State Junior College
Connecticut	The Hartford Graduate Center
	University of Connecticut
	Yale University
Delaware	Delaware State University
	University of Delaware
District of Columbia	George Washington University
	University of the District of Columbia
Florida	Brevard Community College
	Daytona Beach Community College
	Florida Institute of Technology
	Gulf Coast Community College
	Indian River Community College
	Lake City Community College (FT)
	Miami-Dade Community College
	Pensacola Junior College (FT)
	St. Petersburg Junior College
	University of Florida
	University of West Florida
Georgia	Abraham Baldwin Agricultural College (FT)

State or Territory	**School** (FT = Forest Technology)
	Andrew College
	Atlanta Metropolitan College
	Bainbridge College
	Clayton State College
	Coastal Georgia Community College
	Columbus State University
	Dalton College
	Darton College
	Gainesville College
	Mercer University
	Middle Georgia College
	Savannah State University
	University of Georgia (FT)
	Waycross College (FT)
Hawaii	University of Hawaii at Manoa
Guam	Guam Community College
Idaho	College of Southern Idaho
	North Idaho College
	Ricks College
	University of Idaho
Illinois	Elmhurst College
	Illinois Institute of Technology
	Joliet Junior College
	Kishwaukee College
	Lake Land College
	Northwestern University
	Southeastern Illinois College (FT)
	Southern Illinois University at Carbondale
	University of Illinois at Chicago
	University of Illinois at Urbana-Champaign
Indiana	Ball State University
	Huntington College
	Indiana State University
	Indiana University–Purdue University, Indianapolis
	Purdue University (FT)
	Valparaiso University
	Vincennes University
Iowa	Ellsworth Community College

State or Territory	**School** (FT = Forest Technology)
	Iowa Lakes Community College
	Iowa State University of Science and Technology
	Iowa Wesleyan College
	Kirkwood Community College
	Muscatine Community College
	St. Ambrose University
	University of Northern Iowa
	Upper Iowa University
Kansas	Barton County Community College
	Colby Community College
	Dodge City Community College
	Fort Hays State University
	Fort Scott Community College
	Haskell Indian Nations University
	Highland Community College
	Hutchinson Community College and Area Vocational School
	Kansas State University
Kentucky	St. Catharine College
	University of Kentucky
	University of Kentucky, Hazard Community College (FT Only)
	University of Kentucky, Jefferson Community College
	Western Kentucky University
Louisiana	Louisiana State University and A & M College
	Louisiana Tech University
	University of Southwestern Louisiana
Maine	Unity College (FT)
	University of Maine
	University of Maine at Fort Kent
Maryland	Allegany College of Maryland (FT)
	Cecil Community College
	Frostburg State University
	Garrett Community College
	Johns Hopkins University
	University of Maryland College Park
Massachusetts	Berkshire Community College
	Boston University
	Bridgewater State College
	Clark University

State or Territory	**School** (FT = Forest Technology)
	Greenfield Community College
	Hampshire College
	Harvard University
	Springfield College
	Tufts University
	University of Massachusetts Amherst
Michigan	Bay de Noc Community College
	Delta College
	Eastern Michigan University
	Grand Rapids Community College
	Grand Valley State University
	Lake Superior State University
	Michigan State University
	Michigan Technological University (FT)
	Northern Michigan University
	University of Michigan
	University of Michigan, Flint
	Wayne County Community College
Minnesota	College of Saint Benedict
	Itasca Community College (FT)
	Saint John's University
	University of Minnesota, Crookston
	University of Minnesota, Twin Cities Campus (FT)
	Vermilion Community College (FT)
	Winona State University
Mississippi	Copiah-Lincoln Community College, Natchez Campus
	East Mississippi Community College (FT)
	Holmes Community College
	Itawamba Community College (FT Only)
	Jones County Junior College (FT Only)
	Mississippi State University
	Northeast Mississippi Community College (FT)
	Northwest Mississippi Community College
Missouri	Central Missouri State University
	East Central College
	Jefferson College
	Northwest Missouri State University
	Southwest Missouri State University
	University of Missouri

State or Territory	School (FT = Forest Technology)
Montana	Dull Knife Memorial College
	Flathead Valley Community College (FT Only)
	Fort Belknap College
	Fort Peck Community College
	Montana State University, Bozeman
	Montana State University, Northern
	Montana Tech of the University of Montana
	Rocky Mountain College
	Salish Kootenai College (FT)
	University of Montana, Missoula
Nebraska	Nebraska College of Technical Agriculture
	Nebraska Indian Community College
	Peru State College
	University of Nebraska, Lincoln
	Western Nebraska Community College
Nevada	University of Nevada, Las Vegas
	University of Nevada, Reno
New Hampshire	Antioch New England Graduate School
	New Hampshire Technical College
	University of New Hampshire (FT)
New Jersey	Camden County College
	Montclair State University
	New Jersey Institute of Technology
	Rutgers, the State University of New Jersey, Cook College
	Rutgers, the State University of New Jersey, New Brunswick
	Stevens Institute of Technology
	Thomas Edison State College
New Mexico	College of Santa Fe
	Dona Ana Branch Community College
	Northern New Mexico Community College
	Southwestern Indian Polytechnic Institute
	University of New Mexico
New York	Adirondack Community College (FT)
	Alfred University
	Bard College
	Clarkson University
	Cornell University
	Erie Community College, City Campus (FT)

State or Territory	School (FT = Forest Technology)
	Erie Community College, South Campus (FT)
	Finger Lakes Community College
	Fulton-Montgomery Community College (FT)
	Herkimer County Community College
	Jefferson Community College (FT Only)
	Long Island University, C. W. Post Campus
	Long Island University, Southampton, Friends World Prog.
	Monroe Community College
	North Country Community College (FT)
	Paul Smith's College of Arts and Sciences (FT)
	Rensselaer Polytechnic Institute
	Rockland Community College
	State University of New York (SUNY) College at Brockport
	SUNY College at Oneonta
	SUNY College of Agriculture and Technology at Cobleskill (FT Only)
	SUNY College of Agriculture and Technology at Morrisville (FT)
	SUNY College of Environmental Science and Forestry
	SUNY College of Environmental. Science and Forestry, Ranger School (FT Only)
	SUNY College of Technology at Canton (FT Only)
	SUNY College of Technology at Delhi (FT)
	Sullivan County Community College (FT)
	Syracuse University
North Carolina	Brevard College (FT)
	Duke University
	Haywood Community College (FT Only)
	High Point University
	Lees-McRae College
	Montgomery Community College (FT Only)
	North Carolina State University
	Southeastern Community College (FT Only)
	Warren Wilson College
	Wayne Community College (FT Only)
	Western Carolina University
North Dakota	Minot State University, Bottineau (FT)
	North Dakota State University
	Turtle Mountain Community College
Ohio	Central State University
	Heidelberg College
	Hocking College (FT)
	Kent State University

State or Territory	School (FT = Forest Technology)
	Mount Vernon Nazarene College
	Ohio State University
	Ohio State University Agricultural Technical Institute
	Ohio University
	University of Findlay
	University of Toledo
	Wright State University
Oklahoma	Bacone College
	East Central University
	Eastern Oklahoma State College (FT)
	Murray State College
	Northeastern Oklahoma Agricultural and Mechanical College
	Oklahoma State University
	Southeastern Oklahoma State University
	Tulsa Junior College
	University of Oklahoma
Oregon	Central Oregon Community College (FT Only)
	Chemeketa Community College (FT)
	Clackamas Community College
	Eastern Oregon State College
	Linn-Benton Community College
	Mt. Hood Community College (FT Only)
	Oregon State University
	Southwestern Oregon Community College (FT Only)
	Treasure Valley Community College (FT)
	Umpqua Community College
Pennsylvania	Albright College
	California University of Pennsylvania
	Duquesne University
	Keystone College
	Moravian College
	Pennsylvania College of Technology (FT)
	Pennsylvania State University, Mont Alto Campus (FT Only)
	Pennsylvania State University, University Park Campus (FT)
	Saint Joseph's University
	Shippensburg University of Pennsylvania
	Slippery Rock University of Pennsylvania
	Thiel College
Rhode Island	University of Rhode Island

State or Territory	**School** (FT = Forest Technology)
South Carolina	Central Carolina Technical College Clemson University Horry-Georgetown Technical College (FT Only) Orangeburg-Calhoun Technical College (FT Only) University of South Carolina
South Dakota	Oglala Lakota College Sinte Gleska University South Dakota School of Mines and Technology South Dakota State University
Tennessee	Chattanooga State Technical Community College (FT) Hiwassee College (FT) Knoxville College Tennessee Technological University University of Tennessee at Martin University of Tennessee, Knoxville University of the South
Texas	Baylor University Frank Phillips College Grayson County College Hardin-Simmons University Panola College (FT Only) Stephen F. Austin State University Tarleton State University Texas A&M University Texas Tech University University of Houston, Clear Lake
Utah	Brigham Young University Dixie College Snow College Utah State University
Vermont	Goddard College Marlboro College Sterling College University of Vermont Vermont Law School
Virginia	Christopher Newport University Dabney S. Lancaster Community College (FT Only) Lord Fairfax Community College

State or Territory	**School** (FT = Forest Technology)
	Mountain Empire Community College
	Southwest Virginia Community College
	Virginia Polytechnic Institute and State University
	Washington and Lee University
Washington	Central Washington University
	Centralia College (FT Only)
	Evergreen State College
	Grays Harbor College
	Green River Community College (FT)
	Heritage College
	Spokane Community College
	Tacoma Community College
	University of Washington
	Washington State University
West Virginia	Glenville State College (FT Only)
	Potomac State College of West Virginia University (FT)
	West Virginia University
Wisconsin	Fox Valley Technical College (FT)
	Marquette University
	Milwaukee Area Technical College
	Moraine Park Technical College
	Northland College
	University of Wisconsin, Madison
	University of Wisconsin, Milwaukee
	University of Wisconsin, Platteville
	University of Wisconsin, River Falls
	University of Wisconsin, Stevens Point
Wyoming	Casper College
	Central Wyoming College
	Northwest College
	University of Wyoming

APPENDIX IV
Timberland Ownership by State, 1992

(In Thousands of Acres)

Region & State	National Forests	Other Public	Forest Industry	Farmer & Other Private	All Owners
Northeast					
Connecticut	0	216	4	1,549	1,768
Delaware	0	13	31	332	332
Maine	40	487	8,017	8,444	16,987
Maryland	0	246	131	2,048	2,424
Massachusetts	0	430	66	2,463	2,960
New Hampshire	516	196	658	3,390	4,760
New Jersey	0	464	0	1,400	1,864
New York	6	987	1,035	13,717	15,744
Pennsylvania	466	2,925	613	11,847	15,850
Rhode Island	0	45	4	322	371
Vermont	232	238	410	3,549	4,429
West Virginia	920	250	891	9,855	11,916
Total	**2,179**	**6,407**	**11,858**	**58,914**	**79,449**
North Central					
Illinois	226	163	13	3,628	4,030
Indiana	166	369	18	3,743	4,296
Iowa	0	156	0	1,788	1,944
Michigan	2,354	3,803	1,981	9,265	17,442
Minnesota	1,821	5,781	751	6,420	14,773
Missouri	1,328	691	222	11,137	13,377
Ohio	188	331	175	6,874	7,567
Wisconsin	1,245	2,970	1,179	9,527	14,921
Total	**7,366**	**14,263**	**4,340**	**52,380**	**78,350**
Great Plains					
Kansas	0	46	3	1,158	1,208
Nebraska	29	26	0	481	536
North Dakota	0	35	0	304	338
South Dakota	914	91	21	421	1,447
Total	**943**	**198**	**24**	**2,363**	**3,529**

Southeast					
Florida	990	1,444	4,789	7,779	14,983
Georgia	752	894	4,990	16,995	23,631
North Carolina	1,082	868	2,252	14,508	18,710
South Carolina	577	596	2,626	8,380	12,179
Virginia	1,446	507	1,614	11,724	15,292
Total	**4,847**	**4,309**	**16,253**	**59,387**	**84,794**
South Central					
Alabama	615	557	4,795	15,975	21,941
Arkansas	2,338	794	4,386	9,905	17,423
Kentucky	631	329	205	11,196	12,360
Louisiana	568	743	3,937	8,607	13,855
Mississippi	1,144	721	3,267	11,859	16,991
Oklahoma	244	346	1,077	4,455	6,122
Tennessee	565	953	1,122	10,635	13,275
Texas	602	196	3,986	7,763	12,548
Total	**6,707**	**4,639**	**22,774**	**80,395**	**114,515**
Pacific Northwest					
Alaska	3,780	5,102	0	6,185	15,068
Oregon	10,151	2,853	4,926	3,683	21,614
Washington	4,859	2,427	4,108	4,843	16,238
Total	**18,790**	**10,383**	**9,034**	**14,711**	**52,920**
Pacific Southwest					
California	8,370	416	3,280	4,134	16,200
Hawaii	0	338	0	362	700
Total	**8,370**	**754**	**3,280**	**4,497**	**16,900**
Rocky Mountains					
Arizona	2,650	56	0	1,262	3,968
Colorado	7,062	1,401	0	3,277	11,739
Idaho	9,705	1,525	1,239	2,006	14,474
Montana	8,300	1,605	1,618	4,340	15,863
Nevada	102	10	0	112	224
New Mexico	3,321	141	0	1,958	5,420
Utah	2,108	374	0	597	3,078
Wyoming	2,211	677	37	1,407	4,332
Total	**35,459**	**5,789**	**2,894**	**14,959**	**59,099**
Total U.S.	**84,661**	**46,833**	**70,455**	**287,606**	**489,555**

Source: Douglas S. Powell, et al.

APPENDIX V
United States Land Areas

(In Thousands of Acres)

Region & State	Total Land Area	Forested Acres	Forested Percent
Northeast			
Connecticut	3,101	1,819	59%
Delaware	1,251	389	31%
Maine	19,753	17,533	89%
Maryland	6,295	2,700	43%
Massachusetts	5,016	3,203	64%
New Hampshire	5,740	4,981	87%
New Jersey	4,748	2,007	42%
New York	30,223	18,713	62%
Pennsylvania	28,685	16,969	59%
Rhode Island	669	401	60%
Vermont	5,920	4,538	77%
West Virginia	15,415	12,128	79%
Total	**126,816**	**85,380**	**67%**
North Central			
Illinois	35,580	4,266	12%
Indiana	22,957	4,439	19%
Iowa	35,760	2,050	6%
Michigan	36,358	18,253	50%
Minnesota	50,955	16,718	33%
Missouri	44,095	14,007	32%
Ohio	26,210	7,863	30%
Wisconsin	34,761	15,513	45%
Total	**286,764**	**83,108**	**29%**
Pacific Northwest			
Alaska	365,039	129,131	35%
Oregon	61,442	27,997	46%
Washington	42,612	20,483	48%
Total	**469,093**	**177,611**	**38%**

Pacific Southwest

California	99,823	37,263	37%
Hawaii	4,111	1,748	43%
Total	**103,934**	**39,011**	**38%**

Great Plains

Kansas	52,367	1,359	3%
Nebraska	49,202	722	1%
North Dakota	44,156	462	1%
South Dakota	48,575	1,690	3%
Total	**194,299**	**4,232**	**2%**

Southeast

Florida	34,558	16,549	48%
Georgia	37,068	24,137	65%
North Carolina	31,180	19,278	62%
South Carolina	19,271	12,257	64%
Virginia	25,343	15,858	63%
Total	**147,419**	**88,078**	**60%**

South Central

Alabama	32,480	21,974	68%
Arkansas	33,328	17,864	54%
Kentucky	25,429	12,714	50%
Louisiana	27,882	13,864	50%
Mississippi	30,025	17,000	57%
Oklahoma	43,954	7,539	17%
Tennessee	26,380	13,612	52%
Texas	167,625	19,193	11%
Total	**387,104**	**123,760**	**32%**

Rocky Mountains

Arizona	72,731	19,595	27%
Colorado	66,387	21,338	32%
Idaho	52,961	21,621	41%
Montana	93,156	22,512	24%
Nevada	70,276	8,938	13%
New Mexico	77,673	15,296	20%
Utah	52,588	16,234	31%
Wyoming	62,147	9,966	16%
Total	**547,918**	**135,499**	**25%**

Grand Total	**2,263,259**	**736,681**	**33%**

Source: Douglas S. Powell, et al.

APPENDIX VI
The Tree Farm Community, 1994

(In Acres)

State	Tree Farms	Total Acreage
Alabama	2,843	7,595,231
Alaska	32	564,215
Arizona	30	4,208
Arkansas	4,564	5,398,198
California	660	2,728,685
Colorado	236	169,154
Connecticut	459	136,318
Delaware	286	45,410
Florida	4,147	6,967,601
Georgia	4,441	6,980,150
Idaho	754	1,710,372
Illinois	1,095	97,775
Indiana	1,058	113,617
Iowa	898	73,465
Kansas	318	13,951
Kentucky	879	320,195
Louisiana	2,931	4,419,000
Maine	1,938	8,037,567
Maryland	1,469	270,334
Massachusetts	1,365	249,143
Michigan	2,235	2,469,181
Minnesota	2,751	975,525
Mississippi	5,210	4,012,855
Missouri	1,070	325,381
Montana	357	1,202,408
Nebraska	194	8,866
New Hampshire	1,696	966,267
New Jersey	296	86,236
New Mexico	162	361,113
New York	2,447	1,384,975
North Carolina	3,125	2,551,121
North Dakota	295	15,016

Ohio	2,006	533,832
Oklahoma	270	1,223,280
Oregon	1,238	6,045,846
Pennsylvania	1,612	681,744
Rhode Island	204	35,552
South Carolina	1,792	3,341,703
South Dakota	211	27,244
Tennessee	1,370	1,317,150
Texas	3,703	4,425,071
Utah	4	13,896
Vermont	927	514,353
Virginia	2,633	1,890,730
Washington	1,165	6,299,037
West Virginia	765	1,512,326
Wisconsin	3,900	1,688,392
Wyoming	30	20,282
Total U.S.	**72,071**	**88,823,971**

Source: American Forest Foundation

APPENDIX VII

Harvested Volume in the United States by Ownership and Region, 1991

Region	National Forests	Other Public	Forest Industry	Farmer & Other Private	All Owners
Million Cubic Feet					
Northeast	29	83	573	634	1,319
North Central	115	212	151	990	1,468
Great Plains	23	0	0	28	51
Southeast	124	114	1,156	2,777	4,171
South Central	248	90	1,752	2,691	4,781
Pacific Northwest	605	408	1,110	470	2,594
Pacific Southwest	375	15	411	106	908
Rocky Mountains	382	57	173	165	776
Alaska	99	5	0	136	240
Total U.S.	**2,001**	**985**	**5,325**	**7,997**	**16,308**
Million Board Feet					
Northeast	135	387	2,675	2,962	6,160
North Central	537	990	703	4,624	6,854
Great Plains	109	0	0	129	238
Southeast	579	534	5,397	12,970	19,480
South Central	1,160	421	8,180	12,565	22,327
Pacific Northwest	3,239	2,184	5,938	2,516	13,877
Pacific Southwest	1,976	81	2,168	560	4,785
Rocky Mountains	1,782	264	806	769	3,622
Alaska	532	27	0	726	1,285
Total U.S.	**10,049**	**4,889**	**25,868**	**37,821**	**78,628**

Source: Douglas S. Powell, et al.

GLOSSARY

Abiotic disease An unhealthy state of being caused by a nonliving factor or condition.

Acid precipitation A condition that occurs when moisture in the air combines with sulphur or nitrogen compounds to form acid. It is a serious threat to plant and animal life.

Acre A tract of land equal to 43,560 square feet.

Adaptive behavior Changes in the habits of animals that they have learned for the purpose of increasing their chances of survival.

Adenosine triphosphate (ATP) A high-energy molecule that is produced during photosynthesis; source of energy for muscle movement in animals.

Adventitious root A root that grows downward from the main stem of a tree to provide added support to a tree growing in an aquatic environment; also known as a prop root.

Aerial fuel Fuel that is located more than six feet above the ground in the mid to upper canopy of the forest.

Agency An administrative division of government that is assigned specific duties and functions.

Alfisols Slightly acid soils with high mineral contents of calcium, magnesium, sodium, and potassium.

Allelopathic effect A plant response to competition in which live plants on the soil surface release small amounts of poisonous chemicals into the soil around them as protection against invasion by other plants.

Alluvial fan A geological formation of gravel, clay, sand, and silt that has been deposited by water; located where a stream slows down as it enters a plain or where a tributary joins a main stream.

Anaphase The step during the cell division process known as mitosis during which the chromatids are pulled apart by the spindles.

Anatomy The study of the structure of an organism.

Angiosperm A class of tree that produces seeds inside an ovary or fruit; a flowering tree.

Annual ring A new layer of woody xylem tissue that forms each year throughout the life of a tree.

Anther The plant organ in which pollen grains develop and mature.

Apical meristem A rapidly dividing mass of cells located on the ends of stems and roots that cause the stems and roots to elongate or grow longer.

Aquatic Adapted to a water environment.

Arboriculture The scientific care of shrubs and trees in cities and towns.

Are A metric measurement of land area; a land area equal to 100 square meters.

Artificial regeneration Forest renewal that occurs when seeds or seedlings are planted at a harvest site.

Asexual reproduction Propagation of a plant or other organism from its parts, such as seedling production from leaf, root, or stem tissue.

Backfire A fire started along the inside edge of a firebreak to burn the fuel supply back to the wildfire.

Bare-root stock Seedlings that have been removed from the soil in preparation for storage and shipping.

Basal area A measurement of timber volume that may be determined for a single tree or an entire forest.

Baseline A survey line that runs east-west.

Beam A timber that is equal to or greater than 8" × 8" in dimension.

Biltmore stick A measuring instrument that is used to determine the diameter of a tree at breast height.

Biological control A method of controlling undesirable insects and weeds by introducing their natural enemies into the environment.

Biological succession Changes that occur as living organisms replace other lower-order organisms in an environment.

Biological value The relative worth of the life forms that populate an area, taking into account economic values and effects on climate, watersheds, water temperature, soil erosion, wildlife, and so on.

Biomass Vegetation and waste materials containing large amounts of vegetable matter that is rich in cellulose.

Biomass power Electrical power generated from energy obtained from the heat of burning plant materials.

Biotic disease An unhealthy condition that is caused by living agents of infection such as bacteria, fungi, viruses, micoplasmas, parasites, nematodes, and so on.

Bipinnately compound Multiple leaflets attached to side-branches that grow opposite one another on a central leafstalk.

Blade The flat part of a leaf; usually green or red in color.

Block A log or flitch that has been cut to a standard length to fit a lathe in the manufacture of veneer.

Blowup A crown fire that suddenly erupts into an extremely intense fire.

Board A cut of wood that is less than 2" thick and greater than 4" wide.

Board foot (BF) A unit of measurement for lumber equal to a piece of wood 1' long × 12" wide × 1" thick.

Boreal forest A forest that is located in the northern zone of the North American continent; characterized by a wet, cold climate.

British thermal unit (Btu) A measurement of heat.

Brown rots A type of wood rot that occurs in trees as fungi break down the cellulose in the cell walls, especially in heartwood.

Bucking A process by which a felled tree is cut into desirable lengths.

Bunching The process of assembling the logs in an area into small piles.

Butt rot Decay or rotting of the interior wood near the base of a tree.

Callus tissue Plant tissue that is not differentiated into leaf, root, stem, or other specialized tissue.

Calvin cycle A series of chemical reactions that convert carbon dioxide to simple sugars during the process of photosynthesis.

Cambium A layer of meristem tissue located between the wood and the bark of a tree from which new wood and bark are produced.

Canker A fungi infection in a tree that kills the soft tissues of a tree such as the cambium and the bark.

Canopy The highest level of vegetation in a forest, consisting of the branches and foliage of the tallest trees.

Cant A log from which all four slabs have been removed.

Carbonization process A process, also known as destructive distillation, by which wood is heated in the absence of oxygen to produce charcoal and volatile gases.

Carnivore An animal or other organism whose diet consists of meat.

Cell A small structure that contains cytoplasm and a nucleus; the basic unit of life.

Cell membrane The outer flexible membrane that surrounds the contents of a cell.

Cellulose The substance of which the cell walls of plants are composed; wood fiber used in paper manufacturing.

Cellulose xanthate A viscous liquid obtained from the cellulose component of wood that is used to manufacture rayon, cellophane, photographic films, and so on.

Cell wall The rigid outer layer or covering of a plant cell.

Centriole A cell structure found at either end of a cell that anchors the fibers, drawing chromosomes apart during cell division.

Centromere The point of attachment for a pair of chromatids.

Chain The unit of forest measurement of distance; one chain is equal to the 66' length of a surveyor's chain.

Chain saw A gasoline-powered saw used to fell and trim trees as one of the procedures in a logging operation.

Chemical control The use of chemicals to control weeds and insects.

Chemical pulping A processing method that uses chemicals to dissolve the lignin component of wood in the manufacturing process for high-strength paper.

Chlorophyll A green substance found in plant cells that aids the plant in capturing energy from sunlight and converting it to sugars and starches through the process of photosynthesis.

Chloroplast A plant cell structure containing chlorophyll that converts raw materials to sugars and starches using energy obtained from sunlight.

Choker A cable that is fastened around a log for the purpose of transporting it to the yard or landing.

Chromatid Half of a replicated chromosome.

Cleaning operation Removal of forest vegetation that competes with young trees during the seedling to sapling stage of maturity.

Clear-cutting A harvest method in which all of the trees in the stand are cut at the same time.

Climax community The population of plants that occupies an environment when succession of species is complete and plant populations have become stable.

Coastal plain A land area, particularly along the Atlantic or Gulf coasts, consisting of a series of terraces that generally run parallel to the coastline.

Coke A highly combustible wood product that is obtained by heating wood to temperatures in excess of its combustion temperature in large ovens from which oxygen is excluded.

Collenchyma Thick-walled plant cells that add strength to stems and stalks.

Computer A machine that performs mathematical calculations in a programmed sequence. It is capable of cataloguing and integrating large amounts of information.

Conifer A tree or shrub that produces cones containing seeds.

Conk A growth that arises on a tree trunk due to a fungal infection inside the tree, and from which reproductive spores are released into the environment.

Conservation of matter A basic law of physics: matter may change from one form to another, but it cannot be created or destroyed by natural physical or chemical processes.

Containerized seedlings Very young trees that are produced from seeds in individual containers under greenhouse conditions.

Controlled burn The use of fire to burn trash materials from the soil surface in preparation for planting, or as a fire prevention measure in established forests; sometimes known as a prescribed burn.

Coppice method A form of asexual reproduction of a forest in which all of the trees are cut and new forest growth is generated from the stumps of the harvested trees; also known as the sprout method.

Cork A protective outer covering of root and stem tissues consisting of dead plant cells containing a waxy material.

Cork cambium Meristem tissue that produces the cork on the outside of a root.

Cortex Loosely arranged parenchyma cells located in the interior of a root in which sugars and starches are stored.

Cross-banding Orientation of the grain of the wood in each layer of plywood to align the long wood fibers across one another.

Crown fire A surface fire that builds in intensity until it ignites the aerial fuels including the upper foliage of trees.

Cruise The process of estimating the timber yield in a forest.

Cut A category of trees that have been harvested prior to the second forest survey and whose measurements are taken from their stump measurements.

Cuticle A protective coating of waxy material that forms on the outside of plant tissues such as stems and leaves.

Cutting cycle The length of time in years that elapses between timber harvests.

Cuttings Plant parts obtained from leaves, stems, roots, or buds that are treated with a hormone to stimulate rooting and growth of a new plant.

Cytoplasm All of the structures and substances within a cell except for the nucleus.

Database A collection of data in a computer that is organized in a manner that allows it to be retrieved and manipulated with relative ease.

Deadwood A standing tree that has died before it is harvested, such as a tree that has been killed by fire or insects.

Debarking The process of removing the bark from a log by applying pressure and friction to the outer surfaces of the log.

Decibel (dB) A measurement of noise intensity or the loudness of a sound.

Deciduous Shedding leaves annually or at a particular stage of growth.

Dehydration synthesis A process by which a molecule of water is removed each time a single molecule, such as glucose, is added to a larger molecule, such as starch.

Dendrology The scientific study of trees.

Denitrification The process by which some kinds of bacteria break down nitrate compounds, releasing nitrogen gas into the atmosphere.

Department An administrative level of government administered by the executive branch of government.

Depredation hunt A control measure for reducing excessive populations of large game animals by allowing sportsmen to legally hunt them.

Destructive distillation A process by which wood is heated in the absence of oxygen to produce charcoal and volatile gases.

Diameter at breast height (dbh) A forest measurement that is used to estimate the maturity of trees.

Diploid A cell having twice as many chromosomes as it would have during non-reproductive stage.

Direct attack Applying water and fire retardants directly to a fire from aircraft, trucks, and backpacks.

Direct seeding Planting tree seeds to generate new forest growth.

Disease An unhealthy disorder that can be traced to a specific cause, generally with consistent symptoms.

Disk refiner A machine that separates wood fibers between two mechanical disks as the disks rotate in opposite directions.

Dissolving process A process that is used to dissolve the cellulose component of wood to form a viscous liquid called cellulose xanthate in the production of rayon, cellophane, and films.

Dominant use A forest management concept that assigns higher priority to some forest uses than it does to others, and that restricts the forest to high-priority uses.

Dot grid A method for determining the land area on a map by counting the number of dots that fall within the boundary lines.

Doyle's rule A table that is used by the timber industry to estimate the number of board feet in a log.

Draft A supply of fresh air that the convection column draws into the base of a fire.

Duff Ground fuel that is made up of decaying plant material located on or just beneath the surface of the forest floor.

Earth resource satellite A specialized satellite that is placed in an earth orbit that allows it to pass over every part of the earth's surface, and from which data is transmitted to ground bases.

Ecology The branch of biology that describes relationships between living organisms and the environments in which they live.

Electronic mail A computer network service that allows users to communicate by sending letters over the network.

Elemental cycle The recurring circular flow of elements from living organisms to nonliving materials and back again.

Eluviation The loss of soil components from a soil horizon due to downward leaching and movement of water through the soil profile.

E-mail Electronic mail; a computer network service that allows users to communicate by sending letters over the network.

Embryo sac A female gamete in plants consisting of the cell mass located in the ovule that develops into the embryo and the endosperm of a seed after fertilization.

Endodermis The innermost layer of cells in the cortex of a root, surrounded by a waxy waterproof material that restricts the flow of water and some dissolved materials.

Environmental impact statement A science-based study of the harvest area that describes the expected effects of human activities on the environment and the wildlife in the area.

Environmental Protection Agency (EPA) The federal agency charged with protecting and maintaining the environment for future generations.

Epidermis The outer layer of cells that protect plant leaves, stems, flowers, seeds, and roots.

Erosion The loss of topsoil due to the forces of wind or flowing water.

Ethanol An alcohol, used for fuel, that is produced by fermenting carbohydrates followed by distillation of the liquid.

Even-aged stand A population of trees in which most of the trees are approximately the same age.

Evergreen A tree that bears green leaves in all seasons.

Extractive One of several wood products that is dissolved and removed from wood through the use of solvents.

Fermentation A process by which bacteria, yeasts, or enzymes are used to change the chemical makeup of a product, such as producing alcohol from wood.

Fertilization The fusion of male and female gametes; adding plant nutrients to the soil.

Fiberboard A reconstituted board made of wood fibers that become cross-banded in the panel due to the random arrangement of the fibers in the mat from which the board is formed.

Filament The stalk of the stamen in a flower.

Firebreak A zone located in the pathway of a fire from which fuel has been removed for the purpose of stopping the fire from spreading; also called a fire line.

Fire line A zone located in the pathway of a fire from which fuel has been removed for the purpose of stopping the fire from spreading; also called a firebreak.

Firestorm Erratic behavior by a crown fire, such as hurling burning debris beyond the leading edge of the fire and jumping across fire lines and rivers.

Fire suppression All of the activities that are conducted to discover and extinguish a fire.

Flitch A thick piece of high-quality hardwood that is processed into veneer.

Food chain A series of steps through which energy from the sun is transferred to living organisms; members of the food chain feed on lower-ranking members of the community.

Food pyramid An arrangement of organisms in a ranking order according to their dominance in the food web.

Food web A group of interwoven food chains.

Forest A large tract or land area where the most dominant living organisms consist of a population of trees.

Forest floor The layer of decaying plant materials on the soil surface, acting as a mulch to preserve moisture.

Forest growth The increase that occurs in the volume of wood in a forest over a specific period of time.

Forest growth model A computer program that prepares estimates of forest yields by applying all of the factors known to affect forest growth.

Forest inventory A record of the estimated volume of timber of the different varieties that is present in the forest at a given time.

Forestry The science of planting and managing forests for specific purposes such as timber production, conservation, and so on.

Forest type map A map of the locations of different types of trees that is prepared from aerial photographs of the forest with the aid of an instrument called a stereoscope.

Forty A tract of land measuring approximately forty acres.

Forwarder A machine that is used to move logs from the cutting area to the yard or landing area.

Fruiting body A reproductive structure of fungi that grows out of the trunk of an infected tree where it releases reproductive spores into the environment.

Fungi Threadlike plants lacking chlorophyll that obtain their nourishment from other organic materials.

Gamete A haploid reproductive cell.

Gang saw A multiple-bladed lumber processing saw that is capable of making several cuts at the same time.

Gasohol A fuel containing 90% gasoline and 10% ethanol.

Gene A genetic structure that controls the traits and characteristics expressed in an organism.

Genetic engineering The practice of modifying the heredity of an organism by inserting new genes from other organisms into the chromosome structure for the purpose of introducing new hereditary traits.

Geographic information system (GIS) Satellite technology that is used to make observations and photographic images of the earth's surface features and conditions.

Germination The process by which a seed sprouts and begins to grow.

Girdling To cut or remove a ring of bark completely around a tree; a condition often caused by gnawing rodents.

Global positioning system (GPS) The use of satellite technology to accurately and consistently identify exact locations.

Green chain Parallel moving chains that carry boards to new locations in a sawmill.

Gross national product (GNP) A measurement of the strength of the national economy.

Ground fire A fire that burns the decaying fuels on and beneath the surface of the ground.

Ground fuel Combustible materials located beneath the surface that have begun to decay.

Groundwater Water found in or obtained from an aquifer.

Growth habit The natural shape of a mature tree or other plant.

Growth impact A calculation of the extent of insect damage, taking into account the timber losses due to reduced growth rates and deaths of trees.

Guide meridian A survey line that is established running north-south for each interval of slightly less than twenty-four miles (corrected for the curvature of the earth) on either side of the principal meridian.

Gum naval stores Wood products that are refined from a liquid known as oleoresin, obtained by injuring live pine trees and collecting the liquid as it flows from the wounds.

Gymnosperm A class of tree such as pine, spruce, or cedar that bears seeds in cones.

Habitat The living environment of an organism.

Haploid A cell that contains a single chromosome from each homologous chromosome pair.

Hardboard A reconstituted board that is similar to fiberboard except that it is press-bonded between heated steel plates, resulting in a density range for this product of .5 to 1.3 grams/cubic centimeter.

Hardiness A measurement of the tolerance of a particular species of tree or other plant to restrictive environmental factors such as climate, altitude, temperature, and availability of moisture.

Hardware Mechanical and electronic devices and structures that are used to construct a computer.

Hardwood Wood from a broad-leaved tree.

Head rig A large saw that is used to cut logs into marketable dimensions of lumber; also known as a head saw.

Head saw A large saw that is used to cut logs into marketable dimensions of lumber; also known as a head rig.

Heart rot A disease caused by fungi that results in the decay of wood; rotted wood that accumulates at the center of the trunk in a mature tree.

Heartwood Old, dark-colored, nonfunctional xylem located in the center of a tree stem or trunk that has become saturated with tannins, gums, and resins, eliminating the flow of water and dissolved materials.

Hectare A metric measurement for land area that is equal to 10,000 square meters or 100 ares; equivalent to 2.471 acres.

Hemicellulose A polymer that makes up 27–29% of the material found in wood.

Herbicide A chemical that is used to kill plants.

Herbivore An animal or other organism that eats plants.

Herb layer The bottom layer of vegetation in a forest consisting of ferns, grasses, and other low plants that grow on the forest floor beneath the shrub layer.

High forest A forest that has been propagated from seeds.

Home page A menu that advertises and accesses the information and databases that an individual or organization has made available to others who use the Internet system.

Homologous chromosome Each chromosome of an identical pair of chromosomes.

Homologue Each chromosome of an identical pair of chromosomes.

Horizon One of several different layers in a soil profile.

Humus Organic matter in the soil; a source of nutrients for soil organisms and plants.

Hydrapulper A machine that is used to reduce recycled paper to pulp.

Hydrogen ion A hydrogen particle having a positive charge.

Hypsometer A measuring device that uses trigonometry or geometry to calculate the height of a tree without taking a direct measurement.

Illuviation A buildup of translocated soil components that accumulate in the B horizon of a soil; a process by which subsoil is formed.

Incendiarism The willful setting of destructive fires.

Incendiary fire A fire that has been deliberately started by a person.

Increment borer An instrument that is used to extract a core of wood from the cross-section of a living tree to determine its age or the condition of its health.

Indirect attack A method of fighting a fire such as isolating the fire from the fuel supply by establishing firebreaks and lighting backfires.

Industrial waste Any of a variety of materials that are the by-products of a manufacturing process; a variety of harmful chemicals, poisonous metallic compounds, acids, or other caustic materials.

Ingrowth Growth that occurs in a forest that is in excess of the growth accounted for in individual trees; growth due to new trees that were not present or were too small to be tallied when the original timber yield measurements were taken.

Initial point A permanent physical feature that is prominent in the landscape that serves as a beginning point for a rectangular land survey.

Insecticide A chemical used to kill insects.

Integrated pest management (IPM) The use of natural insect enemies and limited chemical applications to control harmful insects while providing some protection for useful insects.

Intermediate cutting A silviculture practice that is used to improve the forest by removing some of the trees to provide enough space for the remaining trees to grow; also known as thinning.

International log rule A table that is used to estimate the number of board feet in a log by calculating the volume of each 4' log section and adding those volumes together.

Internet An international computer network on which information can be communicated to and accessed from the institutions, companies, and individuals who subscribe to the system.

Interphase A resting or nonreproductive stage in the life span of a cell.

IPM Integrated pest management; the use of natural insect enemies and limited chemical applications to control harmful insects while providing some protection for useful insects.

Landing A site to which logs are gathered in preparation for hauling them to a sawmill or processing plant; also known as a yard.

Law Legislation approved by a government that controls specific conduct by its citizens.

Layering A form of vegetative reproduction in which live tree branches are buried in the debris on the forest floor where they generate roots and live stems to regenerate a forest.

Liberation A type of stand improvement in which undesirable older trees are removed from the stand to make sunlight available to young trees.

Light reaction A chemical reaction that occurs during photosynthesis in which energy is captured from the sun.

Lignin A major component making up 25% of the total material in wood; functions to bind wood components together.

Limbing The process of removing limbs from logs during harvesting.

Line-plot cruising A process that is used to gather timber yield data when systematic sampling methods are being used.

Lipid Any one compound of a group of compounds consisting of fats or other similar substances.

Litigation A lawsuit or the process of conducting a lawsuit.

Loading jack A mechanical arm that is raised between a set of rollers to lift lumber to another set of rollers on a higher level.

Lobed simple leaf A leaf from a broadleaf tree with divisions along the leaf margin.

Logging The process of harvesting trees.

Low forest A forest that has been regenerated from the roots, stumps, or branches of other trees.

Lumber Timber that has been sawed into standard dimensions, such as boards, planks, beams, and so on; wood that is cut to dimensions less than 5" × 5".

Margin The distinctively shaped outer edge of a leaf; a useful structure for tree identification purposes.

Mature A tree that is fully developed for a specific purpose; usually 10–24" in diameter.

Mechanical control Pest management using machines or tools to destroy or remove weeds and harmful insects.

Mechanical pulping Separation of wood fibers using abrasive machine action.

Megaspore A haploid plant cell that occurs during the formation of female gametes.

Megaspore mother cell A diploid plant cell from which a female gamete is formed.

Meiosis A cell division process through which the number of chromosomes is reduced by half during the formation of male and female gametes.

Meristem Plant cells, located in a growth zone, that are dividing rapidly to form new plant tissues such as stems and roots.

Metaphase An intermediate step in cell reproduction during which the chromosomes become aligned at the center of the cell.

Metes and bounds A land survey system that uses natural physical features in the landscape as starting points for land measurements.

Micropyle A small opening through which pollen enters the ovule of a flower during fertilization.

Microspore A haploid plant cell that divides to form a pollen grain.

Microspore mother cell A diploid plant cell from which male gametes are formed.

Midrib A reinforced vein in a leaf extending from the leaf stem to the tip of the leaf.

Mitosis A type of cell division that occurs in an animal or plant, resulting in growth.

Model Advanced programming available as a software package that is used to simulate an actual system.

Modem An instrument that makes it possible for computers to exchange information using a telephone line.

Monoculture A population that consists of a single species or variety.

Monomer A simple sugar molecule, such as glucose; also known as a monosaccharide.

Monosaccharide A simple sugar molecule, such as glucose; also known as a monomer.

Mortality A category of timber losses due to the deaths of trees.

Multiple use A form of management in which public lands are put to different uses by several user groups.

NADPH A high-energy molecule that is produced in plant cells during photosynthesis.

Natural regeneration Reforestation of an area from seeds or vegetation without intervention by people.

Naval stores Products that are extracted from wood using organic solvents, such as fatty acids, turpentine, and rosin.

Needle-leaf A uniquely shaped leaf that is narrow in width and relatively long in length.

Network A system consisting of two or more computers that are connected together to allow them to share files and information; a root system consisting of the roots that connect or cross over each other.

Nitrogen cycle The recurring circular flow of nitrogen from living plant and animal tissues to nonliving atmospheric nitrogen and back again.

Nitrogen fixation A process by which certain strains of bacteria convert nitrogen gas from the atmosphere to nitrogen compounds useful to plants.

Nitrogen-fixing bacteria A strain of bacteria that is capable of converting nitrogen gas into nitrates and other nitrogen compounds that are required by plants as nutrients.

Noise duration The length of time that a worker is exposed to a sound.

Noise intensity The amount of energy that is in the sound waves from a noise source; loudness of a noise.

Nonrenewable resource A resource that is formed slowly and cannot be replaced once it is used up.

Nucleoplasm The material that makes up a cell nucleus.

Nucleus A structure in a cell that contains hereditary materials.

Nursery A farm where young trees are raised before they are transplanted to new locations.

Nursery stock Tree seedlings or saplings that are raised for the purpose of transplanting them to forests or other locations.

Oleoresin A mixture of resin and oils obtained from the sap of pine trees.

Omnivore An animal that eats both plants and other animals.

Organ A group of several tissues such as roots and stems that functions as a single unit.

Ovary The part of a flower that produces the ovule or egg cell.

Overmature A tree for which harvest has been delayed until the wood has begun to decay or die.

Oversight Responsibility for, or supervision of, the care of a natural resource such as trees.

Ovule An immature female germ cell located inside the ovary of a flower.

Palmately compound Multiple leaflets arranged on spines or veins that radiate from a single point; shaped like a hand with the fingers spread apart.

Parasite An organism, such as a fungus, that lives in or on plants, harming its host by feeding on fluids or damaging tissues of the host organism.

Parenchyma Thin-walled cells that are loosely packed together to form spongy tissues with air spaces interspersed between them; cells that make up much of the material in plant leaves, roots, stems, and fruits.

Parent material Rocks (not including bedrock) located in the C horizon of a soil.

Parthenogenesis A type of reproduction in certain types of insects, such as aphids, in which mature females produce female offspring without mating with a male.

Particleboard A reconstituted board containing a high percentage of wood shaving in the central core with layers of wood flakes on either side of the core and fine sawdust near the surface.

Particulate matter Tiny particles of dust and waste materials that are suspended in the air.

Peat Plant material found in deep deposits in wet areas such as marshes and bogs where decay is minimal.

Pericycle The outer layer of the vascular cylinder of a root; lateral roots develop from this tissue, growing out through the cortex and epidermis of the primary root.

Permeable The capacity of a membrane to allow fluids to pass through it.

Petal The inner floral leaves of a flower that are often brightly colored.

Petiole A plant leafstalk forming the attachment to the main stem.

Phloem A vascular plant tissue through which sugars and plant foods, manufactured in the leaves, flow to the stems and roots.

Photogrammetry A method for measuring land area by producing aerial photographs of a known scale from which the total area within the boundary lines is determined.

Photosynthesis A process that uses chlorophyll to capture energy from the sun, combining it with carbon dioxide, water, and nutrients to form plant tissues.

Physiology The branch of biology that is concerned with the life functions and processes of living organisms.

Pinnately compound Multiple leaflets arranged along opposite sides of a central leafstalk.

Pioneer or Pioneer species The first plants to grow naturally in an area that has been cleared or burned.

Pistil A female reproductive structure found in a flower.

Pith A structure located in the center of a stem consisting of parenchyma cells that have the primary function of storing plant food.

Planer A machine that reduces rough lumber to standard lumber dimensions by shaving or planing off the outer surfaces.

Planimeter An instrument that accurately measures the land area of a map of known scale by tracing the perimeter boundary.

Plank A piece of wood that measures 1⅞–4" in thickness and more than 11" in width.

Plantation forest A population of trees that is established by planting seeds or seedlings, and that is intensively managed for maximum production.

Plant Health Care (PHC) A program for the prevention of health problems in trees by observing key elements of good tree health.

Plywood A laminated wood product made of several sheets of veneer and lesser-quality wood bonded together by adhesives.

Pneumatic power The use of compressed air to deliver power to a location that is remote to the power source.

Polar nuclei Haploid nuclei that develop during the formation of female gametes.

Pole A young tree that ranges from 4–10" in diameter.

Policy A regulation, based on law, that has been adopted by a government agency.

Pollen grain A male sex cell in plants that fertilizes female flower parts to produce fruits and seeds.

Polymer Any of a large group of compounds consisting of two or more simple sugars such as starch, carbohydrates, or cellulose; also known as a polysaccharide.

Polysaccharide Any of a large group of compounds consisting of two or more simple sugars such as starch, carbohydrates, or cellulose; also known as a polymer.

Pores Specialized vessels, visible in wood, through which dissolved nutrients passed when the tree was living.

Predator An animal that kills and eats other animals.

Prehauler A large machine that is used to load and haul logs to the landing.

Prescribed burn A controlled fire that is purposely started in the forest where it is allowed to burn the trash on the forest floor, reducing the risk of a catastrophic fire due to excessive fuel in the forest; also known as a controlled burn.

Primary consumer An animal that eats plants.

Primary growth Elongation of plant cells, accounting for the lengthwise growth of roots and stems.

Primary succession The development of an ecological community in an area where living organisms were not previously found, such as on a newly formed volcanic island.

Primary tissue The first specialized tissues that form in the area of maturation, such as the epidermis, cortex, and the vascular cylinder tissues in the root.

Principal meridian A survey line that runs north-south.

Producer A plant that converts solar energy and other plant nutrients to starches and sugars.

Program Consists of the organized commands and routines that are used by computers to perform calculations.

Propagation A process by which an organism reproduces.

Prophase The first stage of active cell reproduction.

Protoplasm All of the structures and substances located within a cell.

Pruning Removal of unwanted branches from a tree; shaping a tree by controlling the growth of the limbs and branches.

Quarter section A tract of land measuring approximately 160 acres; one-fourth of a section.

Radial growth Growth resulting in increased diameter in a tree.

Range The east-west location of a township from a principal meridian.

Receptacle The base of a flower.

Recombinant DNA technology The transfer of a desired gene, such as a disease-resistant gene, to the genetic material of a new plant or animal resulting in the expression of the new trait.

Rectangular survey A land survey system that uses an initial point as the beginning point for each survey, and that establishes east-west baselines, north-south meridians, and intermediate survey lines.

Reforestation The return of a population of forest plants to an area from which they have previously been destroyed or removed.

Regeneration The ability of a plant to reproduce itself from seeds or vegetative plant parts; development of new plant tissues to replace missing parts.

Regulation A rule, written by government employees, that defines how a new law is to be implemented.

Renewable resource A resource that is not used up, but continues to produce more of its kind at regular intervals.

Reproductive organ A plant organ in which sexual reproduction occurs; a flower.

Resin Any of a group of thick, sticky liquids consisting of plant juices such as pitch or gum; a wood product that is used to manufacture varnishes, lacquers, naval stores, and so on.

Resin duct A natural channel in a live tree through which pitch flows to different locations in the tree.

Respiration A process in which energy and carbon dioxide are released due to the breakdown of plant tissues during periods of darkness.

Rhizomorphs Thin strands of fungal tissue that enter into the surfaces of tree roots where they infect the tree with new colonies of fungi.

Riparian zone The land adjacent to the bank of a stream, river, or other waterway.

Rodenticide A chemical that is used to kill rodents.

Rodents Small gnawing animals that are identified by the four large incisor teeth located in the front of their mouths.

Root cap A group of specialized cells that develops at the root tip from which a slimy material is produced, helping the root to pass through the soil.

Root graft A connection between the roots of different trees where they have grown together at the point of contact.

Root hair Epidermal cells that have developed into long, thin, threadlike projections to facilitate absorption of water and dissolved nutrients from the soil.

Rotation age The age or stage of maturity at which a stand of trees is harvested.

Rust A class of plant diseases that results in spotted red or brown discoloration of the stems and leaves of a tree.

Saccharification A process by which the polymers that make up cellulose in wood are converted to simple sugars.

Salvage cutting A harvest of trees that have been damaged by fire, wind, insects, and so on.

Sanitation cutting A harvest of trees that are infected with diseases or insects for the purpose of preventing the problem from spreading to healthy trees.

Sapling A young tree that is more than 3' in height and up to 4" in diameter.

Saprophyte Fungus that obtains nutrition from dead organic materials.

Sapwood Light-colored wood through which water and dissolved plant nutrients flow through the tree.

Saw head A large blade mounted on a mechanical tree-feller for the purpose of sawing through the tree at its base after it has been grasped by the mechanical arm.

Sawyer A person who operates the head saw or head rig in a sawmill.

Scale-leaf A leaf that consists of a number of tiny overlapping leaf structures that are thin and flat.

Scaling Measurement of forest products to determine the quantity or amount.

Sclerenchyma A specialized type of plant cell that strengthens tissues by adding fiber to them; cells that form the shells of nuts.

Scribner's rule A table that is used by the timber industry to estimate the number of board feet in a log.

Secondary consumer A carnivorous animal that obtains its nutrition by eating primary consumers and other carnivores.

Secondary growth Growth in the diameter of roots, stems, and branches.

Secondary succession The gradual change in species of plants that live in an area during the time that a damaged ecosystem is returning to its original stage of ecological development.

Secondary tissue The tissue that is responsible for the increase in the diameter of roots, stems, and branches as they grow.

Section A tract of land approximately one mile square with a surface area of 640 acres.

Seedling A young tree that is in the very early stages of development.

Seed tree method A timber harvest method in which mature trees of the desired species are protected from cutting in locations scattered throughout the forest for the purpose of producing seeds.

Selection cutting A harvest method that is used to remove a sustained yield of wood from the forest at regular intervals by cutting only the most mature trees.

Semichemical pulping A process for producing wood pulp in which wood is exposed to a mild chemical treatment to partially separate the wood fibers, followed by processing through a mechanical disk refiner.

Semipermeable A property of a membrane that allows certain kinds of dissolved materials to pass through while restricting the flow of others.

Senescent An advanced stage of maturity in which a tree shows symptoms of heart rot, decay, and other defects due to age.

Sepal A protective leaflike structure that closes over a flower during hours of darkness.

Sexual reproduction The production of male and female gametes and the process by which they join together to produce offspring.

Shelterwood method A timber harvest method in which mature trees are left in the harvested area in sufficient numbers to provide shade and protection for seedlings. The mature trees are harvested once the seedlings have become established.

Short-duration fire cycle The use of prescribed fires on a regular basis to eliminate fuels form the forests.

Short rotation intensive culture (SRIC) Wood crops that are harvested every three to seven years.

Short rotation woody crops (SRWC) Wood crops that are harvested every three to seven years.

Shrub A short woody plant.

Shrub layer Vegetation consisting of short woody plants that occupy the stratum between the herb layer and the understory of the forest.

Sieve element A specialized plant structure that makes up sieve tubes and phloem tissue; important in the flow of plant foods to stems and roots.

Sieve tube A specialized plant structure made up of sieve elements; important component in phloem tissue in facilitating the flow of plant foods to stems and roots.

Silt Tiny soil particles that are easily eroded by becoming suspended in flowing water or blown as dust in the wind.

Siltation The process by which tiny soil particles are suspended in flowing water and deposited in the beds of streams, rivers, lakes, and oceans.

Silt load The amount of eroded soil that is carried in the flowing waters of streams and rivers.

Silvics The study of forests and forest relationships.

Silviculture Management of forests and their environments to establish and cultivate populations of trees, and to promote the growth and harvest of trees for commercial purposes.

Skidder A large machine that is used to drag logs from the harvest site to the landing.

Slab The outer shell of a log that is removed as the log is sawed into lumber.

Smalian's formula A dependable log scaling method that is used to estimate the amount of solid wood in a log by using a formula that takes into account the log length and the measurements from both ends of the log.

Smog High concentrations of chemical pollutants and particulate matter suspended in the atmosphere; atmospheric haze caused by the action of ultraviolet light on atmospheric pollutants.

Smoke jumper A firefighter who jumps from an aircraft and parachutes to a location near a fire zone.

Software Computer programs and routines stored on disks, CD-ROMs, and magnetic tape.

Softwood Wood obtained from a conifer or needle-bearing tree; immature plant materials obtained from new growth trees.

Soil conservation Soil management practices that prevent soil losses and maintain soil fertility.

Soil profile A description of the texture and content of each of the different horizons or layers in a soil.

Soil texture The proportion of mineral particles of different sizes that are found in a sample.

Spiking The illegal act of driving steel spikes into trees with the intention of damaging saws and milling equipment; an act of protest by antilogging demonstrators.

Spindle A bundle of fibers that functions to separate chromosome pairs during cell division.

Spine A leaf structure similar to the midrib that gives structure and shape to the leaf.

Spodosols Light-colored acid soils that are formed from coarse silica parent material in cold, damp climates.

Spore A cell from a fungus or similar organism that grows into a new organism when it is located in a site that is favorable to its growth; a male reproductive cell in a plant.

Spotting The process by which a firestorm hurls burning materials across fire barriers.

Sprout method A form of asexual reproduction of a forest in which all of the trees are cut and new forest growth is generated from the stumps of the harvested trees; also known as coppice method.

Square Any timber that is square-cut with equal dimensions on all four sides.

Stamen The male part of a flower that consists of the anther and the filament.

Stand A population of trees in a given location.

Standard cord A unit of measurement for pulpwood and firewood; the volume of wood in a pile 8' long, 4' high, and 4' wide; 128 cubic feet.

Standard parallel A survey line that is established running east-west for each interval of twenty-four miles on either side of the baseline.

Stand improvement Removal of competing vegetation from a forest that has matured beyond the sapling stage of development.

Starch A molecule that is formed when large numbers of glucose molecules bond together in a long branching chain.

Stereoplotter An instrument that is used to analyze satellite and aerial photographs for the purpose of creating maps.

Stereoscope A mapping instrument that is used to simultaneously view two aerial photographs of the same area from slightly different angles, creating an image in three dimensions.

Stigma The female flower part that functions as a pollen recepter.

Strata Levels or layers of plant growth in a forest or other ecosystem.

Style A female flower part that connects the stigma to the ovary of the flower.

Sulfate naval stores Wood products obtained as by-products from the Kraft pulping process.

Surface fire A fire that burns the dry twigs, dead branches, grass, and leaves that lie on the forest floor.

Surface fuel Fuels such as dry, undecayed leaves and plant materials along with living plants.

Surface water Water flowing in streams, rivers, and lakes as it moves across the land surface to the ocean.

Survival growth The total difference in timber volume of survivor trees between the first and second measurements of a forest growth survey.

Survivor trees Trees that were measured in both surveys for the time interval during which forest growth is evaluated.

Sustained yield The volume of timber products that can be harvested year after year without depleting the timber resource.

Syngas A synthetic gas that is produced from methane and carbon monoxide; useful in the thermochemical liquefaction process for the production of industrial oil.

Systematic sampling A method that is used to estimate the timber yield of a forest by measuring only those trees that have been selected at specific intervals along predetermined lines.

Tannin A wood product that is used to cure animal hides into leather; also known as tannic acid.

Telophase The last stage of cell division during which cell cytoplasm is divided as two new cells are formed.

Tensile strength Strength across the long axis of a tree; the source of strength of wood that is used for timbers and structural beams.

Tensiometer An instrument that indicates the amount of moisture in the soil.

Terminal growth Vertical growth in a tree.

Tetrad A cluster of four haploid cells.

Thermochemical liquefaction A process by which wood chips are heated under high pressure in a hydrogen gas or syngas atmosphere to produce industrial oil products.

Timber A wood piece that measures greater than 5" × 5" and less than 8" × 8".

Timber cruiser A person who estimates the volume of marketable timber by taking sample measurements from sites throughout a stand of trees.

Timber yield The timber volume in a forest on a particular date.

Tissue A group of cells that contribute to a particular life function.

Tissue culture The development of roots, stems, and leaves from callus tissue in a plant medium containing plant nutrients and hormones.

Topographic map A map that shows relative elevations.

Topography The science of creating maps of the surface features of an area; the relative elevations of different features in a landscape.

Township A tract of land measuring approximately six miles on each side.

Tracheid A long, tapered plant cell found in xylem tissue, having pits in the cell walls through which water is conducted.

Transpiration Loss of water to the atmosphere from the leaf surfaces of plants.

Ultisols Soils formed in warm, humid climates, showing evidence of weather action in the illuvial deposits of clay and iron in the subsoil.

Understory Short trees in a forest that fill an intermediate stratum of vegetation beneath the canopy created by the branches and foliage of the tallest trees.

Uneven-aged stand The presence of trees of different ages and stages of maturity in a forest population.

Urban forestry An emerging branch of forestry that is practiced in or near cities.

Vacuole A cell structure that gathers excess water and wastes that are discharged through the cell wall.

Vascular cambium Meristem tissue that forms in a continuous ring between xylem and phloem tissues and that gives rise to new growth of xylem and phloem tissues.

Vascular cylinder The innermost part of a root consisting of xylem and phloem tissues.

Vascular ray A row of parenchyma cells that radiates to the center of the stem or trunk of a tree, and that transports dissolved materials across the woody section of the stem.

Vector An insect or other organism that carries disease organisms in its body from which it spreads the disease to trees or other susceptible life forms.

Vegetative organ A plant organ such as a root, stem, or leaf.

Vegetative reproduction Asexual reproduction of plants from plant parts such as leaf, root, or stem tissue.

Vein Leaf structures that connect to the vascular tissues in spines or midribs, allowing the flow of dissolved materials to the cells of the leaf.

Veneer Thin sheets (¼" thickness or less) of high-quality wood that are glued to the outer surfaces of lesser-quality wood for the manufacture of plywood and the construction of furniture.

Vessel A plant structure that consists of vessel elements that have grown together, end to end, facilitating the flow of large volumes of dissolved nutrients within the plant.

Vessel element A plant structure that becomes hollow after it has died, facilitating the flow of dissolved nutrients in the stems, roots, and leaves of a plant.

Warp Distortion or twisting of lumber products as they lose moisture.

Water cycle The movement of water in the form of vapor from the oceans to the clouds to the earth as precipitation, and back to the oceans through rivers and streams.

Watershed An area bounded by geographic features where precipitation is absorbed in the soil to form groundwater that eventually emerges to become surface water and that ultimately drains to a particular stream, river, or other body of water.

White rots A type of wood rot that occurs in trees as fungi break down the cellulose and lignin components of the cell walls.

Wilderness A land area that is managed and protected as a wild territory.

Wildfire Any fire that burns out of control or that has not been prescribed for a specific purpose.

Wilt A fungal infection in the vessels of the xylem tissue where the flow of water and dissolved nutrients is blocked from the trunk and branches of the tree.

Wood naval stores Wood products that are extracted from chipped or shredded wood obtained from pine stumps or logs by dissolving the wood components in organic solvents.

Xylem A woody tissue that conducts water with dissolved nutrients and plant materials from the roots to the stems and leaves.

Yard A site to which logs are gathered in preparation for hauling them to a sawmill or processing plant; also known as a landing.

Zone map A map that illustrates the severity of climatic conditions that can be expected in each region of the country.

INDEX

495